3rd International Workshop on Statistical Physics and Mathematics for Complex Systems

(SPMCS 2012)

Journal of Physics: Conference Series Volume 394

Kazan, Russian Federation
25-30 August 2012

ISBN: 978-1-62276-858-5
ISSN: 1742-6588

Printed from e-media with permission by:

Curran Associates, Inc.
57 Morehouse Lane
Red Hook, NY 12571

Some format issues inherent in the e-media version may also appear in this print version.

Printed by Curran Associates, Inc. (2013)

Institute of Physics
Dirac House, Temple Back
Bristol BS1 6BE UK

Phone: 44 1 17 929 7481
Fax: 44 1 17 920 0979

techtracking@iop.org

Additional copies of this publication are available from:

Curran Associates, Inc.
57 Morehouse Lane
Red Hook, NY 12571 USA
Phone: 845-758-0400
Fax: 845-758-2634
Email: curran@proceedings.com
Web: www.proceedings.com

TABLE OF CONTENTS

PHYSICAL AND MATHEMATICAL STUDIES

NATURAL COMPLEX SYSTEMS

SOCIAL, ECONOMICAL, POLITICAL AGENT SYSTEMS AND MAN-MADE COMPLEX SYSTEMS

PREFACE

The 3rd International Workshop on Statistical Physics and Mathematics for Complex Systems (SPMCS2012) was held between 25–30 August at Kazan (Volga Region) Federal University, Kazan, Russian Federation. This workshop was jointly organized by Kazan Federal University and Institut Supérieur des Matériaux et Mécaniques Avancées (ISMANS), France. The series of SPMCS workshops was created in 2008 with the aim to be an interdisciplinary incubator for the worldwide exchange of innovative ideas and information about the latest results. The first workshop was held at ISMANS, Le Mans (France) in 2008, and the third at Huazhong Normal University, Wuhan (China) in 2010.

At SPMCS2012, we wished to bring together a broad community of researchers from the different branches of the rapidly developing complexity science to discuss the fundamental theoretical challenges (geometry/topology, number theory, statistical physics, dynamical systems, etc) as well as experimental and applied aspects of many practical problems (condensed matter, disordered systems, financial markets, chemistry, biology, geoscience, etc). The program of SPMCS2012 was prepared based on three categories:

(i) physical and mathematical studies (quantum mechanics, generalized nonequilibrium thermodynamics, nonlinear dynamics, condensed matter physics, nanoscience);
(ii) natural complex systems (physical, geophysical, chemical and biological);
(iii) social, economical, political agent systems and man-made complex systems.

The conference attracted 64 participants from 10 countries. There were 10 invited lectures, 12 invited talks and 28 regular oral talks in the morning and afternoon sessions. The book of Abstracts is available from the conference website (http://www.ksu.ru/conf/spmcs2012/?id=3). A round table was also held, the topic of which was 'Recent and Anticipated Future Progress in Science of Complexity', discussing a variety of questions and opinions important for the understanding of the concept of complexity itself, the behaviours of complex systems as well as for the finding of new theoretical methods.

The papers submitted to this volume were carefully reviewed by referees. We are very grateful to the referees for their very efficient and thoughtful actions. A few submitted papers were unfortunately not included based on the referee reports. As a result, 34 papers are included here. We are very grateful to the members of the international advisory committee for their recommendations of speakers for SPMCS2012. We also appreciate the behind-the-scenes work of the members of the local organizing committee in preparing the conference site, web page, mail correspondence, arrangements for excursions and accommodation, handling the financial support for participants, and so on. Finally, we acknowledge the support from Kazan Federal University.

Sumiyoshi Abe Alain Le Méhauté Dmitrii Tayurskii

Peer review statement

All papers published in this volume of *Journal of Physics: Conference Series* have been peer reviewed through processes administered by the proceedings Editors. Reviews were conducted by expert referees to the professional and scientifi standards expected of a proceedings journal published by IOP Publishing.

Group, geometry and algebra of nonextensive entropies in complex systems

R G Zaripov

Institute of Mechanics and Engineering, Kazan Scientific Center,
Russian Academy of Sciences,
Lobachevsky str. 2/31, Kazan, 420111, Russia

E-mail: zaripov@mail.knc.ru

Abstract. The Abel entropy group and its matrix representation with the general law of nonextensive entropy composition and quadratic nonlinearity are defined. Four types of matrices for the corresponding parametrical entropies are given and geometries for their measures are determined. Algebraic representation of an entropy group is presented for this types of the general entropy classification in statistical thermodynamics of complex systems. The corresponding geometries are global Finsler ones. The well-known relations for the nonextensive entropies follow from the properties of the conformally generalized hypercomplex numbers.

1. The Abel entropy group

In many physical problems, equilibrium and nonequilibrium systems described by non-Gibbs and non-Gauss distributions of nonextensive (nonadditive) statistical mechanics and thermodynamics are investigated [1-3]. These investigations are based on the one-parameter entropy

$$S = S_q = S_q\left[N_{q-1}(p)\right] \tag{1}$$

for classical systems with a real number q.

The seminorm values N and average geometrical of distribution M are determined by expressions

$$\begin{aligned} N = N_{q-1}(p) = \left[\int p^q \mathrm{d}X\right]^{1/(q-1)}, \\ M = \lim_{q\to 1} N_{q-1}(p) = \exp\left(\int p \ln p \mathrm{d}X\right),\ \int p\mathrm{d}X = 1. \end{aligned} \tag{2}$$

Let us consider the Abel entropy group with the commutative law of composition of elements $S_1 \circ S_2 = S_2 \circ S_1$ and group axioms of associativity $(S_1 \circ S_2)\circ S_3 = S_1 \circ (S_2 \circ S_3)$, unit element $S \circ 0 = 0 \circ S = S$, and inverse element $S \circ S^{-1} = S^{-1} \circ S = 0$. The unit group element I corresponds to $S = 0$, and the inverse element is designated by S^{-1}.

Let us present a method of derivation of the law of entropy composition [4]. We assume that the law of composition comprises only S_1 and S_2 and their product $S_1 S_2$ which specifies the quadratic nonlinearity.

Let us write down the law of entropy composition in the general form:

$$S_1 \circ S_2 = g(S_1,S_2)S_1 + w(S_1,S_2)S_2 + h(S_1,S_2)S_1S_2 . \tag{3}$$

From the commutativity property and associativity condition, equate coefficients of arbitrary S_2, S_1S_2, S_2S_3, S_3S_1 and $S_1S_2S_3$, we obtain equalities

$$\begin{aligned} &g(F,S_3)w(S_1,S_2) = w(S_1,G)g(S_2,S_3), \\ &g(F,S_3)h(S_1,S_2) = h(S_1,G)g(S_2,S_3), \\ &h(F,S_3)w(S_1,S_2) = w(S_1,G)h(S_2,S_3), \\ &h(F,S_3)g(S_1,S_2) = h(S_1,G)w(S_2,S_3), \\ &h(F,S_3)h(S_1,S_2) = h(S_1,G)h(S_2,S_3), \end{aligned} \tag{4}$$

where $F = S_1 \circ S_2$ and $G = S_2 \circ S_3$. Using equation (4), we obtain law of composition

$$S = \frac{S_1 + S_2 + \varepsilon S_1 S_2}{1 + \omega S_1 S_2}, \tag{5}$$

with the parameters ε and ω independent of the entropy. From (5) we find the inverse element of the entropy group

$$S^{-1} = -\frac{S}{1+\varepsilon S}, \ \left(1+\omega S S^{-1} \quad 0\right), \tag{6}$$

which is not equal to the opposite element (–*S*).

Let us present the most general properties of the law of entropy composition described by equation (5) that have not yet been considered. Using equation (6), we obtain the following equalities for *S* and S^{-1}:

$$\begin{gathered} \frac{1}{(-S)} - \frac{1}{S^{-1}} = \varepsilon, \ \left(1+\varepsilon S\right)\left(1+\varepsilon S^{-1}\right) = 1, \\ \frac{1}{1+\omega S S^{-1}} = \frac{1+\varepsilon S}{1+\varepsilon S \quad \omega S^2} = \frac{1+\varepsilon S^{-1}}{1+\varepsilon S^{-1} - \omega\left(S^{-1}\right)^2}. \end{gathered} \tag{7}$$

For two entropy types S_1 and S_2, we have

$$\begin{gathered} \left(1+\varepsilon S - \omega S^2\right) \quad \frac{\left(1+\varepsilon S_1 - \omega S_1^{\,2}\right)\left(1+\varepsilon S_2 - \omega S_2^{\,2}\right)}{\left(1+\omega S_1 S_2\right)^2}, \\ \frac{S}{\sqrt{1+\varepsilon S - \omega S^2}} = \frac{S_1 + S_2 + \varepsilon S_1 S_2}{\sqrt{1+\varepsilon S_1 - \omega S_1^2}\sqrt{1+\varepsilon S_2 - \omega S_2^2}}, \\ 1+\varepsilon S_1 - \omega S_1 S = \frac{1+\varepsilon S_1 - \omega S_1^2}{1+\omega S_1 S_2}, \ 1+\varepsilon S_2 - \omega S_2 S \quad \frac{1+\varepsilon S_2 - \omega S_2^2}{1+\omega S_1 S_2}, \\ 1+\varepsilon S - \omega S^2 = \left(1+\varepsilon S_1 - \omega S_1 S\right)\left(1+\varepsilon S_2 - \omega S_2 S\right) \\ \frac{1+\varepsilon S_1 - \omega S_1 S}{1+\varepsilon S_1 - \omega S_1^2} = \frac{1+\varepsilon S_2 - \omega S_2 S}{1+\varepsilon S_2 - \omega S_2^2} \end{gathered} \tag{8}$$

For the three and four entropy types, we have

$$\begin{aligned}
&1+\varepsilon(S_1\circ S_2\circ S_3)-\omega(S_1\circ S_2\circ S_3)^2=\\
&=\frac{(1+\varepsilon S_1-\omega S_1^2)(1+\varepsilon S_2-\omega S_2^2)(1+\varepsilon S_3-\omega S_3^2)}{[1+\omega(S_1S_2+S_2S_3+S_3S_1+\varepsilon S_1S_2S_3)]^2},\\
&(1+\omega S_1S_2)(1+\omega S_3S_4)[1+\omega(S_1\circ S_2)(S_3\circ S_4)]\\
&=(1+\omega S_2S_3)(1+\omega S_4S_1)[1+\omega(S_2\circ S_3)(S_4\circ S_1)]\\
&=(1+\omega S_3S_1)(1+\omega S_4S_2)[1+\omega(S_3\circ S_1)(S_4\circ S_2)].
\end{aligned}\tag{9}$$

Using expression (6) for the inverse element, from equation (5) we obtain S_1 and S_2:

$$S_1=S\circ S_2^{-1}=\frac{S-S_2}{1+\varepsilon S_2-\omega SS_2},\quad S_2=S_1^{-1}\circ S=\frac{S-S_1}{1+\varepsilon S_1-\omega SS_1}.\tag{10}$$

The following law of entropy composition is generally valid:

$$S^{(2n)}=S_1^{(n)}\circ S_2^{(n)}=\frac{S_1^{(n)}+S_2^{(n)}+\varepsilon S_1^{(n)}S_2^{(n)}}{1+\omega S_1^{(n)}S_2^{(n)}},\tag{11}$$

where $S^{(n)}=S\circ S\circ\ldots\circ S$ is the composition of n entropy types S.

The entropies of the common system S and of two independent systems S_1 and S_2 are functions of the seminorm distributions of the examined systems, that is,

$$S=S(N),\ S_1=S_1(N_1),\ S_2=S_2(N_2),,\tag{12}$$

$$N_1=N_{q-1}(p_1)=\left[\int p_1^q\mathrm{d}X_1\right]^{1/(q-1)},\ N_2=N_{q-1}(p_2)=\left[\int p_2^q\mathrm{d}X_2\right]^{1/(q-1)},\tag{13}$$

where $p=p_1p_2$ and $\mathrm{d}X=\mathrm{d}X_1\mathrm{d}X_2$. They have the multiplicativity property: $N=N_1N_2$.

Differentiation of the seminorm of distribution (2) with allowance for equation (5) yields the equation

$$(1+\varepsilon S-\omega S^2)\frac{\mathrm{d}\ln N}{\mathrm{d}S}=\lambda,\tag{14}$$

having the same form for all examined systems. Here λ is an arbitrary constant which can depend on ε and ω.

Upon integration of equation (14) we obtain, depending on the determinant $D=\varepsilon^2+4\omega$ of the quadratic equation $1+\varepsilon S-\omega S^2=0$, four types of functions for four entropies considered in [5] and the corresponding types of matrix representations.

2. A matrix representation entropy group

We now assign the following representation of the Abel entropy group. A matrix **A**(S) from a set of the second-order matrices is juxtaposed to each entropy S of the set of entropies. All matrices juxtaposed to different entropies are different and hence, the group of matrices is isomorphic to the group being represented. The matrices form the Abel group for the binary multiplication operation – the conventional operation of matrix multiplication: $\mathbf{A}(S)=\mathbf{A}(S_1\circ S_2)=\mathbf{A}(S_1)\,\mathbf{A}(S_2)$. The multiplication operation is commutative $\mathbf{A}(S_1)\,\mathbf{A}(S_2)=\mathbf{A}(S_2)\,\mathbf{A}(S_1)$ and associative $[\mathbf{A}(S_1)\,\mathbf{A}(S_2)]\,\mathbf{A}(S_3)=\mathbf{A}(S_1)\,[\mathbf{A}(S_2)\,\mathbf{A}(S_3)]$. The unit element in the condition

$\mathbf{A}(S)\,\mathbf{A}(S^{-1})=\mathbf{A}(S^{-1})\,\mathbf{A}(S)=\mathbf{A}(0)=\mathbf{E}$ is the unit matrix, and the inverse matrix has the form $\mathbf{A}^{-1}(S)=\mathbf{A}(S^{-1})$.

Type I. The equation has different real roots, including the case of $\omega=0$. Considering that $\varepsilon=\varepsilon_1-\varepsilon_2$, $\omega=\varepsilon_1\varepsilon_2$ and $D=(\varepsilon_1+\varepsilon_2)^2$, we obtain matrix and entropy

$$\mathbf{A}(S)=\left(\frac{1-\varepsilon_2 S}{1+\varepsilon_1 S}\right)^{-\frac{\lambda}{\varepsilon_1+\varepsilon_2}}\begin{pmatrix}\dfrac{1+(\varepsilon_1-\varepsilon_2)S}{\sqrt{(1+\varepsilon_1 S)(1-\varepsilon_2 S)}} & \dfrac{\varepsilon_1\varepsilon_2 S}{\sqrt{(1\ \ \varepsilon_1 S)(1\ \ \varepsilon_2 S)}}\\ \dfrac{S}{\sqrt{(1+\varepsilon_1 S)(1-\varepsilon_2 S)}} & \dfrac{1}{\sqrt{(1\ \ \varepsilon_1 S)(1\ \ \varepsilon_2 S)}}\end{pmatrix}, \tag{15}$$

$$S=\frac{1-\left[N_{q-1}(p)\right]^{-(\varepsilon_1+\varepsilon_2)/\lambda}}{\varepsilon_2+\varepsilon_1\left[N_{q-1}(p)\right]^{-(\varepsilon_1+\varepsilon_2)/\lambda}}.$$

Entropy values lying within the limits $-\varepsilon_1^{-1}<S<\varepsilon_2^{-1}$.

From equation (15), the following entropies follow:

a) $\varepsilon_1=k^{-1}(1-q)$, $\varepsilon_2=0$, $\lambda=k^{-1}$,

$$S=\frac{k}{q-1}\left[1-\int p^q \mathrm{d}X\right],\ \text{(Havrda-Charvat, 1967; Daroczy, 1970)},$$

b) $\varepsilon_1=k^{-1}(\beta-1)$, $q=1/\beta$, $\lambda=\ k^{-1}$, -

$$S=\frac{k}{1-\beta}\left[1-\left(\int p^{1/\beta}\mathrm{d}X\right)^{\beta}\right],\ \text{(Arimoto, 1971)},$$

c) $\varepsilon_1=k^{-1}(1-q)$, $\varepsilon_2=\ k^{-1}(1-r)$, $\lambda=\ k^{-1}$,

$$S=\frac{k}{r-1}\left[1-\left(\int p^q\mathrm{d}X\right)^{\frac{r-1}{q-1}}\right],\ \text{(Sharma-Mittal, 1977)}$$

d) $\varepsilon_1=0$, $\varepsilon_2=k^{-1}(1-q)$, $\lambda=\ k^{-1}$, –

$$S=\frac{k}{1-q}\left[1-\left(\int p^q\mathrm{d}X\right)^{-1}\right],\ \text{(Landsberg-Vedral, 1998)},$$

e) $\varepsilon_1=\varepsilon_2=k^{-1}(1-q)$, $\lambda=\ k^{-1}$,

$$S=\frac{k}{1-q}\left[\frac{\int p^q\mathrm{d}X-\left(\int p^q\mathrm{d}X\right)^{-1}}{\int p^q\mathrm{d}X+\left(\int p^q\mathrm{d}X\right)^{-1}}\right]=\frac{k}{1-q}\,\mathrm{th}\left[\ln\left(\int p^q\mathrm{d}X\right)\right],\ \text{(Zaripov, 2005)},$$

f) $\varepsilon_1=k^{-1}(q-1)$, $\varepsilon_1=k^{-1}(r-1)$, $\lambda=\ k^{-1}$, –

$$S=\frac{k}{r-1}\left[\frac{1-\left(\int p^q\mathrm{d}X\right)^{\frac{2(r-1)}{q-1}}}{1+\left(\int p^q\mathrm{d}X\right)^{\frac{2(r-1)}{q-1}}}\right]=\frac{k}{1-r}\,\mathrm{th}\left[\frac{1-r}{1-q}\ln\int p^q\mathrm{d}X\right],\ \text{(Zaripov, 2005)}.$$

It is of interest to note that entropy Landsberg-Verdal is the inverse element in the Havrda-Charvat-Daroczy entropy group given by equation (6).

Type II. The equation has two equal real roots with $\omega=-\varepsilon_1^2$, $\varepsilon=2\varepsilon_1=2k^{-1}(q-1)$, and $D=0$. Then we have the matrix and the entropy of the form

$$\mathbf{A}(S)=\exp\left(-\frac{\lambda S}{1+\varepsilon_1 S}\right)\times\begin{pmatrix}\dfrac{1+2\varepsilon_1 S}{1+\varepsilon_1 S} & -\dfrac{\varepsilon_1^2 S}{1+\varepsilon_1 S}\\ \dfrac{S}{1+\varepsilon_1 S} & \dfrac{1}{1+\varepsilon_1 S}\end{pmatrix},$$

$$S=-\frac{\lambda^{-1}\ln N_{q-1}(p)}{1+\varepsilon_1\lambda^{-1}\ln N_{q-1}(p)}=\frac{k}{1-q}\left(\frac{\ln\int p^q \mathrm{d}X}{1+\ln\int p^q \mathrm{d}X}\right), \text{(Zaripov, 2005)}. \tag{16}$$

Type III. The equation has complex roots with $\omega<0$ and $D<0$, and we obtain

$$\mathbf{A}(S)=\exp\left[\frac{\lambda}{\sqrt{-\omega-\varepsilon^2/4}}\operatorname{arctg}\left(\frac{S\sqrt{-\omega-\varepsilon^2/4}}{1+S\varepsilon/2}\right)\right]\times$$

$$\times\begin{pmatrix}\dfrac{1+\varepsilon S}{\sqrt{1+\varepsilon S-\omega S^2}} & \dfrac{\omega S}{\sqrt{1+\varepsilon S-\omega S^2}}\\ \dfrac{S}{\sqrt{1+\varepsilon S-\omega S^2}} & \dfrac{1}{\sqrt{1+\varepsilon S-\omega S^2}}\end{pmatrix}, \tag{17}$$

$$S=\frac{\operatorname{tg}\left\{\left[\sqrt{-\omega-\varepsilon^2/4}/\lambda\right]\ln\left[N_{q-1}(p)\right]\right\}}{\sqrt{-\omega-\varepsilon^2/4}-\varepsilon\operatorname{tg}\left\{\left[\sqrt{-\omega-\varepsilon^2/4}/\lambda\right]\ln\left[N_{q-1}(p)\right]\right\}/2}.$$

From equation (17) at $\varepsilon=0$ and $\omega=-k^{-2}(q-1)^2$, $\lambda=-k^{-1}$ trigonometric entropy follows

$$S=\frac{k}{1-q}\operatorname{tg}\left[\ln\left(\int p^q \mathrm{d}X\right)\right], \text{(Zaripov, 2005)}. \tag{18}$$

Type IV. The trinomial degenerates into unity at $\varepsilon=\omega=0$, and we obtain the matrix

$$\mathbf{A}(S)=\exp(-\lambda S)\times\begin{pmatrix}1 & 0\\ S & 1\end{pmatrix}, \tag{19}$$

$$S=\frac{k}{1-q}\ln\left(\int p^q \mathrm{d}X\right), \text{(Renyi, 1960)}. \tag{20}$$

In the limit $q\to 1$, we obtain entropy $S=-k\int p\ln p\mathrm{d}X$ (Boltzmann, 1892; Gibbs, 1902; Shannon, 1948) with the additive composition law.

We now assign the following algebraic representation of the Abel entropy group.

3. Algebraic representation of entropy group. Finsler geometry

Let us write down a conformally generalized hypercomplex number

$$R=\varphi\left(S^0,S^1\right)\left(S^0+S^1 e_1\right) \tag{21}$$

with real numbers of the dimensionless functions of entropies S^0 and S^1 (or components), basic elements $e_0 = 1$, e_1 and conformal multiplier $\varphi(S^0, S^1) = \varphi(S^1/S^0)$. The ratio of the functions gives dimensionless entropy: $S = S^1/S^0$. In equation (21), the basic element e_0 by which the number S^0 is multiplied has been omitted.

The law of basic element composition is defined in the general form as [6]

$$e_0 \circ e_1 = e_1 \circ e_0 = e_1,\ e_0^2 = 1,\ e_1^2 = \omega + \varepsilon e_1; \tag{22}$$

it has the properties of commutativity and associativity that represent a special case of the Clifford algebra.

Numbers with $\varphi(S^0, S^1) = 1$ and composition law (22) are called generalized hypercomplex numbers; they were studied in [6], [7], and so on. Depending on the discriminant $D = \varepsilon^2 + 4\omega = (\varepsilon - 2e_1)^2$ for the equation $e_1^2 - \varepsilon e_1 - \omega = 0$, numbers of three **types** can be identified [6]: hyperbolic with $D > 0$, parabolic with $D = 0$, and elliptic with $D < 0$, as well as the nonextensive entropies of three **types**. The fourth **type** of numbers is observed when $\varepsilon = \omega = 0$ and corresponds to the dual numbers with $e_1^2 = 0$.

The canonical form of the number with $\varepsilon = 0$ [6] is well-known form for complex, dual, and double numbers at $\omega = -1, 0, 1$.

According to equation (21), let us write down the composition law for two conformally generalized hypercomplex numbers [8]

$$\begin{aligned} R = R_1 \circ R_2 &= \varphi(S^0, S^1)(S^0 + S^1 e_1) = \\ &= \left[\varphi(S_1^0, S_1^1)(S_1^0 + S_1^1 e_1)\right] \circ \left[\varphi(S_2^0, S_2^1)(S_2^0 + S_2^1 e_1)\right] = \\ &= \varphi(S^0, S^1)\left[S_1^0 S_2^0 + \omega S_1^1 S_2^1 + (S_1^0 S_2^1 + S_1^1 S_2^0 + \varepsilon S_1^1 S_2^1) e_1\right], \end{aligned} \tag{23}$$

where the group properties of the function $\varphi(S^0, S^1)$ have been used:

$$\begin{aligned} \varphi(S) = \varphi(S_1 \circ S_2) = \varphi(S_1)\varphi(S_2), \\ \varphi^{-1}(S) = \varphi(S^{-1}),\ \varphi(S)\varphi(S^{-1}) = 1,\ \varphi(0) = 1. \end{aligned} \tag{24}$$

Then we obtain values of the components

$$\begin{aligned} S^0 &= S_1^0 S_2^0 + \omega S_1^1 S_2^1, \\ S^1 &= S_1^1 S_2^0 + S_1^0 S_2^1 + \varepsilon S_1^1 S_2^1. \end{aligned} \tag{25}$$

As expected, dividing expressions in equation (25) with allowance that $S_1 = S_1^1/S_1^0$ and $S_2 = S_2^1/S_2^0$ yields entropy composition law (5).

From composition law (23), we obtain the conjugate conformally generalized hypercomplex number

$$\bar{R} = \varphi(S^0, S^1)(S^0 + \varepsilon S^1 - S^1 e_1), \tag{26}$$

which for $\varphi(S^0, S^1) = 1$ coincides with the well-known conjugate number [6]. Using conjugate number (26), we obtain the number modulus

$$|R| = (R \circ \bar{R})^{1/2} = \varphi(S^0, S^1)\left[(S^0)^2 + \varepsilon S^0 S^1 - \omega (S^1)^2\right]^{1/2} \tag{27}$$

and the inverse number $R^{-1} = \bar{R}/|R|^2$ with $R \circ R^{-1} = 1$.

For geometrical representation of the entropy group functions, expression (27) with $|R| = 1$ represents a indicatrix or a curve with radius-vector $\mathbf{R}(S^0, S^1)$ and unit length in the global Finsler structure of two-dimensional space of vectors $\mathbf{R}(S^0, S^1)$. The angle in the given space is the additive dimensionless Renyi entropy $\alpha_q = (1-q)^{-1} \ln\left(\int p^q \mathrm{d}X\right)$ [9]. Expression $\left|\mathbf{R}(S^0, S^1)\right|$ is the metric function of the Finsler geometry [10].

The vectors of the examined two-dimensional geometries have the algebraic representation presented above.

According to composition law (25), the components of numbers are calculated in terms of matrices

$$e^{-\nu\alpha_q} \begin{pmatrix} S^0 \\ S^1 \end{pmatrix} = e^{-\nu\alpha_{q1}} \begin{pmatrix} S_1^0 + \varepsilon S_1^1 & \omega S_1^1 \\ S_1^1 & S_1^0 \end{pmatrix} e^{-\nu\alpha_{q2}} \begin{pmatrix} S_2^0 \\ S_2^1 \end{pmatrix}, \quad (28)$$

where $\alpha_q = \alpha_{q1} + \alpha_{q2}$. Matrix

$$\mathbf{A}(S^0, S^1) = e^{-\nu\alpha_q} \begin{pmatrix} S^0 + \varepsilon S^1 & \omega S^1 \\ S^1 & S^0 \end{pmatrix} \quad (29)$$

called the characteristic matrix and obtained by the well-known method according to the composition law of basic elements (22) was also introduced in [6] for the generalized hypercomplex numbers with $\nu = 0$. Transformation (28) between vectors of the product $\mathbf{R}(S^0, S^1) = \mathbf{A}(S_1^0, S_1^1)\mathbf{R}(S_2^0, S_2^1)$ leaves the form-invariant the expression for the indicatrix of independent systems.

Thus, the Finsler geometry of three **types** corresponds to the conformally generalized hypercomplex numbers of three **types**, and the fourth type corresponds to the Galilean geometry. According to the general classification of entropy groups, we have for **type I** and **type III** the corresponding well-known expressions [9] of the function $\varphi(S^0, S^1)$, and as a result, we obtain the conformally generalized hypercomplex numbers

$$R = \left(\frac{S^0 + \varepsilon_1 S^1}{S^0 - \varepsilon_2 S^1} \right)^{\frac{\nu}{\varepsilon_1 + \varepsilon_2}} (S^0 + S^1 e_1),$$
$$e_1^2 = \varepsilon_1 \varepsilon_2 + (\varepsilon_1 - \varepsilon_2) e_1, \quad (\omega = \varepsilon_1 \varepsilon_2, \varepsilon = \varepsilon_1 \quad \varepsilon_2), \quad (30)$$

$$R = \left\{ \exp\left[\frac{2\nu}{\varepsilon_1 + \varepsilon_2} \arctan\left(\frac{S^1(\varepsilon_1 + \varepsilon_2)/2}{S^0 + S^1(\varepsilon_1 \quad \varepsilon_2)/2} \right) \right] \right\} (S^0 + S^1 e_1),$$
$$e_1^2 = -\frac{\varepsilon_1^2 + \varepsilon_2^2}{2} + (\varepsilon_1 - \varepsilon_2) e_1, \quad \left(\omega \quad -\frac{\varepsilon_1^2 + \varepsilon_2^2}{2}, \quad \varepsilon = \varepsilon_1 \quad \varepsilon_2 \right). \quad (31)$$

The Galilean geometry corresponds to the conformally generalized dual number

$$R = \left[\exp\left(\nu S^1 / S^0 \right) \right] (S^0 + S^1 e_1), \; e_1^2 = 0 \; (\omega = \varepsilon = 0) \quad (32)$$

with the additive law of entropy composition

$$S = S_1 + S_2. \quad (33)$$

Finally, after introduction of the angle α_q, the conformally generalized hypercomplex number (21) is written as follows:

$$R=\left[\exp\left(\nu\alpha_q\right)\right]\left(S^0+S^1e_1\right). \tag{34}$$

According to equation (21), the algebraic representation of the entropy group is given by the conformally generalized hypercomplex numbers of the form

$$R=\varphi(S)\frac{(1+Se_1)}{\sqrt{1+\varepsilon S-\omega S^2}}. \tag{35}$$

The law of composition of two numbers is written in the form

$$\begin{aligned} R=R_1\circ R_2&=\left[\varphi(S_1)\frac{(1+S_1e_1)}{\sqrt{1+\varepsilon S_1-\omega S_1^2}}\right]\circ\left[\varphi(S_2)\frac{(1+S_2e_1)}{\sqrt{1+\varepsilon S_2-\omega S_2^2}}\right]\\ &=\varphi(S)\frac{(1+Se_1)}{\sqrt{1+\varepsilon S-\omega S^2}} \end{aligned} \tag{36}$$

and yields the well-known equalities [9]

$$\begin{aligned} \frac{1}{\sqrt{1+\varepsilon S-\omega S^2}}&=\frac{1}{\sqrt{1+\varepsilon S_1-\omega S_1^2}}\cdot\frac{1}{\sqrt{1+\varepsilon S_2-\omega S_2^2}}\\ &+\omega\frac{S_1}{\sqrt{1+\varepsilon S_1-\omega S_1^2}}\cdot\frac{S_2}{\sqrt{1+\varepsilon S_2-\omega S_2^2}},\\ \frac{S}{\sqrt{1+\varepsilon S-\omega S^2}}&=\frac{S_1+S_2+\varepsilon S_1S_2}{\sqrt{1+\varepsilon S_1-\omega S_1^2}\sqrt{1+\varepsilon S_2-\omega S_2^2}}. \end{aligned} \tag{37}$$

The conjugate element and the modulus of the conformally generalized hyperbolic number have the following forms:

$$\begin{aligned} \bar{R}&=\varphi(S)\left(1+\varepsilon S-Se_1\right),\\ |R|&=\varphi(S)\left(1+\varepsilon S-\omega S^2\right)^{1/2}. \end{aligned} \tag{38}$$

4. The functions of hyperbolic entropy and Havrda-Charvat-Daroczy entropy

According to equation (30), we have the algebraic representation of functions of hyperbolic entropy

$$\begin{aligned} &R=\left[\frac{S^0+(1-q)S^1}{S^0-(1-q)S^1}\right]^{\frac{\nu}{2(1-q)}}\left(S^0+S^1e_1\right),\ e_1^2=(1-q)^2,\\ &S^0=k\,\mathrm{ch}\left[(1-q)\alpha_q\right],\ S^1=\frac{k}{1-q}\mathrm{sh}\left[(1-q)\alpha_q\right] \end{aligned} \tag{39}$$

and Havrda-Charvat-Daroczy entropy:

$$\begin{aligned} &R=\left[S^0+(1-q)S^1\right]^{\frac{\nu}{1-q}}\left(S^0+S^1e_1\right),\ e_1^2=(1-q)e_1,\\ &S^0=\exp\left[(q-1)\alpha_q/2\right],\ S^1=\frac{2k}{1-q}\mathrm{sh}\left[(1-q)\alpha_q/2\right] \end{aligned} \tag{40}$$

in **type I** and the corresponding indicatrix of the Finsler geometries for dimensional entropy functions have the following form:

$$\left[\frac{S^0+(1-q)S^1}{S^0-(1-q)S^1}\right]^r\left[\left(S^0\right)^2-(1-q)^2\left(S^1\right)^2\right]=1,\ r=\nu/(1-q), \tag{41}$$

$$\left[S^0+(1-q)S^1\right]^r\left[\left(S^0\right)^2-(1-q)S^0S^1\right]=1,\ r=2\nu/(1-q), \tag{42}$$

where r can have a constant value or depend on q.

Let us consider the case with $\nu=0$ and rewrite equation (42) in the form

$$\left[S^0+(1-q)S^1/2\right]^2-\left[(1-q)/2\right]^2\left(S^1\right)^2=1. \tag{43}$$

Then transition to the oblique-angled system of coordinates described by the system of equations

$$\begin{aligned} S^{0\prime}&=S^0+(1-q)S^1/2=\mathrm{ch}\left[(1-q)\alpha_q/2\right],\\ S^{1\prime}&=S^1=2\,\mathrm{sh}\left[(1-q)\alpha_q/2\right], \end{aligned} \tag{44}$$

leads equation (43) to the canonical form

$$\left(S^{0\prime}\right)^2-\left(\frac{1-q}{2}\right)^2\left(S^{1\prime}\right)^2=1, \tag{45}$$

coinciding with expression (41) for the pseudo-Euclidean geometry. Thereby, the transition with linear transformations of the entropy functions and Havrda-Charvat-Daroczy entropy to form hyperbolic entropy proceeds by the dimensional formula

$$S'=\frac{S}{2\left[1+(1-q)S/2\right]}=\frac{2k}{1-q}\mathrm{th}\left[\frac{1}{2}\ln\left(\int p^q dX\right)\right]. \tag{46}$$

In the limit $q\to 1$, we obtain Boltzmann-Gibbs-Shannon entropy $S'=-k\int p\ln p dX$.

In the algebraic approach, such transition corresponds to the linear transformations of the basic elements

$$\begin{aligned} e_0'&=e_0,\\ e_1'&=-\frac{1}{2}\left[(1-q)e_0+e_1\right]. \end{aligned} \tag{47}$$

In this case, the composition law for new basic elements is written as follows:

$$e_0 \circ e_1=e_1 \circ e_0=e_1,\ e_0^2=1,\ e_1^2=\left(\frac{1-q}{2}\right)^2. \tag{48}$$

References

[1] Tsallis C 2009 *Introduction to Nonextensive Statistical Mechanics. Approaching a Complex World* (New York: Springer) p 382

[2] Zaripov R G 2010 *Principles of Nonextensive Statistical Mechanics and Geometry of Measures of the Disorder and the Order* (Kazan: Kazan A.N. Tupolev State Technical University Press) p 404 (in Russian)

[3] Naudts Jan 2011 *Generalized Thermostatistics* (London: Springer) p 201

[4] Zaripov R G 2006 *Russian Phys. J.* **49** 157
[5] Zaripov R G 2005 *Russian Phys. J.* **48** 267
[6] Lavrent'ev M A and Shabat B V 1973 *Problems of Hydrodynamics and Their Mathematical Models* (Moscow: Nauka) (in Russian)
[7] Catoni F, Cannata R, Catoni V and Zampetti P 2004 *Adv. Appl. Clifford Algebr.* **14** 47
[8] Zaripov R G 2012 *Russian Phys. J.* **55** 17
[9] Zaripov R G 2009 *Russian Phys. J.* **52** 200
[10] Rund H 1959 *The differential geometry of Finsler spaces* (Berlin: Springer)

Journal of Physics: Conference Series **394** (2012) 012002 doi:10.1088/1742-6596/394/1/012002

Nonadditive thermostatistics and thermodynamics

P Ván[1,2], G G Barnaföldi[1], T S Biró[1], and K Ürmössy[1]

[1] Wigner RCP, Hungarian Academy of Sciences H-1525 Budapest, P.O.Box 49, Hungary
[2]Department of Energy Engineering, Budapest University of Technology and Economics, Bertalan Lajos u. 4-6, H-1111, Budapest, Hungary

E-mail: `van.peter@wigner.mta.hu`

Abstract. Nonadditive composition rules for several physical quantities are treated in thermodynamics. It is argued that the zeroth law defines the existence of their additive forms, the formal logarithms. A further principle, the universal thermostat independence leads to a particular formal logarithm, equivalent to Tsallis entropy S_q. We connect q with generalized susceptibilities of the thermostat.

1. Introduction

Neither thermodynamics nor statistical physics includes the other. Near-equilibrium statistical physics fulfills thermodynamic relations particular microscopic models can clarify certain mechanisms behind thermodynamic quantities. On the other hand the general laws of thermodynamics are valid also in those cases when the detailed microscopic treatment is unavailable or hopelessly complicated. The aiming at universality in thermodynamics should be understood from this point of view. Whenever one deduces model relations from thermodynamic principles, the results are by construction independent of particularities of the microscopic model.

The above relation is present in attempts of explaining among others power-law tails in the transverse momentum spectra of high energy collisions based on statistical considerations. These power-law tails can be observed in heavy ion collisions at the RHIC and LHC accelerators [1] cf. Fig. 1, in proton-proton collisions at the LHC, RHIC, Fermilab and SPS [2] cf. Fig. 2 and even in electron-positron collisions [3], cf. Fig. 3. One inspects, that in electron-positron collisions, where the number of outcoming particles is about 10-50, the deviation from Tsallis distribution is salient.

The presented data indicate the universal nature of the cut power law distributions. These fits perform much better also at low p_T than the exponential Boltzmann-Gibbs distribution [4]. Therefore a single microscopic mechanism, like perturbative QCD, cannot explain all observations.

In this paper we propose a universal explanation of the observed Tsallis distributions. This is based on general arguments extending the framework of classical thermostatics by abandoning the additivity property constraint on the extensive quantities by introducing abstract, in a general setting nonadditive composition rules [5, 6]. By doing so the zeroth law of thermodynamics reveals that the very concept of separate physical systems – thermodynamic bodies – restrict those rules. In the following we are going to prove that rules fulfilling the zeroth law are associative and can be given in a special form with the help of the so called formal logarithms. Then the thermodynamic temperature can be introduced as an intensive

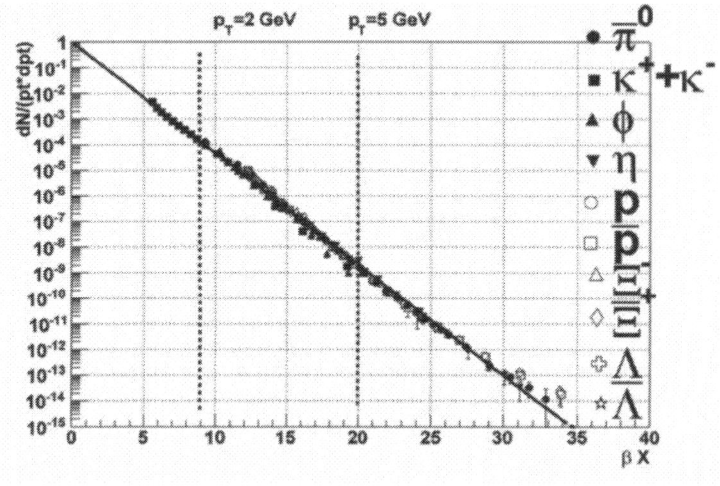

Figure 1. Tsallis fits to transverse momentum distributions of identified particles stemming from AuAu collisions at $\sqrt{s}$ = 200 AGeV collision energy. The expansion of the quark-gluon plasma is taken into account by the blast wave model. X $= \frac{1}{a}ln(1+a\gamma(m_T - vp_T))$ and $\beta = 1/T$. (Fig. 1b in Ref. [1]).

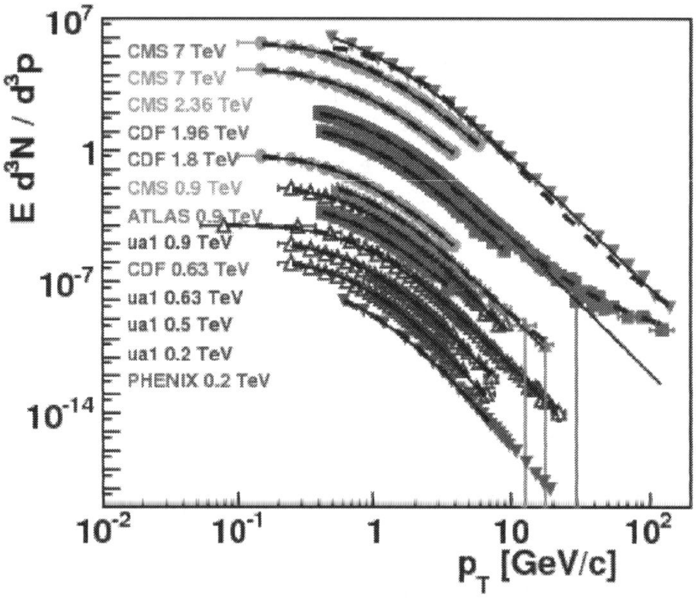

Figure 2. Tsallis fits on transverse momentum spectra of charged hadrons in proton-proton collision on a logarithmic plot (Fig. 1a in Ref. [2]).

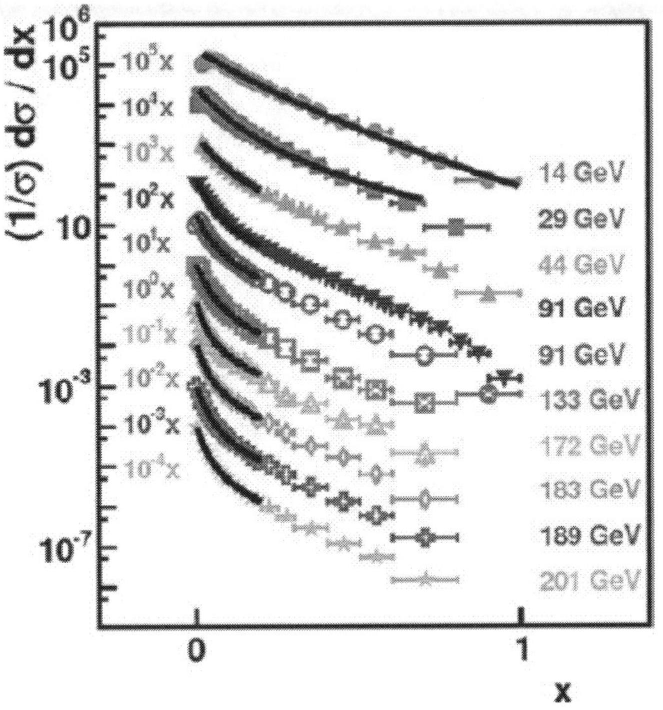

Figure 3. Fragmentation functions from electron-positron collisions measured at various collision energies and fitted one-dimensional Tsallis distributions. The energy fraction x = ϵ/E is the energy of the produced hadron scaled by the maximal acquirable energy in a 2-jet event E = $\sqrt{s}/2$ (Fig. 5 in Ref. [3]).

quantity characterizing the thermodynamic bodies separately. The extension of the maximum entropy principle for nonadditive composition rules becomes straightforward. Nonadditively extended thermodynamics ensures that thermostatistical relations – represented by different statistical entropy formulas – can be interpreted in a thermodynamic framework [7].

This framework can be further refined when dealing with the choice among several different generalizations of thermostatistical entropies (e.g. [8, 9, 10, 11, 12, 13, 14]. One would welcome a reasonable criteria formulated in general thermodynamics [15]. We show that considering a canonical, maximum entropy based approach to an open statistical physical system. The requirement of vanishing first order corrections in the microcanonical treatment of a system including the thermostat, leads to a particular formal logarithm, that transforms the Boltzmann–Gibbs–Shannon entropy density to a Tsallis entropy density. This leads to a physical interpretation of the q parameter in the entropy formula, related to the 'heat capacity' of the reservoir. In this way the Tsallis entropy density and the Rényi total entropy are connected by the selected formal logarithm.

2. The zeroth law of thermodynamics and nonadditive composition rules

2.1. Separability

The basic variables of classical thermodynamics are the so called extensive physical quantities, like the energy (E), particle number of the ith chemical component (N_i) and entropy (S). The adjective 'extensive' is often associated to a different characteristics of these fundamental physical quantities. It is customary to assume that the the extensive quantities are additive if we merge two thermodynamic bodies:

$$X_{12} = X_1 + X_2, \tag{1}$$

where X_i denotes one of the above mentioned quantities of the ith body. This special property of composition is *additivity*. On the other hand traditionally one assumes that these quantities characterize the systems up to the largest meaningful scale operating with finite densities of the extensive quantities [16]:

$$\rho_X = \lim_{N\to\infty} \frac{1}{N}\sum_{i=1}^{N} X_i < \infty. \tag{2}$$

Here one divides the body to N different parts and X_i belongs to the ith of them. This property is the *extensivity*. These two properties are related, but far from being equivalent. If a quantity is additive, then it is necessarily extensive, but there are extensive and yet non-additive quantities [6, 9]. In our treatment the classically extensive quantities are in general nonadditively composed, therefore in the followings we refer to them as *basic quantities*.

Another important class of thermodynamic quantities are the intensives. Their definition is related to thermodynamic equilibrium and to the zeroth and the second laws of thermodynamics. The zeroth law is formulated as the requirement of the *transitivity* of the thermodynamic equilibrium state [17, 18]. Somewhat implicit, but important in this respect, that the thermodynamic bodies are assumed to be separable of each other: in thermodynamic equilibrium their respective equation of states do not depend on the properties of the partner bodies or of the interaction. This *condition of physical separability* is required also in case of nonadditive composition of the extensives. We shall introduce the concept of intensives accordingly. In the following we generalize our reasoning with a nonadditive entropy and nonadditive energy as basic quantities developed in [7] to nonadditive entropy and additional n nonadditive quantities.

Let us consider two thermodynamic bodies, characterized by n basic quantities denoted by X_1^i and X_2^i, $i = 1, ..., n$ for the two bodies respectively and $X_1^0 = S_1$, $X_2^0 = S_2$ are the entropies. The entropy plays a distinguished role, it defines the equation of state of the respective bodies, $S(X^1, ..., X^n)$. In our treatment the basic quantities are not additive, but composable, therefore their value in the two-body system is expressed by the composition rules

$$X_{12}^i(X_1^i, X_2^i) \neq X_1^i + X_2^i, \qquad i = 0, ..., n. \tag{3}$$

This functional relationship is in general nonadditive. We assume that our thermodynamic system of two interacting bodies is isolated, therefore the basic quantities, X_{12}^i are fixed. Their conservation is expressed with the help of differentials:

$$\mathrm{d}X_{12}^i = \frac{\partial X_{12}^i}{\partial X_1^i}\mathrm{d}_1^i + \frac{\partial X_{12}^i}{\partial X_2^i}\mathrm{d}X_2^i = 0, \qquad i = 1, ..., n. \tag{4}$$

The requirement of maximum entropy in isolated systems in equilibrium reads as

$$\mathrm{d}S_{12}(S_1, S_2) = \sum_{i=1}^{n}\left(\frac{\partial S_{12}}{\partial S_1}\frac{\partial S_1}{\partial X_1^i}\mathrm{d}X_1^i + \frac{\partial S_{12}}{\partial S_2}\frac{\partial S_2}{\partial X_2^i}\mathrm{d}X_2^i\right) = 0. \tag{5}$$

The unique condition of the equations (4) and (5) having arbitrary $\mathrm{d}X_1^i$ and $\mathrm{d}X_2^i$ changes in the basic quantities is given by

$$\frac{\partial X_{12}^i}{\partial X_2^i}\frac{\partial S_{12}}{\partial S_1}\frac{\partial S_1}{\partial X_1^i} = \frac{\partial X_{12}^i}{\partial X_1^i}\frac{\partial S_{12}}{\partial S_2}\frac{\partial S_2}{\partial X_2^i}, \quad i = 1, ..., n. \tag{6}$$

The most general form of the partial derivatives now may be written by a multiplicative separation of the variables as

$$\begin{aligned}
\frac{\partial S_{12}}{\partial S_1}(S_1, S_2) &= F_1(S_1)\, G_2(S_2)\, H_1(S_1, S_2),\\
\frac{\partial S_{12}}{\partial S_2}(S_1, S_2) &= F_2(S_2)\, G_1(S_1)\, H_2(S_1, S_2),\\
\frac{\partial X^i_{12}}{\partial X^i_1}(X^i_1, X^i_2) &= A^i_1(X^i_1)\, B^i_2(X^i_2)\, C^i_1(X^i_1, X^i_2),\\
\frac{\partial X^i_{12}}{\partial X^i_2}(X^i_1, X^i_2) &= A^i_2(X^i_2)\, B^i_1(X^i_1)\, C^i_2(X^i_1, X^i_2).
\end{aligned} \tag{7}$$

Then (6) requires, that

$$A^i_2 B^i_1 C^i_2 \cdot F_1 G_2 H_1 \frac{\partial S_1}{\partial X^i_1} = A^i_1 B^i_2 C^i_1 \cdot F_2 G_1 H_2 \frac{\partial S_2}{\partial X^i_2}. \tag{8}$$

This equation factorizes to (X^i_1, S_1) and (X^i_2, S_2) dependent terms only if

$$\frac{C^i_2(X^i_1, X^i_2)}{C^i_1(X^i_1, X^i_2)} = \frac{H_2(S_1, S_2)}{H_1(S_1, S_2)} \quad \text{for} \quad i = 1, ..., n \tag{9}$$

The physical separability of the thermodynamic bodies has to beindependent of any particular form of the equation of state, $S(X^1, ..., X^n)$. Therefore the above ratio can only be constant. Its value can easily be absorbed into one of the factorizing component functions of the entropy F or G. As an immediate consequence one obtains

$$\begin{aligned}
C^i_1(X^i_1, X^i_2) &= C^i_2(X^i_1, X^i_2),\\
H_1(S_1, S_2) &= H_2(S_1, S_2).
\end{aligned} \tag{10}$$

These equalities are the basis for considering formal logarithms for the basic quantities separately. The factorized form of eq. (6)

$$\frac{B^i_1 F_1}{A^i_1 G_1} \frac{\partial S_1}{\partial X^i_1} = \frac{B^i_2 F_2}{A^i_2 G_2} \frac{\partial S_2}{\partial X^i_2}, \tag{11}$$

defines the following generalized intensive quantities Y^i

$$Y^i = \frac{B^i(X^i) F(S)}{A^i(X^i) G(S)} \frac{\partial S}{\partial X^i}(X^1, ... X^n). \tag{12}$$

Finally, using the definitions

$$\begin{aligned}
\hat{L}(S) &:= \int \frac{F(S)}{G(S)} dS,\\
L^i(X^i) &:= \int \frac{A^i(X^i)}{B^i(X^i)} dX^i,
\end{aligned} \tag{13}$$

we arrive at

$$Y^i = \frac{\partial \hat{L}(S)}{\partial L^i(X^i)}. \tag{14}$$

The zeroth law requires that this common value is to be introduced as the generalized entropic intensive quantity.

These functions of the original thermodynamical variables, $\hat{L}(S)$ and $L^i(X^i)$ defined in eq.(13) can be used to map the original composition rules to a simple addition.

Namely considering the conditions (10) one obtains

$$\begin{aligned} \frac{B_1^i}{A_1^i}\frac{\partial X_{12}^i}{\partial X_1^i} &= \frac{B_2^i}{A_2^i}\frac{\partial X_{12}^i}{\partial X_2^i}, \\ \frac{G_1}{F_1}\frac{\partial S_{12}}{\partial S_1} &= \frac{G_2}{F_2}\frac{\partial S_{12}}{\partial S_2}. \end{aligned} \tag{15}$$

Utilizing now the definitions (13) for X_1^i, X_2^i, $i = 1, ..., n$ and and S_1, S_2 separately, the partial derivatives simplify:

$$\begin{aligned} \frac{\partial X_{12}^i}{\partial L_1^i} &= \frac{\partial X_{12}^i}{\partial L_2^i}, \\ \frac{\partial S_{12}}{\partial \hat{L}_1} &= \frac{\partial S_{12}}{\partial \hat{L}_2}. \end{aligned} \tag{16}$$

The general solution of such partial differential equations is an arbitrary function of the sum of variables:

$$\begin{aligned} X_{12}^i &= \Phi^i(L_1^i + L_2^i), \quad \text{for all} \quad i = 1, ...n. \\ S_{12} &= \Psi(\hat{L}_1 + \hat{L}_2). \end{aligned} \tag{17}$$

If the Φ and Ψ functions are invertible we can index their inverse arriving at a more symmetric notation:

$$\begin{aligned} L_{12}^i(X_{12}^i) &= L_1^i(X_1^i) + L_2^i(X_2^i), \\ \hat{L}_{12}(S_{12}) &= \hat{L}_1(S_1) + \hat{L}_2(S_2). \end{aligned} \tag{18}$$

Since the L^i and $\hat{L}$ functions map a non-additive composition rule to an additive one, they are *formal logarithms*.

One recovers the classical additive composition when the formal logarithms are the respective identity functions. The role of the zeroth law in abstract composition rules was investigated by several authors with different generality and purpose [19, 20, 5, 21, 22, 23] and with diverging conclusions.

Composition rules, that can be transformed into the above form with a formal logarithm, are associative. Associative composition rules are also results of an infinite repetitive application of the same rule onto infinitesimal pieces of the same material, as it was proved in [6]. However, in this special case the formal logarithms in (18) are related to the same material and therefore are identical. In the general, heterogeneous case these functions L_1 and L_2 may differ, and the mapping related to the interaction L_{12} can be a further one.

2.2. Transitivity

Up to now we required the separability of the thermodynamic bodies in equilibrium and did not investigate the transitivity aspect of the zeroth law applying nonadditive composition rules. This formulation is not connected to the second law directly. Transitivity universally requires that if bodies 1 and 2 are in thermal equilibrium and independently the bodies 2 and 3, then also the bodies 1 and 3 are in equilibrium [1].

So far we have established additivity of composite functions of the basic quantities of the respective subsystems, $L_I^i(E_I^i)$, $I = 1, 2$. Therefore it is natural to assume that these functions are characteristic to the bodies and only the double-indexed formal logarithms, $L_{IJ}^i(X_{IJ}^i)$, $I, J = 1, 2$ are characteristic to the interaction between bodies in equilibrium. This is valid for the entropy composition, as well. By this construction all subsystems develop the same individual formal logarithm, irrespective to which other system they equilibrate with. Assuming namely the opposite, i.e. a partner-dependent individual formal logarithm, the transitivity would be violated. Let us regard three possible pairings of three subsystems. The composite basic quantities satisfy

$$\begin{aligned} X_{12}^i &= \Phi_{12}^i(L_1^i(X_1^i) + L_2^i(X_2^i)), \\ X_{23}^i &= \Phi_{23}^i(\tilde{L}_2^i(X_2^i) + L_3^i(X_3^i)), \\ X_{13}^i &= \Phi_{13}^i(\tilde{L}_1^i(X_1^i) + \tilde{L}_3^i(X_3^i)). \end{aligned} \tag{19}$$

If $\tilde{L}^i \neq L^i$, then the equilibrium condition is not automatically transitive. The same is valid for the nonadditive entropy, too:

$$\begin{aligned} S_{12} &= \Psi_{12}\left(\hat{L}_1(S_1) + \hat{L}_2(S_2)\right), \\ S_{23} &= \Psi_{23}\left(\check{L}_2(S_2) + \hat{L}_3(S_3)\right), \\ S_{13} &= \Psi_{13}\left(\check{L}_1(S_1) + \check{L}_3(S_3)\right). \end{aligned} \tag{20}$$

Equation (6) requires, that

$$\begin{aligned} \frac{\partial S_{12}}{\partial S_1}\frac{\partial S_1}{\partial X_1^i}\frac{\partial X_{12}^i}{\partial X_2^i} &= \frac{\partial S_{12}}{\partial S_2}\frac{\partial S_2}{\partial X_2^i}\frac{\partial X_{12}^i}{\partial X_1^i} \\ \frac{\partial S_{23}}{\partial S_2}\frac{\partial S_2}{\partial X_2^i}\frac{\partial X_{23}^i}{\partial X_3^i} &= \frac{\partial S_{23}}{\partial S_3}\frac{\partial S_3}{\partial X_3^i}\frac{\partial X_{23}^i}{\partial X_2^i} \\ \frac{\partial S_{13}}{\partial S_1}\frac{\partial S_1}{\partial X_1^i}\frac{\partial X_{13}^i}{\partial X_3^i} &= \frac{\partial S_{13}}{\partial S_3}\frac{\partial S_3}{\partial X_3^i}\frac{\partial X_{13}^i}{\partial X_1^i} \end{aligned} \tag{21}$$

From here the following separation is obtained, also considering (18):

$$\begin{aligned} \frac{\frac{\partial S_{12}}{\partial S_1}\frac{\partial S_1}{\partial X_1^i}}{\frac{\partial S_{12}}{\partial S_2}\frac{\partial S_2}{\partial X_2^i}} &= \frac{\frac{\partial X_{12}^i}{\partial X_1^i}}{\frac{\partial X_{12}^i}{\partial X_2^i}} = \frac{\hat{L}_1' S_1'}{\hat{L}_2' S_2'} = \frac{L_1^{i\prime}}{L_2^{i\prime}}, \\ \frac{\frac{\partial S_{23}}{\partial S_2}\frac{\partial S_2}{\partial X_2^i}}{\frac{\partial S_{23}}{\partial S_3}\frac{\partial S_3}{\partial X_3^i}} &= \frac{\frac{\partial X_{23}^i}{\partial X_2^i}}{\frac{\partial X_{23}^i}{\partial X_3^i}} = \frac{\check{L}_2' S_2'}{\hat{L}_3' S_3'} = \frac{\tilde{L}_2^{i\prime}}{L_3^{i\prime}}, \\ \frac{\frac{\partial S_{13}}{\partial S_1}\frac{\partial S_1}{\partial X_1^i}}{\frac{\partial S_{13}}{\partial S_3}\frac{\partial S_3}{\partial X_3^i}} &= \frac{\frac{\partial X_{13}^i}{\partial X_1^i}}{\frac{\partial X_{13}^i}{\partial X_3^i}} = \frac{\check{L}_1' S_1'}{\check{L}_3' S_3'} = \frac{\tilde{L}_1^{i\prime}}{\tilde{L}_3^{i\prime}}. \end{aligned} \tag{22}$$

[1] J. C. Maxwell was the first who formulated the zeroth law in this form [17]. A representative summary of the history is given here: *http://www.eoht.info/page/Zeroth+law+of+thermodynamics*

This condition can be reduced easily to obtain

$$\frac{\hat{L}_1'\check{L}_2'\check{L}_3'}{\hat{L}_2'\hat{L}_3'\check{L}_1'} = \frac{L_1^{i\prime}\tilde{L}_2^{i\prime}\tilde{L}_3^{i\prime}}{L_2^{i\prime}L_3^{i\prime}\tilde{L}_1^{i\prime}} \tag{23}$$

for $i = 1, ..., n$. The left hand side of the n equalities depends on the body entropies S_1, S_2 and S_3 and the right hand side depends on the ith basic quantities of the bodies X_1^i, X_2^i and X_3^i, respectively. Moreover it is valid for any permuted arrangement of the lowercase body indices 1, 2 and 3. Therefore one concludes that the transitivity of thermal equilibrium can be satisfied only if the

$$\frac{\hat{L}'(S)}{\check{L}'(S)} = const. \quad \text{and} \quad \frac{L^{i\prime}(X^i)}{\tilde{L}^{i\prime}(X^i)} = const. \tag{24}$$

This is a necessary and sufficient condition. Physically sensible composition rules must satisfy two additional simple requirements. The *triviality condition* says, that a composition with zero does not change the value, therefore $L(0) = 0$ is also required. According to the *compatibility condition* for small values of the basic quantities the nonadditive effects are reduced and addition emerges, therefore $L'(0) = 1$. Due to these two conditions, from eq. (24) it follows that the $\tilde{L}^i$ and L^i functions of the basic quantities are identical, as well as the $\hat{L}$ and $\check{L}$ functions of the entropy.

One aspect of the zeroth law – the separability – enforces that composition rules are expressed by formal logarithms. Another aspect of the zeroth law – the transitivity – ensures that $\hat{L}_I$ and L_I^i are characteristics of the thermodynamic body I and only L_{IJ} and $\hat{L}_{IJ}$ depend on the interactions. It is straightforward to see, that zeroth law compatible composition rules are associative if they are *homogeneous*, that is $L_{IJ} = L_I = L_J$.

2.3. First order compositions

A particular non-additive entropy formula, advanced by Tsallis, underlies the following composition rule [9, 8]

$$S_{12}(S_1, S_2) = S_1 + S_2 + aS_1S_2. \tag{25}$$

We have introduced the shorthand notation $\hat{a} = 1 - q$. Additive entropy systems realize $a = 0$, non-additive systems a non-zero value of this parameter. One can get the above formula as a Taylor series expansion of an arbitrary $S_{12}(S_1, S_2)$ up to first order in S_1 and S_2 and requiering the homogeneity of the composition.

According to our previous result a thermodynamic framework requires formal logarithms. The formal logarithm for the above rule is easy to derive from

$$1 + aS_{12} = 1 + aS_1 + aS_2 + a^2S_1S_2 = (1 + aS_1)(1 + aS_2). \tag{26}$$

The product is related to the addition by the logarithm and scaled down to satisfy $\hat{L}'(0) = 1$:

$$\hat{L}(S) = \frac{1}{a}\ln(1 + aS). \tag{27}$$

From the viewpoint of zeroth law compatibility it is strongly advised to consider the additive formal logarithm. In case of the Tsallis entropy,

$$S_T = \frac{1}{a}\left(\sum_i p_i^{1-\hat{a}} - p_i\right), \tag{28}$$

with $\hat{a} = 1 - q$ and the normalization $\sum_i p_i = 1$, its formal logarithm is given by the well known Rényi entropy [24]:

$$S_R = \hat{L}(S_T) = \frac{1}{\hat{a}}\ln(1 + \hat{a}S_T) = \frac{1}{1-q}\ln\sum_i p_i^q. \tag{29}$$

3. Universal Thermostat Independence

The analysis of the previous section shows the conditions of a thermodynamic treatment and also enlightens why the nonadditive composition rule eq. (25) is leading order in a mathematical sense. However, investigating the most researched particular case of nonadditive entropies one can find dozens of different formulas that fulfill either the same, or different but sufficiently simple nonadditive composition rules [12, 25]. They frequently lead to the same equilibrium distributions in a suitable maximum entropy framework. Moreover, accepting one of the suggested formulas one should ask about the origin of the introduced parameters. E.g. how could we determine the $a = 1 - q$ parameter of the Tsallis distribution from microscopic models beyond fitting?

Given a particular microscopic system, such as an ideal gas, a microcanonical treatment assumes that the particles are correlated because the total basic quantities are fixed. A canonical treatment assumes independent distributions with given average basic quantities. The fixed averages are physically provided by suitable reservoirs. A canonical treatment is useful also when the reservoir concept seems purely hypothetic, since its simplicity compared to microcanonical approaches is a great advantage. The Tsallis distribution introduces a particular correlation. It is known that some simple microcanonical distributions are of Tsallis type [26]. If the origin of Tsallis distributions is universal, what is the distinctive property of these correlations?

We seek for a generalized entropy formula first examining the two-body thermodynamics of a subsystem and a reservoir, slightly generalizing the results of [15]. We consider finite reservoirs and the conservation of the basic quantities:

$$X_0^i = X_1^i + X_2^i = const., \quad i = 1, ..., n.$$

We aim at a monotonic function of the Boltzmann – Gibbs entropy $K(S)$ that is maximal in the body-reservoir system and absorbs finite size correlation effects in the Taylor-expansion of the maximum entropy principle. The respective entropy contributions are in general nonadditive and satisfy a composition rule formulated in terms of formal logarithms. We consider homogeneous rules

$$K(S_{12}) = K(S_1) + K(S_2). \tag{30}$$

The maximum entropy principle together with the conservation of the basic quantities reads:

$$K(S(X_1^1, ..., X_1^n)) + K(S(X_0^1 - X_1^1, ..., X_0^n - X_1^n)) = max. \tag{31}$$

The necessary condition of the maximum in case of twice differentiable functions requires vanishing derivatives by the subsystem basic quantities X_1^i:

$$\begin{aligned} Y_1^i &= K'(S(X_1^1, ..., X_1^n))\frac{\partial S}{\partial X_1^i}(X_1^1, ..., X_1^n) = \\ &= K'(S(X_0^1 - X_1^1, ..., X_0^n - X_1^n))\frac{\partial S}{\partial X_1^i}(X_0^1 - X_1^1, ..., X_0^n - X_1^n) = Y_2^i. \end{aligned} \tag{32}$$

Subsystem 1 is much smaller than the reservoir, $X_0^i \gg X_1^i$. This equality in a traditional canonical approach declare the equality of the reservoir and subsystem intensives in the $X_1^i \to 0$ limit. Now, considering effects higher order in X_1^i/X_0^i we request that their leading term vanishes on the right hand side of eq. (32). The Taylor series expansion up to the first order reads as

$$Y_1^i = K'(S_0)\frac{\partial S_0}{\partial X^i} - \left(K''(S_0)\frac{\partial S_0}{\partial X^i}\frac{\partial S_0}{\partial X^j} + K'(S_0)\frac{\partial^2 S_0}{\partial X^i X^j}\right) dX^j + ... \tag{33}$$

where $S_0 = S(X_0^1, ..., X_0^n)$ denotes the function values at the constant basic quantities of the system. It is not obvious that a single function $K(S_0)$ could annul the bracket coefficient of dX^j

in general. One of the reasons is that $\frac{\partial S_0}{\partial X^i}\frac{\partial S_0}{\partial X^j}$ is not invertible. Therefore we restrict ourselves to one additional basic quantity that we denote simply by X and the derivative of the entropy by X is denoted b S'. Then the previous equality reduces to

$$Y = K'(S_0)S_0' - \left(K''(S_0)S_0'^2 + K'(S_0)S_0''\right) dX + ... \tag{34}$$

In this case the vanishing linear term requires for a general $K(S)$ that:

$$\frac{K''(S)}{K'(S)} = -\frac{S''(X)}{S'(X)^2}. \tag{35}$$

Here the right hand side is a function of X, therefore it is a constant solving the above equation for an S dependent K. Therefore

$$\frac{K''(S)}{K'(S)} = a, \tag{36}$$

and the solution of the above equation with the compatibility and triviality conditions $K'(0) = 1$ and $K(0) = 0$ gives

$$K(S) = \frac{e^{aS} - 1}{a}. \tag{37}$$

The derivatives of the $S(X)$ equation of state do have a physical meaning:

$$a = 1/\chi(X_0), \tag{38}$$

where χ characterizes the change of the reservoir entropy per unit change of the intensive $Y = S'$. χ is the *generalized susceptibility*. It is remarkable that function (37) is identical to the inverse of the formal logarithm of the Tsallis additivity composition (27).

Now we generalize the result of the two body analysis to the classical entropy formula $S = -\sum_i p_i \ln p_i$ and recover the Tsallis entropy formula as follows:

$$S_{Tsallis} = \sum_i p_i K(-\ln p_i) = \frac{1}{1-q}\sum_i (p_i^q - p_i), \tag{39}$$

where we have introduced the usual $q = 1 - a$ notation. This can be considered as a modification of the entropy density in the total average, the entropy density representing the limit of the minimal subsystem in an extended state space. According to the previous section the zeroth law requires, that the thermodynamic total entropy of the system must be the Rényi entropy

$$S_{\text{Rényi}} = K^{-1}(S_{\text{Tsallis}}) = \frac{1}{1-q}\ln\sum_i p_i^q. \tag{40}$$

The corresponding maximum entropy principle reads as

$$\frac{1}{1-q}\ln\sum_i p_i^q - \beta\sum_i p_i E_i - \alpha\sum_i p_i = \text{max.}. \tag{41}$$

This is the basis of calculating the equilibrium distribution in a canonical form.

4. Summary

We have derived the requirements of the zeroth law when the basic thermodynamic quantities are nonadditive. We have obtained that then the concept of separate thermodynamic bodies enforces the existence of formal logarithms. The transitivity condition ensure that the formal logarithms represent material properties.

Then we were looking for a universal origin of nonadditive compositions and derived the Tsallis entropy formula. We required that the nonadditive composition cancels linear corrections due to a finite X_0 basic quantity of the reservoir. That requirement is the principle of Universal Thermostat Independence [15]. This derivation explains the particular functional form of the Tsallis and Rényi formulas as generalized entropy expressions.

Acknowledgments

The work was supported by the Hungarian National Science Research Fund OTKA NK778816,NK106119,H07/C 74164,K68108, K81161, K104260, NIH TET-10-2011-0061 and ZA-15/2009. GGB also thanks the János Bolyai Research Scholarship of the HAS.

[1] Biró T S, Ürmössy K and Schram Z 2010 *Journal of Physics G* **37** 094027
[2] Barnaföldi G G, Ürmössy K and Biró T S 2011 *Journal of Physics: Conference Series* **270** 012008
[3] Ürmössy K, Barnaföldi G and Biró T S 2011 *Physics Letters B* **701** 111–116
[4] CMS collaboration 2010 *Journal of High Energy Physics* **1002** 41
[5] Abe S 2002 *Physical Review E* **66** 046134
[6] Biro T S 2008 *EPL* **84** 56003
[7] Biró T S and Ván P 2011 *Physical Review E* **83** 061147
[8] Tsallis C 1988 *Journal of Statistical Physics* **52** 479–487
[9] Tsallis C 2009 *Introduction to Nonextensive Statistical Mechanics* (Springer)
[10] Nauenberg M 2003 *Physical Review E* **67** 036114
[11] Arndt C 2001 *Information Measures (Information and its description in science and engineering)* (Berlin-Heidelberg-New York: Springer)
[12] Taneja I 20019 *Generalized Information Measures and Their Applications* (SpringerOnline book, http://www.mtm.ufsc.br/ taneja/book/book.html)
[13] Kaniadakis G 2001 *Physica A* **296** 405–425
[14] Landsberg P T and Vedral V 1998 *Physics Letters A* **247** 211–
[15] Biró T S, Barnaföldi G G and Ván P 2012 *Preprint* ArXiv:0531803
[16] Mackey M C 1992 *Time's Arrow: The Origins of Thermodynamic Behaviour* (Springer)
[17] Maxwell J C 1902 *Theory of Heat* 10th ed (London: Longmans, Green and Co.)
[18] Fowler R and Guggenheim E 1939 *Statistical Thermodynamics: A Version of Statistical Mechanics for Students of Physics and Chemistry* (Cambridge University Press (printed by W. Lewis))
[19] Abe S 2001 *Physica A* **300** 417–423
[20] Abe S 2001 *Physical Review E* **63** 061105
[21] Johal R S 2003 *Physics Letters A* **318** 48–52
[22] Johal R S 2004 *Physics Letters A* **332** 345–349
[23] Scarfone A 2010 *Physics Letters A* **374** 2701–2706
[24] Rényi A 1961 On measures of entropy and information *Proceedings of the 4th Berkeley Symposium on Mathematical Statistics and Probability, 1960* vol I ed Neyman J (Berkeley: University of California Press) pp 547–561
[25] Zaripov R 2010 *Principles of nonextensive statistical mechanics and geometry of measures of the disorder and the order* (Kazan: Kazan A. N. Tupolev State Technical University Press) (*in Russian*)
[26] Naudts J 2011 *Generalized Thermostatistics* (Springer)

Conceptual difficulties with the q-averages in non-extensive statistical mechanics

Sumiyoshi Abe

Department of Physical Engineering, Mie University, Mie 514-8507, Japan

suabe@sf6.so-net.ne.jp

Abstract. The q-average formalism of nonextensive statistical mechanics proposed in the literature is critically examined by considerations of several pedagogical examples. It is shown that there exist a number of difficulties with the concept of q-averages.

1. Introduction

In the last 1.5 decades, a lot of works have been done on so-called nonextensive statistical mechanics [1], which had been expected to be a generalization of Boltzmann-Gibbs statistical mechanics for describing certain complex systems. This theory is primarily based on a kind of the maximum entropy method, where the form of entropy and the definition of averages are changed respectively as follows [1,2]:

$$S_q[p] = \frac{1}{1-q}\left[\sum_{n=1}^{W} (p_n)^q - 1\right], \tag{1}$$

$$\langle Q \rangle_q = \sum_{n=1}^{W} Q_n \, P_n^{(q)}, \tag{2}$$

where $\{P_n^{(q)}\}_{n=1,2,\ldots,W}$ is the *escort distribution* defined in terms of the original distribution $\{p_n\}_{n=1,2,\ldots,W}$ as follows:

$$P_n^{(q)} = \frac{(p_n)^q}{\sum_{m=1}^{W} (p_m)^q}. \tag{3}$$

Here, q is a positive constant, W the number of accessible states and $\{Q_n\}_{n=1,2,\ldots,W}$ a certain quantity of interest. Equation (2) is termed the q-average. S_q in equation (1) is referred to as the q-entropy. Together with the normalization constraint, $\sum_{n=1}^{W} p_n = 1$, the maximum-S_q-distribution is found to be given as follows:

$$p_n = \frac{1}{Z_q(\beta)} e_q\left(-(\beta / c_q)(Q_n - \tilde{Q}_q)\right), \tag{4}$$

where β is a Lagrange multiplier for the constraint on the q-average in equation (2), $e_q(x) = [1+(1-q)x]_+^{1/(1-q)}$ with the notation $[a]_+ = \max\{0, a\}$,

$Z_q(\beta) = \sum_{n=1}^{W} e_q\left(-(\beta / c_q)(Q_n - \tilde{Q}_q)\right)$, and $c_q = \sum_{n=1}^{W} (p_n)^q$. Equation (4) is termed the q-exponential distribution. $\tilde{Q}_q$ has to be calculated in terms of p_n in equation (4) itself in a self-referential manner. A basic requirement is that this scheme tends to the ordinary maximum entropy method [3] in the limit $q \to 1$. In such a limit, equations (1) and (2) in fact converge to the Boltzmann-Gibbs-Shannon entropy, $S[p] = -\sum_{n=1}^{W} p_n \ln p_n$ (with the Boltzmann constant being set equal to unity), and the ordinary average, $\langle Q \rangle = \sum_{n=1}^{W} Q_n p_n$, respectively. Accordingly, equation (4) tends to the exponential distribution familiar in equilibrium statistical mechanics.

There is a naive question why q in equations (2) and (3) is the same as that in Eq. (1). In other words, why not

$$\langle Q \rangle_{f(q)} = \frac{\sum_{n=1}^{W} Q_n (p_n)^{f(q)}}{\sum_{m=1}^{W} (p_m)^{f(q)}} \tag{5}$$

with any $f(q)$ satisfying $f(q) \to 1$ $(q \to 1)$?

In this article, we focus our attention on the notion of q-averages. We show that it does suffer from a series of conceptual difficulties and therefore cannot be employed in nonextensive statistical mechanics.

2. Lottery

This example can be seen as the simplest/trivial one, which shows that it is unreasonable to employ the q-average for a q-exponential distribution.

Suppose you are going to sell 869 lots. And among these 869, there are 100 first prizes with prize money 1000 Yen each, 144 second prizes with 500 Yen, 225 third prizes with 100 Yen and finally 400 fourth prizes with 0 Yen. The total amount of money you have to prepare is $0 \times 400 + 100 \times 225 + 500 \times 144 + 1000 \times 100 = 197500$ Yen. Therefore, if you sell each piece of lot more expensive than $197500 / 869 = 227.272727...$ Yen, then your pocket will be in safety.

The corresponding probability distribution is given by

$$p_4 = N/9 \text{ (first prize)}, \; p_3 = 4N/25 \text{ (second prize)},$$
$$p_2 = N/4 \text{ (third prize)}, \; p_1 = 4N/9 \text{ (fourth prize)}.$$

where $N = 900 / 869$ is the normalization constant. You see that using this probability distribution you obtain the following normal average number of the prize money, $\langle M \rangle$:

$$\langle M \rangle = 0 \times p_1 + 100 \times p_2 + 500 \times p_3 + 1000 \times p_4,$$

which is precisely your previously obtained average value, 227.272727... Yen.

Now, note that the above probability distribution has the form

$$p_n = \frac{N}{(1 + n/2)^2} \qquad (n = 1, 2, 3, 4),$$

which is simply rewritten in the form of the q-exponential distribution

$$p_n = N e_q(-n)$$

with $q = 3/2$. If you use the q-average with this q-exponential distribution as in [2], then you have

$$\langle M \rangle_{q=3/2} = \frac{1}{\sum_{n=1}^{4} (p_n)^{3/2}} \left[0 \times \left(\frac{400}{869} \right)^{3/2} + 100 \times \left(\frac{225}{869} \right)^{3/2} \right.$$

$$\left. +500 \times \left(\frac{144}{869} \right)^{3/2} + 1000 \times \left(\frac{100}{869} \right)^{3/2} \right],$$

which is $\langle M \rangle_{3/2} = 156.1015387 \ldots$ Yen.

A question is if you sell the lots with reference to $\langle M \rangle_{3/2} = 156.1015387 \ldots$. Yen. If you do so, then your pocket will seriously be damaged.

So far, you considered only the average. There is another important quantity, which is the variance or its square root, i.e., the standard deviation. Using the normal average, you obtain for the standard deviation, $\sigma \equiv \sqrt{\langle M^2 \rangle - \langle M \rangle^2} = 327.7773887 \ldots$ Yen, whereas using the q-average, $\sigma_{q=3/2} \equiv \sqrt{\langle M^2 \rangle_{3/2} - \langle M \rangle_{3/2}^{\ 2}} = 282.0710854 \ldots$ Yen. Clearly, the latter is smaller than the former. This is natural, since it is nothing but a general feature of equation (5): the larger the value of $f(q)\,(>1)$ is, the smaller the generalized standard deviation, $\sigma_{f(q)}$, is. That is, *the q-average makes fluctuation small.*

The above toy example indicates that the use of the q-average for the q-exponential distribution is in doubt.

3. Experiment

Imagine an experimentalist, who does not know/need nonextensive statistical mechanics. He/She does measurements of a certain physical quantity of a statistical-mechanical system and obtains a probability distribution of a positive random variable, x, that turns out to be well fitted by the Zipf-Mandelbrot distribution

$$f(x) = \frac{A}{(1+\lambda x)^{\alpha}} \tag{6}$$

over a finite range, say $[a,b]$. Here, α and λ are observed positive constants, and A is a normalization factor. He/She may naively calculate the average value of a physical quantity, $Q(x)$, using the normal definition

$$\langle Q \rangle = \int_a^b dx\, Q(x)\, f(x) . \tag{7}$$

There is nothing wrong here.

On the other hand, the distribution in equation (6) can be rewritten as a q-exponential distribution with the following identifications:

$$\alpha = \frac{1}{q-1}, \qquad \lambda = (q-1)\beta . \tag{8}$$

In fact, with these identifications, equation (6) is expressed as follows:

$$f(x) = Ae_q(-\beta x). \tag{9}$$

Then, according to [2], one has to use, instead of equation (7), the following q-average:

$$\langle Q \rangle_q = \frac{\int_a^b dx\, Q(x)\, [f(x)]^q}{\int_a^b dx'\, [f(x')]^q}. \tag{10}$$

The experimentalist will never understand why equation (7) cannot be employed, and the expression

$$\langle Q \rangle_q = \frac{\int_a^b dx\, Q(x)\, [f(x)]^{1+1/\alpha}}{\int_a^b dx'\, [f(x')]^{1+1/\alpha}} \tag{11}$$

has to be used.

Furthermore, suppose that the experimentalist does another experiment and obtains this time the following distribution of a positive random variable, x:

$$F(x) = \frac{Be^{-x_0/x}}{(1+\kappa x)^\gamma}, \tag{12}$$

where x_0, κ, and γ are observed positive constants, and B is a normalization factor. Equation (12) asymptotically tends to the q-exponential distribution for large x but contains the extra exponential factor, here. Should the q-average with $q = 1 + 1/\gamma$ be used following [2]? One may see no reasonable answers to this question for supporting the q-average.

4. Engineering

Suppose that an engineer, who does not know anything about nonextensive statistical mechanics, is analyzing diffusion in a structured medium. He/She is trying to model it by using the following nonlinear Fokker-Planck equation for a *physical* distribution $p(x,t)$ of diffusing particles:

$$\frac{\partial p(x,t)}{\partial t} = -\frac{\partial}{\partial x}[k\, x\, p(x,t)] + \frac{D}{2}\frac{\partial^2 p^\nu(x,t)}{\partial x^2}, \tag{13}$$

where k and D are positive constant and a generalized diffusion coefficient, respectively. He/She may find an analytic solution of this equation, which has the form

$$p(x,t) = \frac{1}{Z(t)}\left[1-(\nu-1)\, B(t)\, (x-\xi(t))^2\right]_+^{1/(\nu-1)}, \tag{14}$$

where $Z(t)$, $B(t)$, and $\xi(t)$ are somewhat involved. He/She is interested especially in the case when $\nu < 1$, because equation (14) is an asymptotically power-law distribution in this case. Then, he/she may try to evaluate the shifted variance as follows:

$$\left\langle (x-\xi)^2 \right\rangle(t) = \int_{-\infty}^{\infty} dx\,(x-\xi)^2\, p(x,t). \tag{15}$$

The engineer will soon recognize that the variance in equation (15) is finite if and only if $\nu > 1/3$. But, after this recognition, he/she may be informed that the solution in equation (14) has already been obtained in [4], in which the following identification is made:

$$\nu = 2 - q. \tag{16}$$

Then, the solution is written as

$$p(x,t) = \frac{1}{Z(t)} e_q\left(-B(t)\left(x-\xi(t)\right)^2\right). \tag{17}$$

Now, imagine that the engineer may look at the literature [1,2] and find that, in the field of nonextensive statistical mechanics, the distribution in equation (17) is termed the q-Gaussian distribution, which may/might maximize the q-entropy

$$S_q[p] = \frac{1}{1-q}\left[\int_{-\infty}^{\infty} dx\,[p(x,t)]^q - 1\right] \tag{18}$$

under the constraints on the normalization condition and the shifted q-variance

$$\left\langle (x-\xi)^2 \right\rangle_q(t) = \frac{\int_{-\infty}^{\infty} dx\,(x-\xi)^2 [p(x,t)]^q}{\int_{-\infty}^{\infty} dx'\,[p(x',t)]^q}, \tag{19}$$

where the time-dependence is assumed to be mild.

Of course, the engineer understands that the range of ν for finite $\left\langle (x-\xi)^2 \right\rangle_q$ becomes widened from $1/3 < \nu < 1$ to $-1 < \nu < 1$, if equation (19) is used. However, he/she will never understand why he/she cannot use equation (15) but must calculate

$$\frac{\int_{-\infty}^{\infty} dx\,(x-\xi)^2 [p(x,t)]^{2-\nu}}{\int_{-\infty}^{\infty} dx'\,[p(x',t)]^{2-\nu}}$$

from his/her data, $p(x,t)$, because basically his/her work is perfectly fine with the nonlinear Fokker-Planck equation (13) and does not need any entropic approach.

5. Quantum theory

Consider an isolated quantum system. Its state is represented by $|\psi\rangle$, which can be expanded in terms of the complete set of eigenstates $\{|u_n\rangle\}_n$ of some observable as $|\psi\rangle = \sum_n a_n |u_n\rangle$, where the expansion coefficients $\{a_n\}_n$ satisfy the normalization condition: $\sum_n a_n^* a_n = 1$. A state of the system is described by the expansion coefficients, largely. For example, consider two states, $|\psi\rangle$ and

$|\psi'\rangle$. They have the following expansions: $|\psi\rangle = \sum_n a_n |u_n\rangle$ and $|\psi'\rangle = \sum_n a'_n |u_n\rangle$. The difference between $|\psi\rangle$ and $|\psi'\rangle$ is reflected in the difference between the expansion coefficients.

The quantum-mechanical average of some observable, Q, in the state $|\psi\rangle$ is given by $\langle Q\rangle = \sum_{m,n} a_n^* a_m \langle u_n|Q|u_m\rangle$. This bilinear structure with respect to the expansion coefficients is radical for the probabilistic interpretation of quantum theory, i.e., the Born rule. One may also consider a pure-state density matrix, $\rho_0 = |\psi\rangle\langle\psi| = \sum_{m,n} a_n^* a_m |u_m\rangle\langle u_n|$. Using this, one has $\langle Q\rangle = \mathrm{Tr}(Q\rho_0)$.

Now, a crucial question is the following. Is the q-average, $\langle Q\rangle_q = \mathrm{Tr}(Q\rho^q)/\,\mathrm{Tr}(\rho^q)$, consistent with quantum theory?

The point is that in the probabilistic interpretation of quantum theory the bilinear form $a_n^* a_m$ is essential as mentioned above, and this bilinear structure must be respected in any quantum-mechanical calculation. Let us construct a mixed state using the most general linear race-preserving positive quantum operation

$$\rho_0 = |\psi\rangle\langle\psi| \to \rho = \Phi(\rho_0) = \sum_k V_k \rho_0 V_k^+ = \sum_{m,n} a_n^* a_m \sum_k V_k |u_m\rangle\langle u_n| V_k^+ , \tag{20}$$

where the condition, $\sum_k V_k^+ V_k = I$, is imposed. The quantum-mechanical average is still in the bilinear form: $\langle Q\rangle = \mathrm{Tr}(Q\rho_0) \to \mathrm{Tr}(Q\rho) = \sum_{m,n} a_n^* a_m \langle u_n|\Phi(Q)|u_m\rangle$.

On the other hand, $\langle Q\rangle_q$ is obviously not in a bilinear form unless $q=1$ and therefore is inconsistent with the Born rule in quantum mechanics.

6. Comment on the escort-distribution representation

The above stories certainly cause a serious problem for the researchers working on nonextensive statistical mechanics, since they want to establish a statistical-mechanical bridge between q-exponential distributions and the maximum q-entropy principle.

If the q-average is still to be employed, then the one and only way of resolving the difficulties may be to identify the observed q-exponential distribution with the escort distribution.

However, the situation turns out to remain hopeless. If an observed distribution is the escort distribution, $P_n^{(q)} = (p_n)^q / \sum_{m=1}^{W} (p_m)^q$, then the value of the q-entropy should be calculated using it. The result for this is

$$S_q = \frac{1}{1-q}\left\{\left[\sum_{n=1}^{W} \left(P_n^{(q)}\right)^{1/q}\right]^{-q} - 1\right\}. \tag{21}$$

Unfortunately, this is not an entropic quantity any more, since it fails to be concave with respect to $\left\{P_n^{(q)}\right\}_{n=1,2,\ldots,W}$ [5].

7. Farewell to the *q*-averages

These considerations lead to the conclusion that the *q*-average formalism suffers from fundamental conceptual difficulties and is not acceptable. One need employ the normal definition for averages and identify an observed physical distribution with the original distribution, $\{p_n\}_{n=1,2,\ldots,W}$, not its associated escort distribution.

This conclusion is actually supported strongly by another recent physical discussion based on the generalized *H*-theorem [6].

8. Concluding remarks

We have shown the concept of the *q*-averages has fundamental difficulties. Once it has been discussed [7] that the Shore-Johnson theorem [8-10] leads to necessity of using the *q*-averages in nonextensive statistical mechanics. A point, which became clear today, is that the requirement of "subset independence" in the theorem is not realized in nonextensive statistical mechanics. Therefore, the basic premise in [7] is actually not fulfilled. Thus, the *q*-average formalism has lost its basis.

In addition, it has recently been shown [11-13] that the *q*-average is unstable (i.e., not uniformly continuous). In other words, its value can drastically changes even for very small changes of an observed distribution. Thus, it fails to possess experimental robustness

Acknowledgement

The author was supported in part by a Grant-in-Aid for Scientific Research from the Japan Society for the Promotion of Science.

References

[1] Tsallis C 2010 *Introduction to Nonextensive Statistical Mechanics: Approaching a Complex World* (New York: Springer)
[2] Tsallis C, Mendes R S and Plastino A R 1998 *Physica* A **261** 534
[3] Rosenkrantz R D (ed.) 1989 *E. T. Jaynes: Papers on Probability, Statistics and Statistical Physics* (Dortrecht: Kluwer)
[4] Tsallis C and Bukman D J 1996 *Phys. Rev.* E **54** R2197
[5] Abe S 2000 *Phys. Lett.* A **275** 250
[6] Abe S 2009 *Phys. Rev.* E **79** 041116
[7] Abe S and Bagci G B 2005 *Phys. Rev.* E **71** 016139
[8] Shore J E and Johnson R W 1980 *IEEE Trans. Inf. Theory* **IT-26** 26
[9] Shore J E and Johnson R W 1981 *IEEE Trans. Inf. Theory* **IT-27** 472
[10] Shore J E and Johnson R W 1983 *IEEE Trans. Inf. Theory* **IT-29** 942
[11] Abe S 2008 *EPL* **84**, 60006
[12] Abe S 2009 *J. Stat. Mech.* P07027
[13] El Kaabouchi A and Abe S 2011 *J. Appl. Math.* 804836

Journal of Physics: Conference Series **394** (2012) 012004 doi:10.1088/1742-6596/394/1/012004

Superfluid hydrodynamic in fractal dimension space

D A Tayurskii and Yu V Lysogorskiy

Institute of Physics, Kazan Federal University, 18 Kremlevskaya st., Kazan, 420018, Russia

E-mail: dtayursk@gmail.com

Abstract. The complex behavior of such quantum fluids like liquid ^{4}He and liquid ^{3}He in nanoporous media is determined by influence of randomly distributed geometrical confinement as well as by significant contribution from the surface atoms. In the present paper Fractional Schrodinger equation has been used for deriving two-fluid hydrodynamical equations for describing the motion of superfluid helium in the fractal dimension space. Nonlinear equations for oscillations of pressure and temperature are obtained and coupling of pressure and temperature oscillations is observed. Moreover coupling should disappear at very low temperatures which provide an experimental test for this theory.

1. Introduction

Liquid of stable isotopes of helium ^{4}He and ^{3}He at very low temperature belong to the class of quantum fluids with strong correlations between atoms. The first one represents a Bose-system and shows superfluid transition at λ-point $T_\lambda = 2.17$ K (below this temperature it is called He-II). In recent years it has been recognized that the quantum fluids in the confined geometry at nanoscale length can be considered as a new state of quantum matter due to close values between characteristic lengths for these quantum liquids and the size of geometrical confinement and significant contribution from the surface atoms.

So the question of influence of geometrical factors (confinement dimension, dimension of nanopores space, etc) has been emerged[1–6]. And one has to apply new physics to describe such systems with taking into account their complex nature. For example, last two years the attempts to develop the fractionalized two-fluid hydrodynamics for nanoporous media with fractal dimensions have been made[7, 8].

In the present paper Fractional Schrodinger equation has been used for deriving two-fluid hydrodynamical equations for describing the motion of superfluid helium in the fractal dimension space (like aerogel) and nonlinear equations for oscillations of pressure and temperature are obtained and coupling of pressure and temperature oscillations is observed.

2. Two-fluid hydrodynamical model

The usual two-fluid hydrodynamic model [9], [10] of superfluid helium ^{4}He is described by the system of four differential equations (1), (2), (3), (4) which represent continuity equation, entropy conservation law and Eiler equations for superfluid and normal components respectively

$$\frac{\partial \rho}{\partial t} + \nabla \mathbf{j} = 0 \tag{1}$$

$$\frac{\partial \rho\sigma}{\partial t} + \nabla(\rho\sigma \mathbf{v}_n) = 0 \tag{2}$$

$$\rho_s \frac{\partial \mathbf{v}_s}{\partial t} = -\frac{\rho_s}{\rho} \nabla p + \rho_s \sigma \nabla T \quad (3)$$

$$\rho_n \frac{\partial \mathbf{v}_n}{\partial t} = -\frac{\rho_n}{\rho} \nabla p - \rho_s \sigma \nabla T, \quad (4)$$

where ρ – He-II density, ρ_s, ρ_n – superfluid and normal component density, $\mathbf{v}_s, \mathbf{v}_n$ – superfluid and normal component velocity, $\mathbf{j} = \rho_s \mathbf{v}_s + \rho_n \mathbf{v}_n$ – He-II flow density. Interaction between the fluid components, forces of viscous friction and energy dissipation are neglected because of the small velocities of both components. This model provide two type of oscillations - oscillations of pressure (first sound, both component move in phase) and oscillations of temperature (second sound, components move in antiphase). Neglecting the anomalously small thermal expansion coefficient of He-II, oscillations of pressure and temperature in bulk sample of superfluid helium are independent. But in superfluid helium in aerogel there is experimentally proven coupling of these two type of oscillations[11].

3. Fractional quantum mechanics

For taking into account complex nature of internal geometry of confinement (aerogel) one can propose to use formalism of fractional quantum mechanics. One of the point of view on quantum mechanical motion is the so-called Feynman formalism of quantum mechanics, which is focused on the concept of trajectory, and the particle can move along any possible trajectories. To move a particle from point A to point B one have to take into account the contribution from all possible trajectories with the corresponding weight (complex factor). Possible trajectories resemble Brownian trajectory of a free particle and have a fractal dimension of $\alpha = 2$.

De Broglie thermal wavelength for the helium atom inside the aerogel at temperatures of about 1 K is about 10 Å, which is in proper relation with the characteristic length scale of the fractal structures formed aerogel. Thus, some quantum-mechanical trajectory of the helium atom will be forbidden due to the influence of the structure of aerogel. Realized trajectory will resemble the motion of a Brownian particle in a porous medium, where the mean square displacement depends on time as $< x^2 > \propto t^\alpha$. This phenomenon is called subdiffusion. To describe this type of diffusion an equation with fractional Riesz derivative is used. Probability density function for such case is written in terms of Levy function, which is a generalization of the Gaussian distribution.

Laskin proposed to generalize the Feynman's path integrals to an arbitrary fractal dimension of trajectories α [12]. From this type of path integrals one can obtain fractional Schrödinger equation which is written as

$$i\hbar \frac{\partial \psi(\mathbf{r}, t)}{\partial t} = D_\alpha \left(\hbar \nabla\right)^\alpha \psi(\mathbf{r}, t) + V(\mathbf{r}, t)\psi(\mathbf{r}, t). \quad (5)$$

Here Riesz fractional derivative is introduced as

$$\left(\hbar \nabla\right)^\alpha \psi(\mathbf{r}, t) = -\frac{1}{(2\pi\hbar)^3} \int d^3 p e^{ipr/\hbar} \left|\mathbf{p}\right|^\alpha \varphi(\mathbf{p}, t), \quad (6)$$

$$\varphi(\mathbf{p}, t) = \int d\mathbf{r} e^{-ipr/\hbar} \psi(\mathbf{r}, t), \quad (7)$$

and D_α is the generalized coefficient, the physical dimension of which is $[D_\alpha] = \mathrm{erg}^{1-\alpha} \cdot \mathrm{cm}^\alpha \cdot \mathrm{sec}^{-\alpha}$. Thus fractional Hamiltonian can be written in the form

$$\hat{H}_\alpha = D_\alpha \left(\hbar \nabla\right)^\alpha + V\left(\mathbf{r}, t\right) = D_\alpha \left|\hat{\mathbf{p}}\right|^\alpha + V\left(\mathbf{r}, t\right). \quad (8)$$

Hamiltonian (8) is hermitian operator and provide probability conservation law. Also it obeys parity conservation law, so one divide particles at the fermoins and bosons, as in usual case.

This type of Hamiltonian (8) already has been used to describe the specific heat of noncrystalline solids (glasses) associated with the unusual structure of these materials[13].

4. Galilean noninvariance

On other hand, using of nonlinearity in Hamiltonian results in breaking some other symmetries in equations of dynamics. One of them is Galilean invariance. In usual nonfractional case, transition from one inertial system to another will conserve general view of dynamical equations. If we consider Schrödinger equation in two inertial reference frame K (nonprimed variables) and K' (primed variables), which move relative to each other with velocity V, then by substitution[14]

$$\Psi(x,t) = \varphi(x',t)e^{i\left(\frac{mV}{\hbar}x' + \frac{mV^2}{2m\hbar}t\right)}$$

one can obtain Schrödinger equation in another inertial reference frame

$$i\hbar\frac{\partial\Psi}{\partial t} = -\frac{\hbar^2}{2m}\nabla^2\Psi \Rightarrow i\hbar\frac{\partial\varphi}{\partial t} = -\frac{\hbar^2}{2m}\nabla'^2\varphi,$$

On other hand, by making similar substitution in the form

$$\Psi(x,t) = \varphi(x',t)e^{i\frac{mVx'}{\hbar} + i\frac{D_\alpha|p|^\alpha t}{\hbar}},$$

for fractional Schrödinger equation, one can obtain dynamical equation in new inertial reference frame

$$i\hbar\frac{\partial\varphi}{\partial t} - \varphi D_\alpha|mV|^\alpha - i\hbar V\frac{\partial\varphi}{\partial x} + \varphi mV^2 = \frac{D_\alpha}{2\pi\hbar}\int_{-\infty}^{\infty} dp e^{ipx/\hbar}\tilde{\varphi}(p)|p+mV|^\alpha. \tag{9}$$

The only possibility to reduce eqn. (9) to the form (5) is to set $V = 0$, i.e. do not change the intertial reference frame. As a consequence there is some special reference frame where, fractional Schrödinger equation has form (5). One can identify this reference frame with some reference frame where, for example, nanoporous media is in a rest.

5. Fractional two-fluid hydrodynamical model

By using Heisenberg representation one can obtaian dynamical equations for operators $\hat{\mathbf{r}}$ and $\hat{\mathbf{v}} = \frac{d\hat{\mathbf{r}}}{dt}$ by means of general dynamical equation for arbitrary operator $\hat{O}$

$$\frac{d\hat{O}(t)}{dt} = \frac{\partial\hat{O}(t)}{\partial t} + \frac{i}{\hbar}\left[\hat{H}_\alpha, \hat{O}\right]. \tag{10}$$

Then for velocity and acceleration we will obtain next equations

$$\hat{\mathbf{v}} = \alpha D_\alpha \left|\hat{\mathbf{p}}\right|^{\alpha-2}\hat{\mathbf{p}}, \tag{11}$$

$$\frac{d\hat{\mathbf{v}}}{dt} = -\frac{i}{\hbar}\alpha D_\alpha \sum_{l=1}^{\infty}\frac{(-i\hbar)^l}{l!}\nabla^l V(\hat{\mathbf{r}})\frac{(\alpha-1)!}{(\alpha-l-1)!}\hat{\mathbf{p}}^{\alpha-1-l}. \tag{12}$$

Suppose that velocity is more general quantity than momentum [15], one can rewrite (12) in the form

$$\frac{d\hat{\mathbf{v}}}{dt} = -\alpha D_\alpha\frac{(\alpha-1)}{(\alpha D_\alpha)^{(\alpha-2)/(\alpha-1)}}\left|\hat{\mathbf{v}}\right|^{\frac{\alpha-2}{\alpha-1}}\nabla V(\hat{\mathbf{r}}), \tag{13}$$

where we left only first spatial derivation from potential V. From eqn. (5) one can obtain mass conservation law in the form

$$\frac{\partial\rho(\mathbf{r},t)}{\partial t} + \nabla\mathbf{J}(\mathbf{r},t) + K(\mathbf{r},t) = 0, \tag{14}$$

where $\rho = \Psi^*\Psi$ is density, $\mathbf{J} = \frac{1}{\alpha}\left(\Psi^*\hat{\mathbf{v}}\Psi + \Psi\hat{\mathbf{v}}\Psi^*\right)$ is density flow and

$$K = \frac{i}{\alpha\hbar}\left(\hat{\mathbf{p}}\Psi\hat{\mathbf{v}}\Psi^* - \hat{\mathbf{p}}\Psi^*\hat{\mathbf{v}}\Psi\right) \tag{15}$$

is new term, which represent additional sources in fractal space. If we decompose wave function on the basis of plane waves $\Psi = \sum_{\mathbf{p}} C_{\mathbf{p}} \exp(i\mathbf{pr}/\hbar)$ then substitution it in the eqn. (15) results in

$$K = \frac{iD_\alpha}{\hbar}\sum_{\mathbf{p_1}\neq\mathbf{p_2}} \mathbf{p_1 p_2} C^*_{p_1} C_{p_2} e^{\frac{(\mathbf{p_2}-\mathbf{p_1})\mathbf{r}}{\hbar}}\left(|p_2|^{\alpha-2} - |p_1|^{\alpha-2}\right).$$

If we assume that density of superfluid helium is almost homogeneous, i.e. we have only set of plane waves with close values of $\mathbf{p}$ in wave function and helium atoms is strongly delocalized, then one can suppose $K \approx 0$. In that case in fractional two-fluid hydrodynamical model one can left continuity equations in the form (1) and (2).

Instead of constant D_α one can introduce new constant with dimension of velocity as $\alpha D_\alpha = v_0^{2-\alpha} m^{1-\alpha}$. Then dynamical equation (13) for the superfluid component with taking into account thermodynamical relations [9] can be rewritten as

$$\rho_s \frac{d\mathbf{v}_s}{dt} = (\alpha-1)\left|\frac{\mathbf{v}_s}{v_0}\right|^{\frac{\alpha-2}{\alpha-1}}\left(-\frac{\rho_s}{\rho}\nabla p + \rho_s\sigma\nabla T\right). \tag{16}$$

Also we can suppose the similar form for dynamical equation of density flow and write

$$\frac{d\mathbf{j}}{dt} = -(\alpha-1) v_0^{\frac{2-\alpha}{\alpha-1}}\left|\frac{\mathbf{j}}{\rho}\right|^{\frac{\alpha-2}{\alpha-1}}\nabla p. \tag{17}$$

Thus equations (1), (2), (16) and (17) form *fractional two-fluid hydrodynamical model* of superfluid helium in nanoporous space. These equations set result in two oscillation equations

$$\frac{\partial^2 p}{\partial t^2} = (\alpha-1)u_1^2\nabla\left(\left|\frac{\mathbf{v}}{v_0}\right|^{\frac{\alpha-2}{\alpha-1}}\nabla p\right), \tag{18}$$

$$\frac{\partial^2 T}{\partial t^2} = (\alpha-1)u_2^2\nabla\left(\left|\frac{v_s}{v_0}\right|^{\frac{\alpha-2}{\alpha-1}}\nabla T\right) - (\alpha-1)\frac{u_2^2}{\rho\sigma}\nabla\left(\left(\left|\frac{v_s}{v_0}\right|^{\frac{\alpha-2}{\alpha-1}} - \left|\frac{v}{v_0}\right|^{\frac{\alpha-2}{\alpha-1}}\right)\nabla p\right), \tag{19}$$

where u_1 and u_2 is first and second sound velocity for bulk superfluid helium.

Let us consider approximate solution of equations (18) and (19). By using "weak fractality" approximation $\alpha - 2 \ll 1$ in low temperature region $T < 0.5$K when $\rho_n/\rho_s \ll 1$ and $v_n/v_s \ll 1$ one can rewrite eqn.(19) as

$$\frac{\partial^2 T}{\partial t^2} = (\alpha-1)u_2^2\left(1 + \gamma\ln\left|\frac{v_s}{v_0}\right|\right)\nabla^2 T + (\alpha-2)u_2^2\frac{\rho_n}{\rho\sigma\rho_s}\left(\frac{v_n}{v_s} - 1\right)\nabla^2 p, \tag{20}$$

where $\gamma = (\alpha-2)/(\alpha-1) \ll 1$. We will find solutions in the form

$$p = p_0 + p' e^{i(ukt - kr)},$$
$$T = T_0 + T' e^{i(ukt - kr)}$$

and suppose that velocity has dependence $v = v^0 \exp(i(ukt - kr))$. As a result one can obtain two type of oscillations: temperature oscillations with sound speed $u \approx u_2\left(1 + (\alpha - 2)(1 + \frac{1}{2}\ln\left|\frac{v_s^0}{v_0}\right|)\right)$ and pressure-temperature oscillations with sound speed $u \approx u_1\left(1 + \frac{\alpha-2}{2}(1 + \ln\left|\frac{v^0}{v_0}\right|)\right)$. The latter one has relation between pressure and temperature amplitude $T' = \beta p'$, where coupling coefficient has form

$$\beta = ((\alpha - 2)\frac{u_2^2 \rho_n}{\rho \sigma \rho_s}\left(\frac{v_n^0}{v_s^0} - 1\right))/(u_2^2 - u_1^2).$$

At low temperatures, when $u_1^2 \approx \text{const}$, $u_2^2 \approx \text{const}$ and $\sigma \propto T^n$, this coupling coefficient should be linear on temperature $\beta \propto T$.

6. Conclusion

It was proposed that for the microscopical description of superfluid in nanoporous media with complex fractal structure one can use fractional Schrödinger equation. But it is necessary to keep in mind that the fractal geometry of nanoporous media leads to the Galilean noninvariance of this equation and as a consequence one needs to choose the special frame of reference where, for example, nanoporous media is in a rest. One can interpret dynamical equations of fractional quantum operator in Heisnberg representations as a classical dynamical equations in fractal media and generalize them to obtain fractional hydrodynamical set of equations. From this two-fluid fractional hydrodynamical model one can obtain equations of pressure-temperature and pure temperature oscillation. It was shown that the pressure-temperature coupling constant has linear dependence on temparature at low temperature region which provide us with possibility of experimental proof of given model.

Acknowledgments

This work is supported in part by the Ministry of Education and Science of Russian Federation.

References

[1] Naletov V V, Tagirov M S and Tayurskii D A 2002 *Low Temp. Phys.* **28**(5) 431-448
[2] Minnillin A R and Tayurskii D A 2000 *JETP Letters* **72**(12) 616-620
[3] Wong G, Crowell P and Cho H *et al.* 1990 *Phys. Rev. Lett.*, **65**(19) 2410-2413
[4] Matsumoto K, Tsuboya H and Yoshino K *et al.* 2007 *J. of Low Temp. Phys.* **148**(5-6) 615-620
[5] Vasquez C and Paredes R 2006 *Cond. Matt. Phys.* **9**(2) 305-317
[6] Azuah R T, Glyde H R and Scherm R *et al.* 2003 *J. of Low Temp. Phys.* **130** (5) 557-567
[7] Tayurskii D, Lysogorskii Y and Zvezdov D 2009 *J. Phys.: Conf. series* **150**(3) 032110
[8] Tayurskii D A and Lysogorskii Y V *J. of Low Temp. Phys.* **158** (1-2) 237-243
[9] Tilley D and Tilley J 1977 *Superfluidity and superconductivity* (Moscow: Mir)
[10] Khalatnikov I M 1988 *An Introduction to the Theory of Superfluidity* (Redwood City: Addison-Wesley)
[11] McKenna M J, Slawecki T and Maynard J D 1991 *Phys. Rev. Lett.* **66** (14) 1878-1881
[12] Laskin N 2002 *Phys. Rev.* E **66** (5) 56108
[13] Lenzi E K, Oliveira B F and Astrath N G *et al.* 2008 *The European Physical Journal* B **62**(2) 155-158
[14] Greenberger, D 2001 *Phys. Rev. Lett.* **87**(10) 100405
[15] Dong J and Xu M 2008 *J. of Math. Physics.* **49**(5) 052105

Critical fluctuations in the one-dimensional Bak-Sneppen model

W Y Qian and C B Yang

Institute of Particle Physics, Central China Normal University, Wuhan 430079, P. R. China
Key Laboratory of Quark and Lepton Physics (CCNU), Ministry of Education, P. R. China

E-mail: cbyang@mail.ccnu.edu.cn

Abstract. The critical fluctuation properties of fitness distribution in the one-dimensional Bak-Sneppen model are studied in terms of the normalized factorial moments, erraticity moments and the factorial correlators. For a fitness window below the gap intermittent behaviors are observed. The scaling exponent for the BS model is different from that in two dimensional Ising model for second order phase transition. There is no correlation between fluctuations in two windows separated by the critical gap.

1. Introduction

Last a few decades witnessed the rapid development of theories for complex systems. It has been found that some complex systems can evolve into a critical state without tuning any external parameters. System at such self-organized critical states, such as in the sand-pile model [1] and the Bak-Sneppen (BS) model [2], have as fingerprints for criticality power-law distributions [3] for the size and lifetime of avalanches.

As is well-known, systems at traditional critical point exhibit nontrivial critical fluctuations and long-range correlations. A typical example is the critical fluctuations in the two dimensional (2D) Ising model [4]. For the purpose of comparing properties among system under normal and self-organized critical states, one can investigate the fluctuation properties of systems in the self-organized critical state. To achieve this end, we study the dynamical fluctuations in a typical self-organized critical model, the one-dimensional BS model. In such a model, sites in a lattice are initially assigned random numbers uniformly drawn from (0, 1). Those random numbers are called fitness of the sites. At each time step of the evolution, a site with the minimum fitness is found and the fitness on that site and its nearest neighbors are replaced by new random numbers also uniformly drawn from (0, 1). A variable named gap at time t is defined as the maximum of the minimum fitness before that time t. With the update of the system, it is found that the system's gap will saturate to a critical value f_c and a critical state will be reached.

In this paper, we study the critical fluctuations of fitness distribution and try to find a characteristic quantity for those fluctuations for the one-dimensional BS model.

2. Factorial moments and dynamical fluctuations

Consider a system in which each site or particle has a property of d components which fluctuates because of the interactions or other effects. One can represent a particle as a point in a d-dimensional space. To study the fluctuations quantitatively, one can first choose a fixed region in the d-dimensional space and divide it into M^d bins of equal size. Then one counts the number of points in each bin and study the fluctuations of the point numbers in each bin. This process can be repeated for different numbers of bins or different bin sizes.

A useful variable used for analyzing the fluctuations is the factorial moments. The q-th order factorial moment is defined as

$$f_q = \langle n(n-1)\cdots(n-q+1)\rangle \ , \tag{1}$$

where n is the number of points in a bin and $\langle\cdots\rangle$ represents average over all bins and different realizations of the system's configuration. The normalized factorial moments are defined as ratios of the factorial moments by

$$F_q = f_q/f_1^q \ . \tag{2}$$

An important advantage of F_q is that they can filter out the statistical fluctuations. One can see this easily by considering a case in which the probability of having n points in a bin is given by a Poisson or binomial distribution. In this case, the normalized factorial moments are all fixed numbers of about 1, independent of the bin size. Specifically, if the fluctuations in the system has some kind of self-similarity, as in the random cascading model [5], F_q have a power-law dependence on the bin size δ, $F_q(\delta) \propto \delta^{-\varphi_q}$ with $\varphi_q > 0$, a phenomenon known as intermittency [6].

Considering the above, one may ask whether the normalized factorial moments can be used for complex system and tell us some features of the critical fluctuations in those systems. An application [7] has been appeared along this line for the financial fluctuations. For the BS model we are considering now, each site on the lattice has a fitness in (0, 1). So it is natural to represent the fitness state of a lattice by points in a one dimensional space from 0 to 1. In this paper, we only investigate the critical fluctuations in the one-dimensional BS model. Generalization to higher dimensional BS models is straightforward.

Since we are interested only in the critical fluctuations of the fitness distribution in the BS model, we start our simulation of the system's evolution from the critical state when fitness on all sites are uniformly distributed from a critical gap f_c to 1. For a one dimensional lattice we are considering, $f_c = 0.667$ [8]. In our simulations, we fix the lattice size $L = 500$. Let the system evolve $5L$ times from the initial critical state, then we put fitness on all sites in a one-dimensional space from 0 to 1. To calculate the factorial moments configuration by configuration, we focus on a fixed fitness window Y from $f_{\min}$ to $f_{\max}$ and divide the window into M bins. Then it is straightforward to count the number of points in each bin and calculate the factorial moments $f_q(q = 1, 2, \cdots)$ for each bin in the given configuration. The factorial moments for a given configuration of the system can be obtained by averaging f_q over all bins in the window. Then one gets the normalized factorial moments F_q from the configurational averaged moments f_q. The calculation is repeated for different numbers of bins or different bin sizes. To eliminate such statistical fluctuations, the above process must be repeated for a lot of configurations, and each configuration is separated by $5L$ evolution steps. We use 2×10^7 configurations from the same critical state, and average the normalized factorial moments F_q over those configurations. Suppose the normalized factorial moments F_q^e for a configuration has a distribution $P_q(F_q^e)$ for each q in those configurations, one can easily see that the normalized factorial moments after average over the configurations is in fact

$$F_q \equiv \langle F_q^e\rangle = \int dF_q^e F_q^e P_q(F_q^e) \ , \tag{3}$$

where $\langle\cdots\rangle$ represents average over configurations, thus very little information about the distribution $P_q(F_q^e)$ is contained in the averaged moments. To get more information about the fluctuations of the moments from configuration to configuration, a new quantity has been proposed in Ref. [9], which is defined as

$$C_{p,q} = \frac{\int dF_q^e P_q(F_q^e)(F_q^e)^p}{(\int dF_q^e P_q(F_q^e) F_q^e)^p} = \frac{\langle (F_q^e)^p \rangle}{\langle F_q^e \rangle^p} . \tag{4}$$

By definition, $C_{0,q} = C_{1,q} = 1$. Please note that p needs not be an integer. For $p > 1$, $C_{p,q}$ emphasizes on contributions of large F_q^e, while for $p < 1$ $C_{p,q}$ probes the lower F_q^e behavior of $P_q(F_q^e)$. So with different choices of p, a lot of more information on the fluctuations can be obtained from $C_{p,q}$. For multi-fractal fluctuations, one expects a generalized scaling behavior, called erraticity, [10]

$$C_{p,q} \propto h(M)^{\psi_{p,q}} \text{ , with } \psi_{p,q} > 0 \tag{5}$$

where $h(M)$ is a properly chosen function of the number of bins in the given window, common for all p and q. Here we only investigate the behavior of $C_{p,q}$ near $p = 1$ and calculate a new set of moments

$$\Sigma_q = \left. \frac{\partial C_{p,q}}{\partial p} \right|_{p=1} = \left\langle \frac{F_q^e}{\langle F_q^e \rangle} \ln \frac{F_q^e}{\langle F_q^e \rangle} \right\rangle . \tag{6}$$

When Eq. (5) holds, one can see that

$$\Sigma_q \sim \left. \frac{\partial \psi_{p,q}}{\partial p} \right|_{p=1} \ln h(M) = \mu_q \ln h(M). \tag{7}$$

So Σ_q is a linear function of $\ln h(M)$ for all q, and therefore scaling properties of Σ_q are described by slope parameters $\mu_q = \partial \psi_{p,q} / \partial p|_{p=1}$. μ_q are sometimes called entropy indices.

3. Results of the moments

In the one dimensional fitness space, there is a special point, $f = f_c$. For $f > f_c$, the distribution is on average uniform because of the building up of the BS model. For $f < f_c$, the distribution is very complicated. For this reason, we investigate the factorial moments in two separate windows, one above f_c, another below f_c. The window above f_c is chosen to be (0.68, 0.98). Our calculated results for F_q are constants, independent of the bin size, thus not very interesting.

Now we study the fluctuations of fitness distribution in a fitness window below f_c. There is some difference in the average fitness densities for different bins in this window, because the minimum fitness in the lattice can be from 0 to f_c and only the sites involved in the evolution may have fitness larger than the minimum fitness but less than f_c, thus contribute to the fitness distribution in the new window. Because of the fact that the minimum fitness fluctuates from time step to time step in the evolution, fitness distribution in our new window range cannot be flat after average over a long evolution history or over many configurations. To reduce the effect from such a bin to bin difference, we choose a fitness window close to the critical gap f_c and not too big. We choose the window to be from 0.45 to 0.65 and perform exactly the same calculations for the moments as for the last case we just studied. We suppose that the discussed bin to bin difference will not lead to very significant modification to our results, as claimed in Ref. [11] that the bin averaging done also washes out the nonstationary effects. Because the number of fitness points in the window is a small fraction of the lattice size, one has to work with smaller bin number M. We start with $M = 2$ and increase M by fifty percent successively six times. The obtained normalized factorial moments F_q as functions of number of bins M are shown in Fig. 1 in log-log plot.

Power-law dependence of F_q on the bin size or number of bins M or intermittent behavior can be seen clearly. The scale invariance of F_q is a signature for critical fluctuations, as for the 2D Ising model. So we have

$$F_q \propto M^{\varphi_q} \quad \text{with} \quad \varphi_q > 0 . \tag{8}$$

One can display the scaling results of F_q as functions of F_2, which can exhibits scaling behaviors even when F_q does not scale with M. This more general scaling behavior reads

$$F_q \propto F_2^{\beta_q} \text{ ,with } \beta_q = \varphi_q/\varphi_2 \ . \tag{9}$$

Such a scaling behavior is shown in Fig. 2. Power-law fitting results are also shown in the figure.

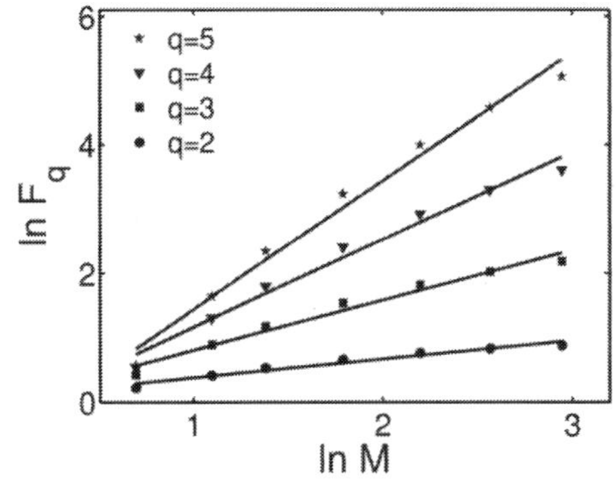

Figure 1. F_q as functions of number of bins for the fitness window below the critical gap f_c.

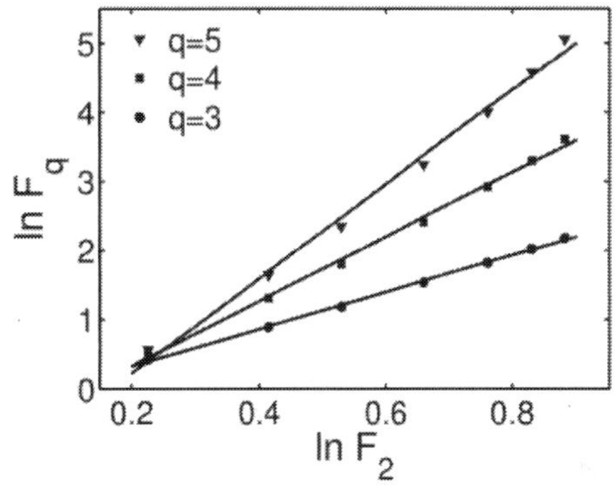

Figure 2. Scaling behavior of F_q vs F_2 in log-log plot for the results shown in Fig. 1. The straight lines are from the power-law fit.

The exponents β_q in Eq. (9) are shown in Fig. 3 as a function of $q-1$ in log-log plot. From the power-law fitting curve $\beta_q \propto (q-1)^\nu$, one gets the scaling exponent $\nu = 1.35$ for the scaling behavior, which is very close to that obtained from the 2D Ising model, where $\nu \simeq 1.3$.

For describing the fluctuation properties of the normalized factorial moments F_q, Σ_q can also be calculated in the simulation of the system evolution. For the fitness window below f_c we found that $\ln h(M) = (\ln M)^2$ for the scaling of Σ_q. The results for Σ_q are shown in Fig. 4 as functions of $(\ln M)^2$.

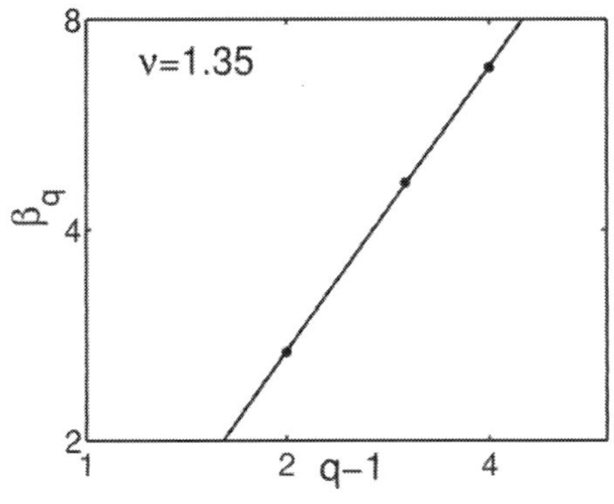

Figure 3. Exponents β_q for the curves in Fig. 2 as a function of $q-1$.

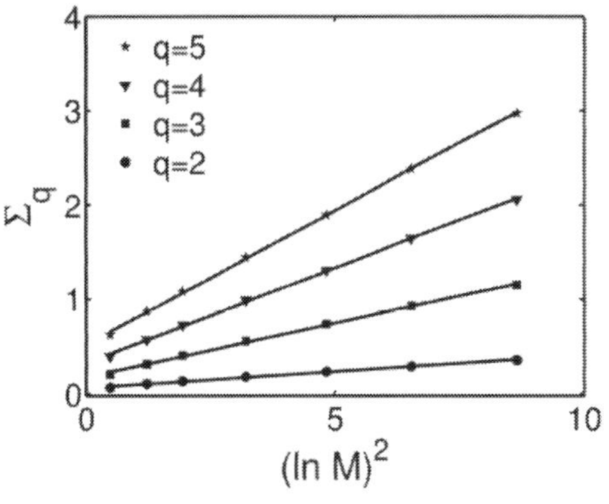

Figure 4. Σ_q as functions of bin number $h(M) = (\ln M)^2$ for the window below f_c.

With the scaling of Σ_q at hand, one can get the slopes μ_q of the lines in Fig.4 and the results are shown in Fig. 5 as a function of $q-1$ in log-log plot. Again a scaling behavior is found

$$\mu_q = (q-1)^{\nu'} \text{ ,with } \nu' = 1.53. \tag{10}$$

In Ref.[12] a scaling behavior of Σ_3 vs Σ_2 is shown for the 2D Ising model. They obtained the slope $\omega_3 = 3.24$, which corresponds to an exponent $\nu' = \ln(\omega_3)/\ln(3-1) = 1.696$ from the above equation. Thus the value of ν' we just obtained is a little bit smaller than that from 2D Ising model.

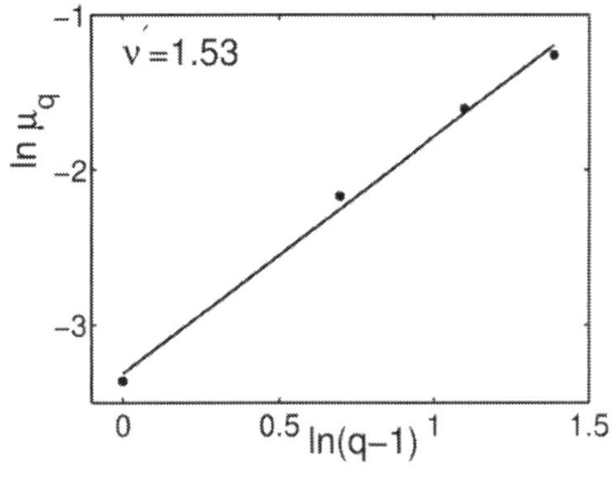

Figure 5. Exponents ω_q for the curves in Fig. 4 as a function of $q-1$.

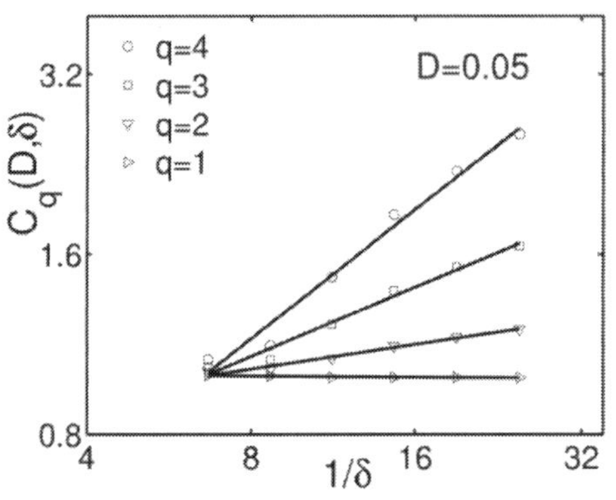

Figure 6. C_q as functions of D and δ for $q = 1, 2, 3$ and 4 when D is fixed at 0.05.

4. Factorial correlators

The critical fluctuations can be investigated from the so-called factorial correlators

$$f_{q_1,q_2} \equiv \langle n_1(n_1-1)\cdots(n_1-q_1+1)n_2(n_2-1)\cdots(n_2-q_2+1)\rangle \ , \tag{11}$$

where n_1 and n_2 are multiplicities in two different bins in some phase space region. As in F_q, $\langle\cdots\rangle$ denotes average over many bins and different configurations. The scalednormalized factorial correlators can be defined as

$$C_{q_1,q_2} = f_{q_1,q_2}/\left((f_{1,0})^{q_1}(f_{0,1})^{q_2}\right) \ . \tag{12}$$

C_{q_1,q_2} depends on the separation D between the two bins and the bin widths. Now we only investigate behaviors of C_{q_1,q_2} for $q_1 = q_2 = q$ and we choose equal size for the two bins. Let one bin lie on the left of f_c and the other symmetrically on the right of f_c. The distance between the centers of the two bins is D and each bin has a width δ. As for investigating F_q, we perform Monte Carlo simulation and calculate $C_q \equiv C_{q,q}$ as functions of D and δ for $q = 2, 3, 4$. The results are shown in Fig. 6 for an arbitrary $D = 0.05$ and in Fig. 7 for $D = 0.2$. Similar results can be obtained for other values of D. Power-law behaviors can be seen as for the normalized factorial moments. One can get the scaling exponents φ_q for C_q as functions of q for three different values of D as shown in Fig. 8. One can see that the scaling exponent can also be expressed as $\varphi_q \propto (q-1)^\nu$ with ν almost the same as from F_q, independent of value of D. Thus there is no correlation between multiplicity fluctuations in those two bins.

5. Conclusion and discussion

We studied the normalized factorial moments, the entropy indices and the factorial correlators for the fluctuations of fitness distribution in the one-dimensional BS model for two fitness windows, one above and the other below the critical gap f_c. For the window above f_c, F_q do not show any anomaly scaling in the bin size or bin number. But for the window below f_c, intermittent

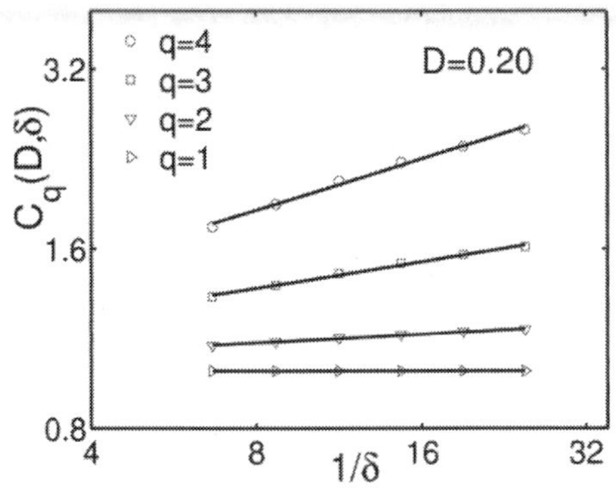

Figure 7. C_q as functions of D and δ for $q = 1, 2, 3$ and 4 when D is fixed at 0.2.

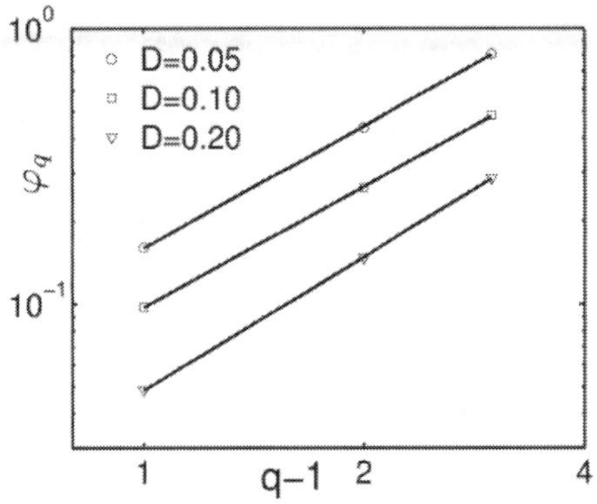

Figure 8. Scaling exponents φ_q of C_q as a function of $q - 1$ for three values of D.

behaviors are observed with the scaling exponent for the normalized factorial moments very close to that for the 2D Ising model, while the critical exponent ν' for the erraticity is found a little bit smaller than that from 2D Ising model. The results show that there is no correlation between multiplicity fluctuations in those two bins separated by the critical gap f_c.

Acknowledgments

This work was supported in part by the National Natural Science Foundation of China under Grant No. 11075061, by the programme of Introducing Talents of Discipline to Universities under No. B08033.

6. References

[1] Bak P, Tang C and Wiesenfeld K 1988 *Phys. Rev.* A **38** 364
[2] Bak P and Sneppen K 1993 *Phys. Rev. Lett.* **71** 4083
Flyvbjerg H, Sneppen K and Bak P 1993 *Phys. Rev. Lett.* **71** 4087
[3] Butenberg B and Richter C F 1956 *Ann. Geofis.* **9** 1
Paczuski M, Maslov S and Bak P 1996 *Phys. Rev.* E **53** 414
[4] For the use of Ising model to study particle production, see for examples, Satz H 1989 *Nucl. Phys.* B **326** 613
Chau L L and Huang D W 1992 *Phys. Lett.* B **283** 1
Burda Z, Zalewski K, Peschanski R and Wosiek J 1993 *Phys. Lett.* B **314** 74
[5] Bialas A and Peschanski R 1988 *Phys. Lett.* B **207** 59
[6] Bialas A and Peschanski R 1986 *Nucl. Phys.* B **273** 703
ibid 1988 *Nucl. Phys.* B **308** 867
[7] Schoeffel L *Preprint* arXiv:1108.5946v1
[8] Paczuski M, Maslov S, and Bak P 1996 *Phys. Rev.* E **53** 414
Grassberger P 1995 *Phys. Lett.* A **200** 277
[9] Hwa R C 1996 *Acta Physica Polonica* B **27** 1789
[10] Ochs W, Wosiek J 1988 *Phys. Lett.* B **214** 617
1989 *Phys. Lett.* B **232** 271
[11] Eggers H C, Carruthers P, Lipa P and Sarcevic I 1991 *Phys. Rev.* D **44** 1975
[12] Cao Z and Hwa R C 2000 *Phys. Rev.* D **61** 074011

Study of energy fluctuation effect on the statistical mechanics of equilibrium systems

Yu V Lysogorskiy[1,2], **Q A Wang**[1,3] **and D A Tayurskii**[2]
[1]LUNAM Université, ISMANS, Laboratoire de Physique Statistique et Systèmes Complexes, 44, Avenue F.A. Bartholdi, Le Mans 72000, France
[2]Institute of Physics, Kazan Federal University, Kazan 420008, Russia
[3]LPEC, Université du Maine, Ave. O. Messiaen, Le Mans 72035, France

E-mail: yura.lysogorskii@gmail.com

Abstract. This work is devoted to the modeling of energy fluctuation effect on the behavior of small classical thermodynamic systems. It is known that when an equilibrium system gets smaller and smaller, one of the major quantities that becomes more and more uncertain is its internal energy. These increasing fluctuations can considerably modify the original statistics. The present model considers the effect of such energy fluctuations and is based on an overlapping between the Boltzmann-Gibbs statistics and the statistics of the fluctuation. Within this "overlap statistics", we studied the effects of several types of energy fluctuations on the probability distribution, internal energy and heat capacity. It was shown that the fluctuations can considerably change the temperature dependence of internal energy and heat capacity in the low energy range and at low temperatures. Particularly, it was found that, due to the lower energy limit of the systems, the fluctuations reduce the probability for the low energy states close to the lowest energy and increase the total average energy. This energy increasing is larger for lower temperatures, making negative heat capacity possible for this case.

1. Introduction

A fluctuation is a universal phenomenon in nature. So many and diverse physical phenomena in macroscopic systems are determined directly by or influenced by fluctuations and the importance of a right taking into account of them does not cease to increase following the development of the physics and technology of small size systems[1, 2]. When a thermodynamic systems becomes smaller, not only the surface and size enter into play in the thermodynamic properties[1–3], the increasing thermal fluctuations can also affect the behavior of the thermodynamic quantities and may be responsible for many peculiar properties such as negative specific heat, compressibilities and susceptibilities[3, 4]. There are already interesting works on the relationship between fluctuation and negative specific heat using relations between energy variance and heat capacities[5, 6]. The goal of the present work is to propose a model for the effects of energy fluctuations on the base of their statistical distributions.

The increase of fluctuations with decreasing size can be traced by observing many quantities such as particle density, particle number, and internal energy which is proportional to the inverse square root of the number of particles $1/\sqrt{N}$ [7]. For macroscopic system, N is very large (of order 10^{23}), hence the fluctuations of these quantities around the statistical means are often negligible. Their thermodynamic properties can be tackled as if they were deterministic and smoothly variable quantities. However, for small size systems, the things are much more uncertain due to the nondeterministic and random variations of these quantities which fluctuate. Hence one can expect important impacts of fluctuation in the functionality and performance of, for instance, small size molecular engines or biological engines[1, 2].

Suppose we have a given statistics for the macroscopic system when the fluctuations are small. With increasing fluctuations this statistics should be modified by large fluctuating motion of the elements of the system. The result can be a very complex statistics with respect to the original one. A mathematical approach to tackle this complexity is to decompose the resulted statistics into the original statistical distribution and the distribution of fluctuations, both being well known, and to take the averaging of the original statistics over fluctuations. The basic physics of this procedure is the following. Imagine a thermodynamic system undergoing large fluctuations in time, or an ensemble of identical statistical systems all distributed in a fluctuating environment. The long time or ensemble behavior of the systems can then be estimated by the time or ensemble averaging of the fluctuating statistical means of the quantities of interest. An example of such approach is the so-called superstatistics[8] which considers the temperature fluctuations in nonequilibrium systems and estimates the modification to the Boltzmann-Gibbs statistics by superposing the distribution of fluctuating temperature over the conventional Gibbs distribution functions for canonical ensembles.

In this work we consider the fluctuating systems in equilibrium. When a system fluctuates in equilibrium, one of the most important fluctuating quantities is its internal energy for any statistical ensemble under consideration (micro-canonical, canonical or grand-canonical). We are interested in the studying the results of superimposing or overlapping of an energy fluctuations distribution on the Boltzmann-Gibbs statistics when the fluctuations gradually become dominating. We used the Gaussian, stretched exponential and q-Gaussian distributions of energy fluctuations. In a previous work[9], we have investigated the fluctuation influence on the quantum statistical distributions. Here we are focused on classical systems in the canonical ensemble. In what follows, we present the details of the model and the calculation results for the probability distributions and the temperature behaviors of internal energy and heat capacity.

2. The overlapping

Imagine a large system governed by the Boltzmann distribution $\propto e^{-e_i/\theta}$ (θ is temperature) becomes smaller and smaller, so that a given state i (i=1,2 ... w) possesses more and more uncertain energy due to the fluctuations, with increasing magnitude of the fluctuations $\Delta e_i/\varepsilon_i$ where ε_i is the mean of the energy e_i over the fluctuations with the distribution $f(e_i - \varepsilon_i)$, i.e., $\varepsilon_i = \int_{e_{\min}}^{\infty} e_i f(e_i - \varepsilon_i) de_i$. The lower limit of integration is determined by the lowest possible energy $e_{\min}$ of the system. When fluctuations are small, $f(e_i - \varepsilon_i) \to \delta(e_i - \varepsilon_i)$. We have an ensemble of systems in contact with a thermostat at temperature θ. The systems are distributed over all the w states of fluctuating energy states e_i. Hence this ensemble can be separated into w sub-ensembles, each one containing a large number of systems distributed over the fluctuating energy e_i around the mean ε_i for a given state i. On the other hand, the number of systems of each sub-ensemble i is proportional to $p(\varepsilon_i)$. Our model supposes that the new overlapped distribution is determined by the following way:

$$p(\varepsilon_i) = \frac{1}{Z} \int_{e_{\min}}^{\infty} e^{-x/\theta} f(x - \varepsilon_i) dx, \tag{1}$$

where $x = e_i$ and Z is the normalization constant calculated with $Z = \sum_{i=1}^{w} \int_{e_{\min}}^{\infty} e^{-x/\theta} f(x - \varepsilon_i)dx$ or $Z = \int_{\varepsilon_{\min}}^{\infty} \int_{e_{\min}}^{\infty} e^{-x/\theta} f(x - \varepsilon_i) dx g(\varepsilon) d\varepsilon$ in case of continuous states, where $g(\varepsilon)$ is the density of states. It should be mentioned that we have to normalize overlapped distribution even when fluctuations distribution and Boltzmann distribution are both normalized. The usual canonical distribution can be restored by setting fluctuations distribution as δ-function $f(x) = \delta(x)$.

Below we will consider different types of fluctuations (Gaussian, stretched exponential and q-Gaussian) and their influences on the total energy and heat capacity of the system of interest. The density of states $g(\varepsilon)$ is assumed to be the same as that of the system when the fluctuations are small.

2.1. Gaussian fluctuations

The Gaussian fluctuations distribution is taken in the usual form $f(x) = \exp(-x^2/2\sigma^2)/\left(\sqrt{2\pi}\sigma\right)$, where σ is the standard deviation. It should be mentioned that energy ε, ε_{min}, temperature θ and standard deviation σ are measured in some energy units. For the case of simplicity we consider the constant density of states (i.e. a particles with linear dispersion law in one dimension box of length L). The density of state for this system is $g(\varepsilon) = L/(\pi\hbar c) = 1/\varepsilon_0 = \text{const}$, where c denotes sound velocity and we can perform the integration analytically. We introduce here the unit of energy $\varepsilon_0 = \pi\hbar c/L$ and will measure all energies and probability distribution parameters with respect to this unit, i.e. $\widetilde{\varepsilon} = \varepsilon/\varepsilon_0$, $\widetilde{\sigma} = \sigma/\varepsilon_0$, $\widetilde{\theta} = \theta/\varepsilon_0$, $\widetilde{p} = \varepsilon_0 p$ will denote dimensionless quantities, but hereafter we will omit symbol $\widetilde{}$ for the clarity. The probability distribution function after integration has the form

$$p(\varepsilon) = \frac{1}{\theta} e^{-\frac{\varepsilon-\varepsilon_{\text{min}}}{\theta}} \frac{\operatorname{erfc}\left(-\frac{\varepsilon-\varepsilon_{\text{min}}}{\sqrt{2}\sigma} + \frac{\sigma}{\sqrt{2}\theta}\right)}{\operatorname{erfc}\left(\frac{\sigma}{\sqrt{2}\theta}\right) + e^{-\frac{\sigma^2}{2\theta^2}}}, \qquad (2)$$

where $\operatorname{erfc}(z) = 1 - \operatorname{erf}(z)$ denotes the complementary error function. This probability distribution is invariant under the uniform energy shift, i.e. it depends only on the difference $\varepsilon - \varepsilon_{\text{min}}$. On Fig. 1(a) this probability function is plotted for three different values of σ. The fluctuations effect is especially visible in the low energy region with considerable decrease of probability with increasing fluctuations.

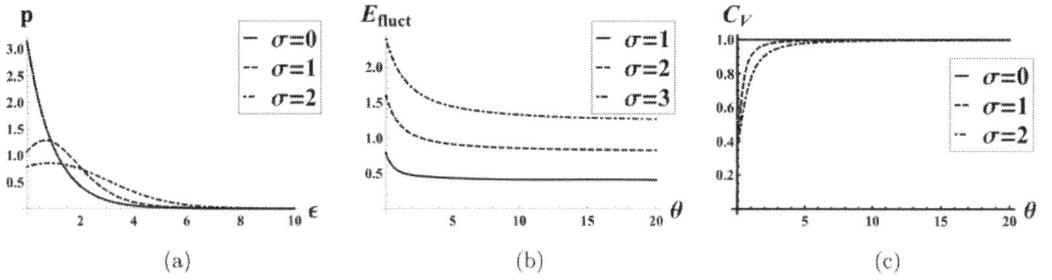

Figure 1. (a)Probability distribution at the temperature $\theta = 1$; (b) fluctuation energy; (c) total heat capacity dependence for different values of σ for Gaussian fluctuations

By knowing the probability distribution function one can calculate the total energy $E = \int_{\varepsilon_{\text{min}}}^{\infty} \varepsilon p(\varepsilon) g(\varepsilon) d\varepsilon$. The heat capacity C is the first derivative of E with respect to temperature, that is $C = \partial E/\partial\theta$. The obtained expression for energy is

$$E = \underbrace{\varepsilon_{\text{min}}}_{\text{energy shift}} + \underbrace{\theta}_{\text{Boltzmann energy}} + \underbrace{\sqrt{\frac{2}{\pi}} \frac{\sigma}{1 + e^{\frac{\sigma^2}{2\theta^2}} \operatorname{erfc}\left(\frac{\sigma}{\sqrt{2}\theta}\right)}}_{\text{fluctuation energy } E_{\text{fluct}}}. \qquad (3)$$

It can be seen from Eq. (3) that the total energy "starts" from the minimal energy $\varepsilon_{\min}$. For the sake of simplicity hereafter we set $\varepsilon_{\min} = 0$. It is follows from Eq. (3) that the energy remains finite even at the zero temperature. It could be shown that the fluctuations energy at zero temperature is equal to $\sqrt{2/\pi}\sigma$ and is proportional to the fluctuations strength. At high-temperatures ($\sigma \ll \theta$) the fluctuations energy is $\sigma/\sqrt{2\pi}$ and also is proportional to the fluctuations strength but does not depend on the temperature. In Fig. 1(b) the dependence of the fluctuation energy versus temperature for two different values of σ is depicted.

The temperature dependence of the heat capacity is shown in the Fig. 1(c). An exact calculation of the heat capacity at the zero temperature results in $C(\theta = 0) = 1 - 2/\pi \approx 0.36$ independently on the fluctuations strength σ. Similar calculations for two-dimensional and three-dimensional systems with linear dispersion law give $C_V^{2D}(\theta = 0, \sigma) = 2 - \pi/2 \approx 0.43$, $C_V^{3D}(\theta = 0, \sigma) = 3 - 8/\pi \approx 0.45$. It should be indicated here that the positive heat capacity at zero temperature is related to the well known fact that when the classical statistical mechanics has been used, the quantum correction should be considered for low temperature cases[7]. Considering the universal vanishing heat capacity at zero temperature and the negative fluctuations effect at low temperatures, the negative heat capacity is not possible with Gaussian fluctuations. In what follows we will investigate the influence of other fluctuations.

2.2. *Stretched exponential fluctuations*

Now we will consider a deformation of the Gaussian fluctuations distribution - the stretched exponential distribution given as

$$f(x;\alpha) = \frac{1}{(2^{1+\frac{1}{\alpha}})\sigma\Gamma\left(1+\frac{1}{\alpha}\right)} \exp\left(-\frac{|x|^\alpha}{2\sigma^\alpha}\right),$$

where $\alpha \in (0, \infty)$. This family of distributions is frequently encountered in nature[10] whose probabilistic uncertainty measure has been well studied[11, 12]. The first moment $< x >$ of this distribution is zero, whereas second moment is $< x^2 >= 4^{\frac{1}{\alpha}}\Gamma\left(\frac{3}{\alpha}\right)\sigma^2/\Gamma\left(\frac{1}{\alpha}\right)$. The most interesting case for us is the case $\alpha < 2$, when the distribution is wider than for the Gaussian case ($\alpha = 2$) and as a result, the influence of fluctuations is stronger. When $\alpha = 1$ the standard deviation is $< x^2 >= 8\sigma^2$, while for $\alpha = 1/2$ the standard deviation is very large $< x^2 >= 1920\sigma^2$.

Hereafter we will consider two different type of systems - free particle with quadratic spectrum $\varepsilon = \frac{\hbar^2 k^2}{2m}$ and particles with linear spectrum $\varepsilon = \hbar ck$ both in 3D space. Characteristic energy of free particle with mass m in box with side length L is $E_0 = \frac{\hbar^2}{mL^2}$. For a particle with linear dispersion and velocity c characteristic energy is $E_0 = \frac{\hbar c}{L}$. The corresponding densities of states are $g(\varepsilon) = \frac{\sqrt{\epsilon}}{\sqrt{2}\pi^2 E_0^{3/2}}$ and $g(\varepsilon) = \frac{\epsilon^2}{E_0^3 2\pi^2}$.

In the case if $\alpha = 1$ one can analytically integrate the probability distribution function $p(\varepsilon)$ and the total energy $E(\theta)$ and thus obtain the analytical expression for the heat capacity. The plots of heat capacities are depicted in the Fig. 2(a) and Fig. 2(b) correspondingly. The asymptotic behaviors of heat capacity close to zero temperature are $C(\theta) \approx 15/2^{5/2}(\theta/\sigma)^{\frac{3}{2}} \geq 0$ and $C(\theta) \approx 3(\theta/\sigma)^3 \geq 0$ respectively. As it can be seen, if fluctuations have the stretched exponential distribution with $\alpha = 1$, the heat capacity at zero temperature is always zero. One can notice that the low temperature behavior of heat capacity of particles with linear dispersion has a cubic temperature dependence, which could be only obtained with Debye model[7] when fluctuations are weak. But it should be underlined again that the quantum effects are not considered here. In view of the negative effect of fluctuations on the heat capacity at low temperatures, the negative heat capacity can be expected if the quantum statistics is considered.

In case of the stretched exponential fluctuation distribution with $\alpha = 1/2$ the only analytically integrable system is the case of particles with linear dispersion. On the Fig. 3 one can find the

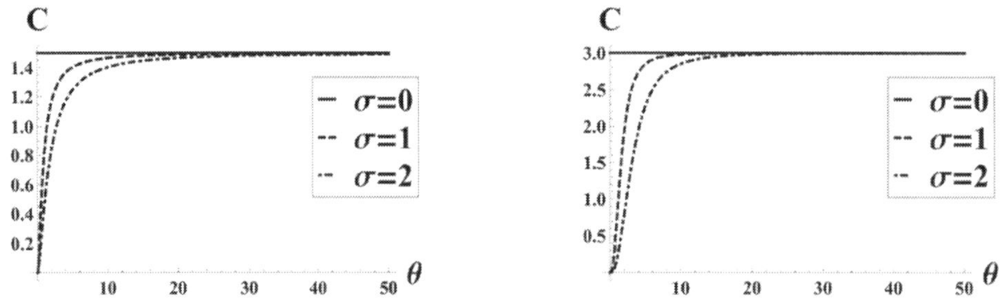

(a) (b)

Figure 2. Heat capacity for (a) free particle ; (b) particle with linear dispersion in the 3D space for the case of the stretched exponential ($\alpha = 1$) distribution of fluctuations.

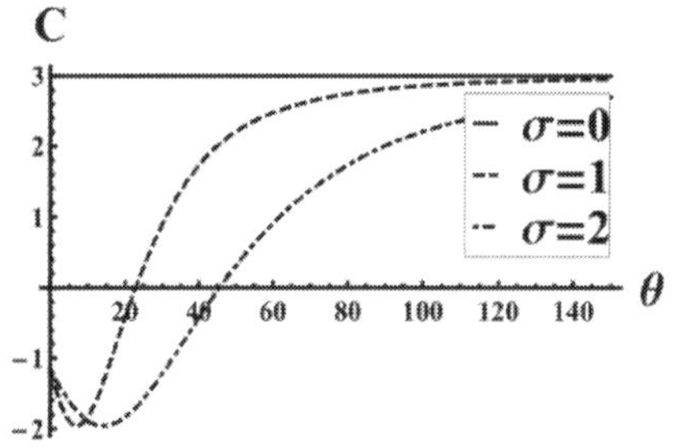

Figure 3. Heat capacity for particles with linear dispersion in 3D space for the case of the stretched exponential ($\alpha = 1/2$) distribution of fluctuations.

heat capacity for such particles in 3D space. It was calculated, that for the particles with linear dispersion one could obtain the negative heat capacity in the temperature range from zero temperature up to some T_z, and that the larger the fluctuations (value of σ), the higher T_z we have.

The low temperature behaviors of the heat capacities around zero temperature for the spaces with different dimensionality are $C_V^{(1D)}(\theta) \approx -2 + \frac{9\sqrt{\pi}}{8}\left(\frac{\theta}{\sigma}\right)^{1/2}$, $C_V^{(2D)}(\theta) \approx -\frac{4}{3} - \frac{5}{9}\frac{\theta}{\sigma}$, $C_V^{(3D)}(\theta) \approx -\frac{6}{5} - \frac{7}{50}\frac{\theta}{\sigma}$. Notice that, with increasing of dimensionality of space, the negative heat capacity at zero temperature increases.

2.3. q-Gaussian fluctuations

Now let us consider another type of the deformed Gaussian distribution, namely the q-Gaussian distribution given by

$$f(x) = \frac{\sqrt{q-1}\,\Gamma\left(\frac{1}{q-1}\right)}{\sqrt{2\pi}\sigma\Gamma\left(-\frac{1}{2}+\frac{1}{q-1}\right)}\left(1 - \frac{(1-q)x^2}{(3-q)\sigma^2}\right)^{\frac{1}{1-q}} \tag{4}$$

which approaches power law distribution at large values of x. This function has been introduced in the context of nonextensive statistics[13]. There are quite large number of works focusing on the properties of this q-exponential functions[15] and its application to modeling of different phenomena especially in complex situations[14]. In the limit case of $q \to 1$, this q-Gaussian recovers the usual Gaussian function. q-Gaussian is a class of deformation of usual Gaussian which is easy to treat mathematically.

We will exploit the expression for the total energy $E = \int_0^\infty \varepsilon\ p(\varepsilon)\ d\varepsilon$, where for the sake of simplicity, we take the DOS as a constant $g(\varepsilon) =$ const. With the help of Eq. (1), it can be rewritten as $E = \int_0^\infty dx\ e^{-\frac{x}{\theta}}\ 1/Z\ \int_0^\infty d\varepsilon\ \varepsilon\ f(x-\varepsilon)$. The inner integral has to converge at infinity, which implies the condition $\varepsilon\ f(x-\varepsilon) \sim \varepsilon^{-\gamma}$, with $\gamma > 1$. This condition yields limits $1 \leq q < 2$. The total energy is not analytically integrable, so all further calculations should be made numerically. The dependence of heat capacity on temperature is show in Fig. 4. As one can see from this figure, the negative heat capacity at low temperatures can be found for $q \geq 7/4$. The high temperature limit for the heat capacities in all cases is 1, which restores the classical limit.

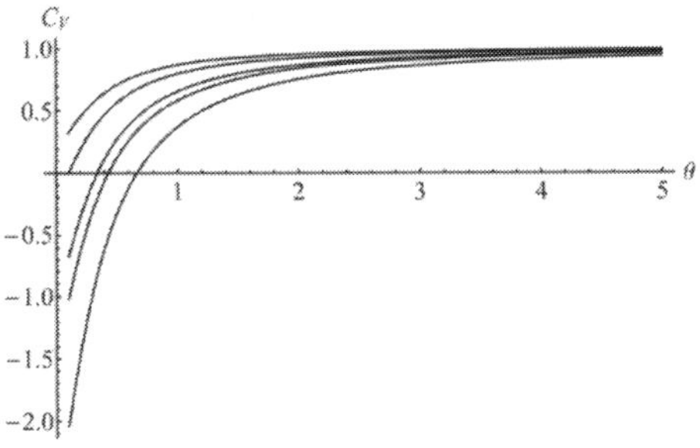

Figure 4. Heat capacity of 2D free particle in case of q-Gaussian fluctuation distribution for different values of $q = \frac{3}{2}, \frac{7}{4}, \frac{15}{8}, \frac{19}{10}, \frac{31}{16}$ from top to bottom ($\sigma = 1$).

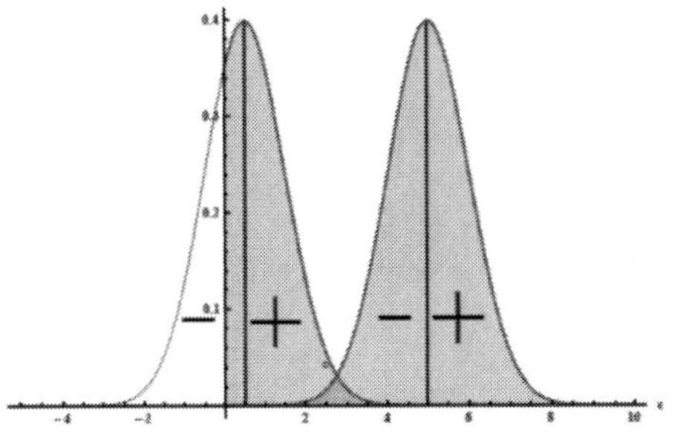

Figure 5. Scheme of the energy fluctuation, see comments in the text.

3. Concluding remarks

In the present work we have studied the effect of energy fluctuations in small classical thermodynamic systems by considering the overlaps of the classical Boltzmann distribution and the fluctuations distribution. Several fluctuations distributions were considered in this work in the framework of classical statistical mechanics including Gaussian, the stretched exponential and q-Gaussian distributions. The results show that this overlapping can modify considerably the Boltzmann statistics, the temperature behavior of the total energy and heat capacity. Within the fluctuation range considered in this work, the influence is visible especially for the low energy states and at low temperatures. The fluctuations effect must mainly reduce the probability of the low energy states close to the lower energy bound, increase the energy of these states and reduce the heat capacity at low temperatures, so the negative heat capacity becomes possible. It can be expected that larger fluctuations (or σ) will extend their influence to more high energy states with more complex effects.

The reason why the behavior of probability distribution, energy and heat capacity is especially affected by fluctuations for low energy energy states and for low temperatures can be explained as follows. The fluctuations distributions used here suppose that, for a given state, the energy fluctuations are symmetrical, i.e., positive and negative energy deviations are equally probable for sufficiently high energy state, as shown in Fig. 5. But for low energy states, this symmetry is broken by existing of the lowest energy - energies lower than this boundary are forbidden (see Fig. 5). Hence the probability of the states close to the boundary are diminished (see for instance Fig. 1(a)) due to the Boltzmann factor $e^{-e_i/\theta}$, the average energy increases (see Fig. 1(b)).

On the other hand, the contribution of these states is reduced by increasing temperature due to Boltzmann factor. Consequently, the increase of average energy gets smaller for higher temperatures when $\theta \ll \sigma$ but constant when $\theta \gg \sigma$, making the negative heat capacity possible (see Fig. 3) at low temperatures (sufficiently high temperature cancels the effect of this lack of energy symmetry). Hence fluctuations effect on the heat capacity is negligible at high temperatures.

Finally it should be stressed again that quantum effects and Debye model for heat capacity are not considered at this stage, hence the heat capacity at low temperatures is the same as at high temperatures when there is no fluctuation. It can be expected that the heat capacity will be lowered more at low temperatures with the consideration of these effects.

References

[1] Gemmer J, Michel M and Mahler G 2004 *Quantum thermodynamics* (Berlin: Springer)
[2] Blickle V and Bechinger C 2012 *Nature Physics* **8** 143-6
[3] Carignano M A and Gladich I 2010 *Euro Phys. Lett.* **90** 63001
[4] Chomaz P and Gulminelli F 2006 *Eur. Phys. J.* A **30**, 317-31
[5] D'Agostino M, Gulminelli F, Chomaz Ph, Bruno M, Cannata F, Bougault R, Colonna N, Gramegna F, Iori I, N Le Neindre *et al.* 2002 *Phys. Lett.* B **473** 219-25;
D'Agostino M, Bougault R, Gulminelli F, Bruno M *et al.* 2002 *Nuclear Physics* A **699** 795-818
[6] Campi X, Krivine H, Plagnol E and Sator N 2005 *Phys. Rev.* C **71** 041601R
[7] Terletskii Ya P 1971 *Statistical Physics* (North Hollande)
[8] Beck C, Cohen E G D and Swinney H L 2005 *Phys. Rev.* E **72** 056133
Beck C and Cohen E G D Superstatistics 2003 *Physica* A **322** 267-75
[9] Petrova A , Nedopekin O, Tayurskii D and Wang Q A 2011 *J. Phys.: Conf. series* **324** 012032
[10] Laherrere J and Sornette D 1998 *Eur. Phys. J.* B **2** 525-39
[11] Ou C, El Kaabouchi A, Nivanen L, Chen J, Tsobnang F and Wang Q A 2010 *International Journal of Modern Physics* B **24** 3461-8
[12] Wang Q A 2008 *J. Phys.* A **41** 065004
[13] Tsallis C 2009 *Braz. J. Phys.* **39** 337-56
[14] Douglas P, Bergamini S and Renzoni F 2006 *Phys. Rev. Lett.* **96** 110601
[15] Nivanen L, Le Mehaute A and Wang Q A 2003 *Reports on Mathematical Physics* **52** 437 (Preprint cond-mat/0303061)

A dynamical model for hierarchy and modular organization: The trajectories en route to the attractor at the transition to chaos

Alberto Robledo
Instituto de Física y Centro de Ciencias de la Complejidad,
Universidad Nacional Autónoma de México,
Apartado Postal 20-364, México 01000 D.F., Mexico.

E-mail: robledo@fisica.unam.mx

Abstract. We show that the full features of the dynamics towards the Feigenbaum attractor, present in all low-dimensional maps with a unimodal leading component, form a hierarchical construction with modular organization that leads to a clear-cut emergent property. This well-known nonlinear model system combines a simple and precise definition, an intricate nested hierarchical dynamical structure, and emergence of a power-law dynamical property absent in the exponential-law that governs the dynamics within the modules. This classic nonlinear system is put forward as a working example for complex collective behavior.

1. Introduction

Prototypical complex systems are inherently hierarchical [1]-[3]. Hierarchical systems consist of interrelated subsystems or modules, each of which has also a hierarchical constitution. These modular structures form levels down to a lowest one made of elementary units. Examples of complex systems made of smaller complex systems are widespread: astrophysical objects, social organizations, biological organisms, molecules and atoms themselves, etc. In these kinds of systems there can be many subsystems. For example, tissues put individual living beings together and tissues consist of cells, a cell has nuclei and organelles and these are made of macromolecules such as proteins. The proteins in turn are made of many atoms, and electrons and nucleons, which are comprised of hadrons, quarks, etc, form the atoms. In describing a hierarchy it is necessary to define how the subsystems interact and how these interactions associate with the dynamical processes within and between the structures. This applies to physical, biological as well as to social complex systems.

The dynamical properties of hierarchies, within subsystems and between those that interconnect them, are essential for the understanding of complex systems. A purely static description is insufficient for the comprehension of the modular organization and functioning of the entire system, and, most importantly, for the appreciation of the properties of the system called emergent properties, those that result in capabilities and behaviors of the whole that are not attributable to any of the separate subsystems or modules. Emergence 'arises' from the dynamic interactions of parts within the whole. Experience accumulated in studying complex systems have established that the theoretical generation of dynamical hierarchies from a formal system is not trivial and that the quest for a possible general explanation for hierarchical structures is at least elusive. However, it is believed that genuine emergence is strongly associated with nonlinearity, and that a hierarchy of building blocks operating at the "Edge of Chaos" is a characteristic of complex systems.

Here we take the view that it is possible to generate hierarchical systems with dynamical organization from a formal starting point in the form of a simple closed-form nonlinear dynamical system. And also that nonlinear dynamics tuned at the transition to chaos can lead to an emergent property. We illustrate this standpoint by resorting to the properties of the simplest well-known nonlinear iterated map, the iconic one-dimensional unimodal map, as represented by the quadratic logistic map [4], [5].

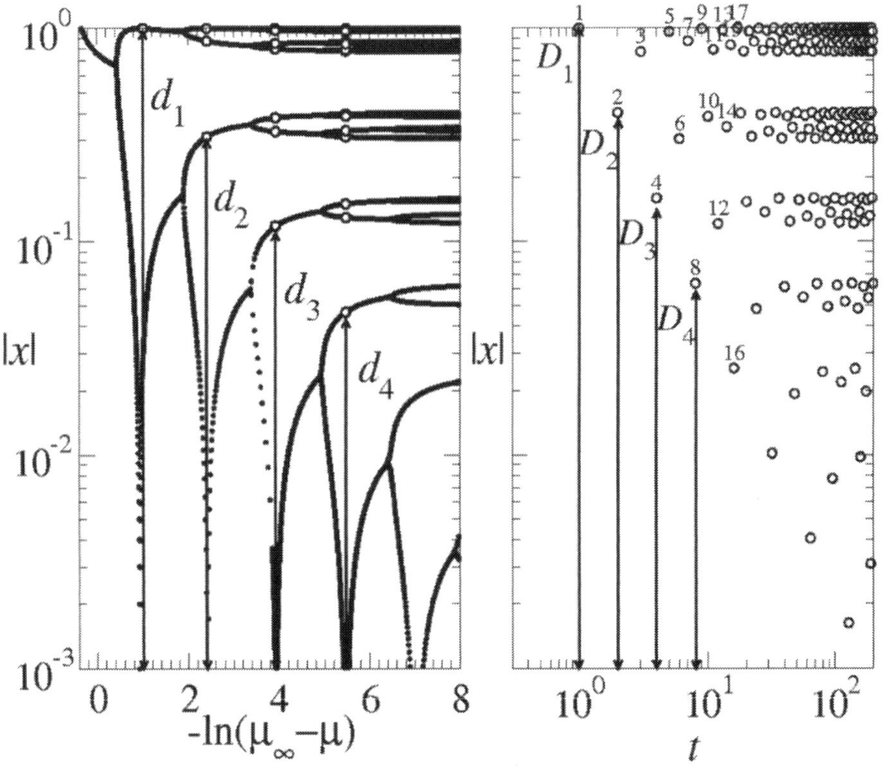

Figure 1. Left panel: Absolute value of attractor positions for the logistic map $f_\mu(x)$ in logarithmic scale as a function of $-\ln(\mu_\infty - \mu)$. Right panel: Absolute value of trajectory positions for $f_\mu(x)$ at μ_∞ with initial condition $x_0 = 0$ in logarithmic scale as a function of the logarithm of time t, also shown by the numbers close to the circles. The arrows indicate the equivalence between the diameters $d_{n,0}$ in the left panel, and position differences D_n with respect to $x_0 = 0$ in the right panel.

Elements in our analysis are the following: i) The preimages of the periodic attractors along the bifurcation cascade that leads to the transition to chaos appear entrenched in a fractal hierarchical structure of increasing complexity as period doubling develops. ii) The limiting form of this rank structure results in an infinite number of families of well-defined phase-space gaps in the positions of the attractor at the transition to chaos, the so-called Feigenbaum attractor. iii) The gaps in each of these families can be ordered with decreasing width in accordance to power laws and are seen to appear sequentially in the dynamics generated by uniform distributions of initial conditions. iv) The emergent power law with log-periodic modulation associated with the rate of approach of trajectories towards the Feigenbaum attractor is explained in terms of the progression of gap formation. v) The relationship between the law of rate of convergence to the Feigenbaum attractor and the inexhaustible hierarchy feature of the preimage structure is elucidated. A detailed account of these properties is given in Refs. [6]-[8]

2. Brief recollection of the dynamics towards the Feigenbaum attractor

We start by recalling that unimodal maps are dissipative due to their one-dimensionality and consequently their dynamical properties are associated with sets of positions called attractors as they capture at sufficiently long times most trajectories. Other trajectories evolve towards related sets of positions called repellors. The trajectories associated to the period-doubling route to chaos in unimodal maps exhibit intricate dynamical properties that follow concerted patterns. At the period-doubling accumulation points, the transitions to chaos, periodic attractors become multifractal before turning chaotic. We call the attractor at the transition to chaos the Feigenbaum attractor when the Lyapunov exponent vanishes as it changes sign [4], [5]. There are two sets of properties associated with the attractors involved in these dissipative systems: those of the dynamics inside the attractors and those of the dynamics towards the attractors. These properties have been characterized in detail, the organization of trajectories and of the sensitivity to initial conditions at the attractor are described in Ref. [6], while the features of the rate of approach of an ensemble of trajectories to the attractor has been explained in Ref. [7].

We recall now some of the basic features of the bifurcation forks that form the period-doubling cascade sequence in unimodal maps, often illustrated by the logistic map $f_\mu(x) = 1 - \mu x^2$, $-1 \leq x \leq 1$, $0 \leq \mu \leq 2$ [4], [5]. The dynamics towards a particular family of periodic attractors, the superstable attractors [2], makes possible the understanding of the rate of approach to the Feigenbaum attractor, located at $\mu = \mu_\infty = 1.401155189...$, and highlights the source of its discrete scale invariant property [8]. The family of trajectories associated with these attractors - called supercycles - of periods 2^N, $N = 1, 2, 3, ...$, are located along the bifurcation forks. The positions (or phases) of the 2^N- attractor are given by $x_m = f^{(m)}_{\overline{\mu}_N}(0)$, $m = 1, 2, \ldots, 2^N$. Notice that infinitely many other sequences of superstable attractors appear at the period-doubling cascades within the windows of periodic attractors for values of $\mu > \mu_\infty$. Associated to the 2^N-attractor at $\mu = \overline{\mu}_N$ there is a $(2^N - 1)$-repellor consisting of $2^N - 1$ positions y_m, $m = 1, 2, \ldots, 2^N - 1$. These positions are the unstable solutions, $\left|df^{(2^{n-1})}_{\overline{\mu}_N}(y)/dy\right| < 1$, of $y = f^{(2^{n-1})}_{\overline{\mu}_N}(y)$, $n = 1, 2, \ldots, N$. The first, $n = 1$, originates at the initial period-doubling bifurcation, the next two, $n = 2$, start at the second bifurcation, and so on, with the last group of 2^{N-1}, $n = N$, setting out from the $N - 1$ bifurcation. See Fig. 1.

Basic to the understanding of the dynamical properties of unimodal maps is the following far-reaching property: Time evolution at μ_∞ from $t = 0$ up to $t \to \infty$ traces the period-doubling cascade progression from $\mu = 0$ up to μ_∞. There is an underlying quantitative relationship between the two developments. Specifically, the trajectory inside the Feigenbaum attractor with initial condition $x_0 = 0$, the 2^∞-supercycle orbit, takes positions x_t such that the distances between appropriate pairs of them reproduce the diameters $d_{N,m}$ defined from the supercycle orbits with $\overline{\mu}_N < \mu_\infty$. See Fig. 1, where the absolute value of positions and logarithmic scales are used to illustrate the equivalence. This property has been central in obtaining rigorous results for the sensitivity to initial conditions for the Feigenbaum attractor [6]. Other families of periodic attractors share most of the properties of supercycles. We consider explicitly the case of a map with quadratic maximum but the results are easily extended to general nonlinearity $z > 1$.

The organization of the total set of trajectories as generated by all possible initial conditions as they flow towards a period 2^N attractor has been determined in detail [8]. It was found

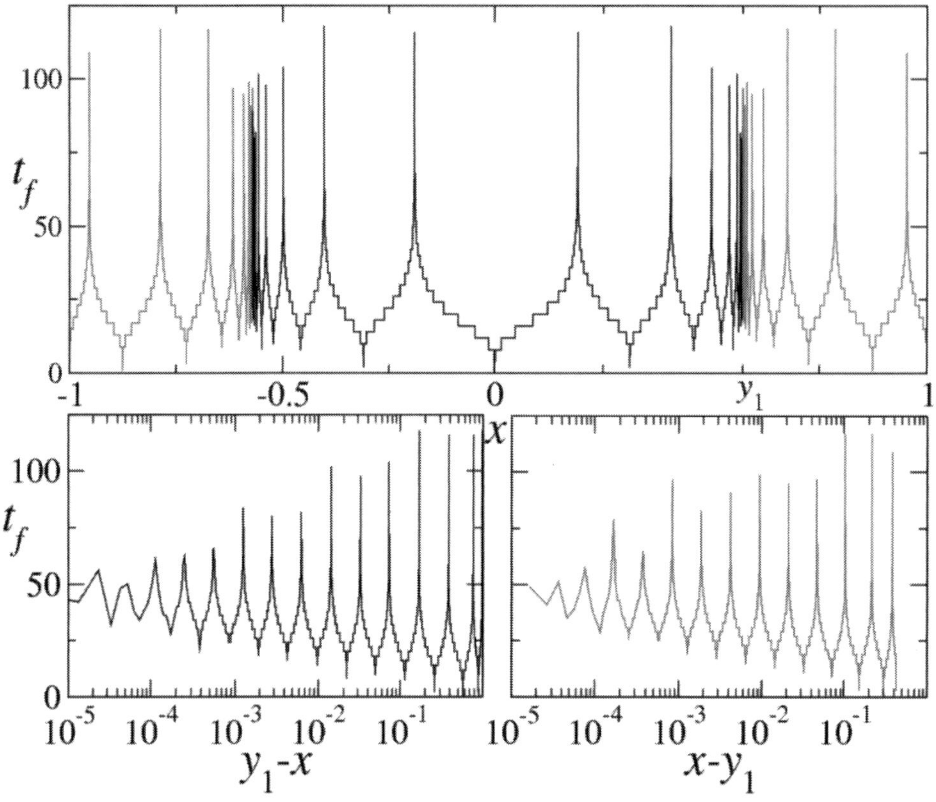

Figure 2. Top panel: Time of flight $t_f(x)$ for $N = 2$, the black lines correspond to initial conditions that terminate at the attractor positions $x = 0$ and $x \simeq -0.310703$, while the grey lines to trajectories ending at $x = 1$ and $x \simeq 0.8734$. Right (left) bottom panel: Same as top panel, but plotted against the logarithm of $x - y_1$ $(y_1 - x)$. It is evident that the peaks are arranged exponentially around the old repellor position y_1, i.e., they appear equidistant in a logarithmic scale.

that the paths taken by the full set of trajectories in their way to the supercycle attractors (or to their complementary repellors) are exceptionally structured. We define the preimage $x^{(k)}$ of order k of position x to satisfy $x = h^{(k)}(x^{(k)})$ where $h^{(k)}(x)$ is the k-th composition of the map $h(x) \equiv f_{\overline{\mu}_N}^{(2^{N-1})}(x)$. The preimages of the attractor of period 2^N, $N = 1, 2, 3, ...$ are distributed into different basins of attraction, one for each of the 2^N phases (positions) that compose the cycle. When $N \geq 2$ these basins are separated by fractal boundaries whose complexity increases with increasing N. The boundaries consist of the preimages of the corresponding repellor and their positions cluster around the $2^N - 1$ repellor positions according to an exponential law. As N increases the structure of the basin boundaries becomes more involved. Namely, the boundaries for the 2^N cycle develops new features around those of the previous 2^{N-1}cycle boundaries, with the outcome that a hierarchical structure arises, leading to embedded clusters of clusters of boundary positions, and so forth. The dynamics associated to families of trajectories always displays a characteristically concerted order in which positions are visited, which in turn reflects the repellor preimage boundary structure of the basins of attraction. That is, each trajectory has an initial position that is identified as a preimage of a given order of an attractor (or repellor) position, and this trajectory necessarily follows the steps of other trajectories with initial conditions of lower preimage order belonging to a given chain or pathway to the attractor

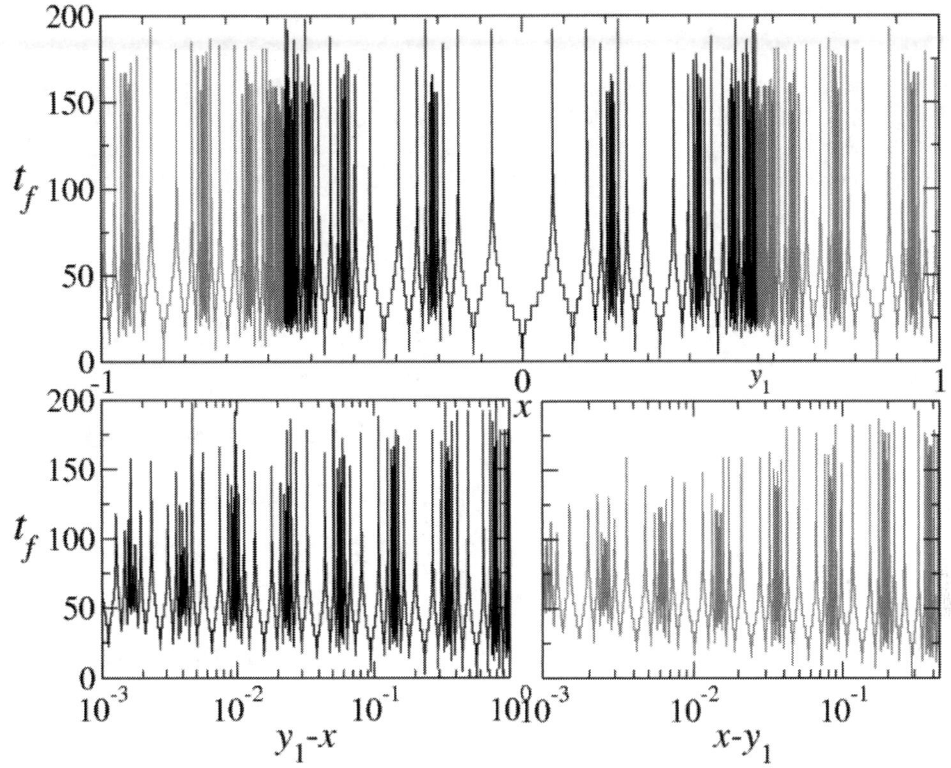

Figure 3. Same as Fig. 2 for $N = 3$. The black lines correspond to initial conditions that terminate at any of the four attractor positions close or equal to $x = 0$, while the grey lines to trajectories ending at any of the other four attractor positions close or equal to $x = 1$. As the bottom panels show, in logarithmic scale, in this case there are (infinitely) many clusters of peaks (repellor preimages) equidistant from each other.

(or repellor). When the period 2^N of the cycle increases the dynamics becomes more involved with increasingly more complex stages that reflect the hierarchical structure of preimages. See Figs. 2 and 3, and see Ref. [8] for details. The fractal features of the boundaries between the basins of attraction of the positions of the periodic orbits develop a structure with hierarchy, and this in turn reflects on the properties of the trajectories. The set of trajectories produce an ordered flow towards the attractor or towards the repellor that reflect the ladder structure of the sub-basins that constitute the mentioned boundaries.

Another way by which the preimage structure described above manifests in the dynamics of the supercycles of periods 2^N is via the successive formation of gaps in phase space (the interval $-1 \leq x \leq 1$) that finally give rise to the attractor and repellor multifractal sets. To observe explicitly this process we consider an ensemble of initial conditions x_0 distributed uniformly across phase space and record their positions at subsequent times. This set of gaps develops in time beginning with the largest one containing the $k = 0$ repellor, then followed by a set of two gaps associated with the $k = 1$ repellor, next a set of four gaps associated with the $k = 2$ repellor, and so forth. This process stops when the order of the gaps k reaches $N - 1$. To facilitate a visual comparison between the process of gap formation at $\mu = \mu_\infty$ and the dynamics inside the Feigenbaum attractor—as illustrated by the trajectory in Fig. 1 right panel—we plot in Fig. 4 the time evolution of an ensemble composed of 10^4 trajectories. We use logarithmic scales for both $|x_t|$ and t and then superpose on the evolution of the ensemble the positions for the trajectory starting at $x_0 = 0$. It is clear from this figure that the gaps that form consecutively all have the same width in the logarithmic scale of the plot and therefore their actual widths decrease as a power law, the same power law followed, for instance, by the position sequence $x_t = \alpha^{-N}$, $t = 2^N$, $N = 0, 1, 2, ...$, for the trajectory inside the attractor starting at $x_0 = 0$ (and where α is Feigenbaum's universal constant). The locations of this specific family of consecutive gaps (the largest gaps for each value of k) advance monotonically toward the sparsest region of the multifractal attractor located at $x = 0$. Other gaps cannot be observed in the scale of the figure due to the way the data is plotted. See Refs. [6]-[8] for more details.

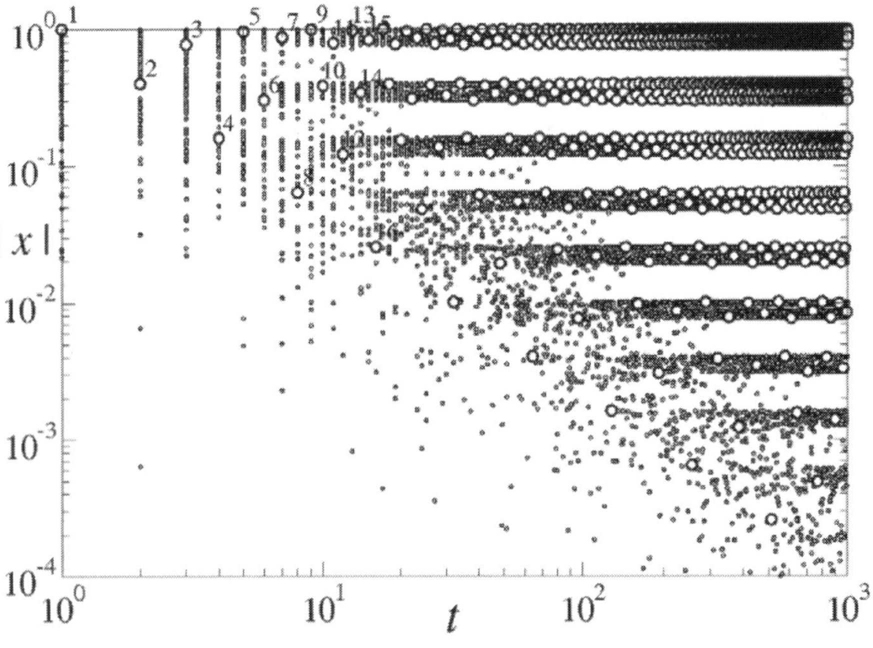

Figure 4. Phase-space gap formation for $\mu = \mu_\infty$. The black dots correspond to time evolution of a uniform ensemble of 10000 trajectories as a function of $|x|$ vs t, both in logarithmic scales. The open circles are the positions, labeled by the times at which they are reached, for the trajectory inside the Feigenbaum attractor with initial condition x_0= 0, same as right panel in Fig. 1.

There is straightforward quantitative way to measure the rate of convergence of an ensemble of trajectories to the attractor and to the repellor that consists in evaluating a single time-dependent quantity. A partition of phase space is made of N_b equally sized boxes or bins and a uniform distribution of N_c initial conditions placed, as above, along the interval $-1 \leq x \leq 1$. The number r of trajectories per box is $r = N_c/N_b$. The quantity of interest is the number of boxes W_t that contain trajectories at time t. This is shown in Fig. 5 on logarithmic scales for the first five supercycles of periods 2^1 to 2^5 where we can observe the following features: In all cases W_t shows a similar initial nearly-constant plateau, and a final well-defined decay to zero. In between these two features there are $N - 1$ oscillations in the logarithmic scales of the figure. As it can be observed in the left panel of Fig. 5, the duration of the overall decay grows approximately proportionally to the period 2^N of the supercycle. We are now in a position to appreciate the dynamical mechanism at work behind the features of the decay rate W_t. From our previous discussion we know that, every time the period of a supercycle increases from 2^{N-1} to 2^N by a shift in the control parameter value from $\overline{\mu}_{N-1}$ to $\overline{\mu}_N$ the preimage structure advances one stage of complication in its hierarchy. Along with this, and in relation to the time evolution of the ensemble of trajectories, an additional set of 2^N smaller phase-space gaps develops and also a further oscillation takes place in the corresponding rate W_t for finite period attractors. At μ_∞ the time evolution tracks the period-doubling cascade progression, and every time t increases from 2^{N-1} to 2^N the flow of trajectories undergoes equivalent passages across stages in the itinerary through the preimage ladder structure, in the development of phase-space gaps, and in logarithmic oscillations in W_t.

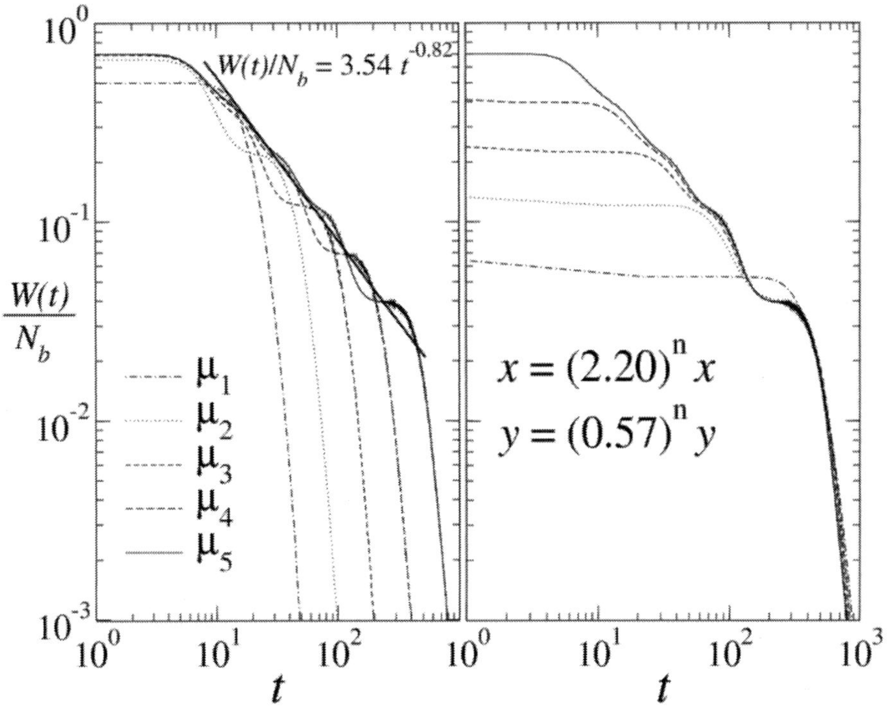

Figure 5. Left panel: The rate $W(t)$, divided by the number of boxes N_b employed, of approach to the attractor for the supercycles of periods 2^N, $N = 1, 2, 3, 4$ and 5 in logarithmic scales. The expression shown corresponds to the power-law decay of the developing logarithmic oscillations. Right panel: Superposition of the five curves for $W(t)$ in the left panel via n-times repeated rescaling factors shown for the horizontal x and vertical y axis.

In summary, each doubling of the period introduces additional modules or building blocks in the hierarchy of the preimage structure, such that the complexity of these added modules is similar to that of the total period 2^N system. As shown in Ref. [7], each period doubling adds also new components in the family of sequentially-formed phase space gaps, and moreover increases in one unit the number of undulations in the transitory log-periodic power-law decay displayed by the fraction W_t of ensemble trajectories still away at a given time t from the attractor (and the repellor). As a consequence of this we have obtained detailed understanding of the mechanism by means of which the discrete scale invariance implied by the log-periodic property [7] in the rate of approach W_t arises.

3. The dynamics towards the Feigenbaum attractor as a dynamical hierarchy with modular organization

We proceed now to the identification of the features of the dynamics towards the Feigenbaum attractor as those of a bona fide model of dynamical hierarchy with modular organization. These are: i) Elementary degrees of freedom and the elementary events associated with them.

Journal of Physics: Conference Series **394** (2012) 012007 doi:10.1088/1742-6596/394/1/012007

ii) Building blocks and the dynamics that takes place within them and through adjacent levels of blocks. iii) Self-similarity characterized by coarse graining and renormalization group (RG) operations. iv) Emerging property of the entire hierarchy absent in the embedded building blocks.

The elementary degrees of freedom at the bottom of the hierarchy are the preimages of the attractors of period 2^2. These are assigned an order k, according to the number k of map iterations they require to reach the attractor. The preimages are also distinguished in relation to the position, $x_1, \ldots, x_{2^2}$, of the attractor they reach first. The preimages of each attractor position appear grouped in basins with fractal boundaries. See Fig. 2. The elementary dynamical event is the reduction of the order k of a preimage by one unit. This event is generated by a single iteration of the map for an initial position placed in a given basin. The result is the translation of the position to a neighboring basin of the same attractor position. The (operative) degrees of freedom higher up in the hierarchy are infinite sets of preimages of attractors of period 2^N, $N > 2$, for which boundary basins are replaced by clusters of boundary basins. See Fig. 3.

The modules or building blocks are clusters of clusters formed by families of boundary basins. The boundary basins of attraction of the 2^N positions of the periodic attractors cluster exponentially and have an alternating structure [8]. In turn, these clusters cluster exponentially themselves with their own alternating structure. Furthermore, there are clusters of these clusters of clusters with similar arrangements, and so on. The dynamics associated with the building blocks consists of the flow of preimages through and out of a cluster or block. Sets of preimages 'evenly' distributed (say one per boundary basin) across a cluster of order N (generated by an attractor of period 2^N) flow orderly throughout the structure. If there is one preimage in each boundary basin of the cluster, each map iteration produces a migration from a boundary basin to a neighboring boundary basin in such a way that at all times there is one preimage per boundary basin, except for the inner ones in the cluster that are gradually emptied.

The self-similar feature in the hierarchy is demonstrated by the coarse-graining property amongst building blocks of the hierarchy that leaves the hierarchy invariant when $\mu = \mu_\infty$. That is, clusters of order N can be simplified into clusters of order $N-1$. There is a self-similar structure for the clusters of any order and a coarse graining can be performed on clusters of order N such that these can be reduced to clusters of lower order, the basic coarse graining is to transform order N into order $N-1$. Also coarse graining can be carried out effectively via the RG functional composition and rescaling, as this transformation reduces the order N of the periodic attractor. Automatically the building-block structure simplifies into that of the next lower order and the clusters of boundary basins of preimages are reduced in one unit of involvedness. Dynamically, the coarse-graining property appears as flow within a cluster of order N simplified into flow within a cluster of order $N-1$. As coarse graining is performed in a given cluster structure of order N, the flow of trajectories through it is correspondingly coarse-grained. e.g., flow out of a cluster of clusters is simplified as flow out of a single cluster. The RG transformation via functional composition and rescaling of the cluster flow is displayed dynamically since by definition functional composition establishes the dynamics of iterates, and the RG transformation $Rf(x) = -\alpha f(f(-x/\alpha)$ leads for N finite to a trivial fixed point that represents the simplest dynamical behavior, that of the period one attractor. For $N \to \infty$ the RG transformation leads to the self-similar dynamics of the non-trivial fixed point, the period-doubling accumulation point.

There is a modular structure of embedded clusters of all orders. The building blocks, clusters of order N, form well-defined sets that are embedded into larger building blocks, sets of clusters of order $N+1$. As the period 2^N of the attractor that generates this structure increases the hierarchy extends and as N diverges a fully self-similar structure develops. The RG transformation for $N \to \infty$ no longer reduces the order of the clusters and a nontrivial fixed point arises. The trajectories consist of embedded flows within clusters of all orders.

The entire flow towards the 2^N-period attractor generated by an ensemble of initial conditions (distributed uniformly across the phase space interval of the map) follows methodically a pattern predetermined by the hierarchical structure of embedded clusters of the preimages.

Each module exhibits a basic kind of flow property. This is the exponential emptying of trajectories within a cluster of order N. Trajectories initiated in the boundary basins that form a cluster of order N flow out of it with an iteration time exponential law. The flow is transferred into a cluster of order $N+1$. This flow is a dynamical module from which a structure of flows is composed. The emerging property that appears when $N \to \infty$ is that there is a power-law emptying of trajectories for the entire hierarchy. The flow of trajectories towards an attractor of period 2^N proceeds via a sequence of stage or step flows each within a cluster of a given order. Thus the first is through a cluster of order 1, then through a cluster of order 2, etc., until the last stage is through a cluster of order N. The sequence evolves in time via a power law decay that is modulated by logarithmic oscillations. This is the emerging property of the model.

4. Summary and discussion

The remarkable properties of unimodal maps already known for a long time [4], [5] have contributed significantly to the historical development of nonlinear dynamics. The two sets of properties that correspond to the dynamics within and towards the Feigenbaum attractor have been studied in detail a few years ago, in the former case the intricate structure of the sensibility to initial conditions was determined [6], while in the latter case a surprisingly rich dynamical organization was discovered [7]. Additionally, a relationship of statistical-mechanical nature between these two types of dynamical properties was established [7].

Here we have made use of the results from these studies to identify the features of the dynamics of an ensemble of trajectories that approach the Feigenbaum attractor as a genuine model hierarchy where all the elements are clearly defined and evaluated, including the observation of an emergent property, the power-law decay with logarithmic oscillations that accompanies the formation of the band structure of the multifractal attractor, and that is in fact an integral part of a basic statistical-mechanical expression that relates an entropy to a partition function [7].This statistical mechanics is of a one-parameter deformed type [7].

We have demonstrated that a dynamically-organized hierarchy with well-defined modules arises directly from a simple formal expression with an iteration rule without the need of any other kind of ingredients. We have chosen a classical one-dimensional nonlinear system, a unimodal map as exemplified by the logistic map, to demonstrate that the intricate dynamical behavior at the transition to chaos fulfills the stipulations of such hierarchical system. This may serve as a basic idea to build or design different types of workable hierarchical models appropriate to different purposes.

Acknowledgements. Support by DGAPA-UNAM-IN100311 and CONACyT-CB-2011-167978 (Mexican Agencies) is acknowledged.

5. References

[1] D. Pumain 2006 *Hierarchy in Natural and Social Sciences* (New York: Springer-Verlag)
[2] V. Ahl and T. F. H. Allen 1996 *Hierarchy theory, a vision, vocabulary and epistemology* (New York: Columbia University Press)
[3] A. Koestler 1967 *The ghost in the machine* (Macmillan, New York)
[4] H.G. Schuster 1988 *Deterministic Chaos. An Introduction* 2nd Revised Edition (Weinheim: VCH Publishers)
[5] R.C. Hilborn 2000 *Chaos and nonlinear dynamics* 2nd Revised Edition (New York: Oxford University Press)
[6] E. Mayoral and A. Robledo 2005 *Phys. Rev. E* **72** 026209
[7] A. Robledo and L.G. Moyano 2008 *Phys. Rev. E* **77** 036213 ; *2009 Braz. J. Phys.* **39** 364
[8] L.G. Moyano, D. Silva and A. Robledo 2009 *Cent. Eur. J. Phys.* **7** 591

Log-periodic oscillations in the specific heat behaviour for self-similar Ising type spin systems

A A Khamzin[1], R R Nigmatullin[1], I I Popov[1] and M P Zhelifonov[2]

[1]Kazan (Volga Region) Federal University, 420008, Kremlevskaya 18, Kazan, Russia

[2]Kazan State Power-Engineering University, 420066, Krasnosel'skaya 51, Kazan, Russia

E-mail: airat.khamzin@rambler.ru

Abstract. The self-similar model of spin-system of the Ising type is formulated. The thermodynamic properties of this model are considered. Analytically and numerically the specific heat of this system is calculated in the nearest neighbor approximation (only the influence of two neighboring spins was taken into account). It is shown that in temperature dependence of the specific heat the log-periodic oscillations are appeared. These oscillations are imposed on the expected power-law dependence.

1. Introduction

Log-periodic oscillations phenomenon can be considered as a general property belonging to a wide class of self-similar systems having discrete scaling invariance [1]. These oscillations have been found in different systems with discrete scale invariance, for example, on lattices (which determine geometrical fractals [2, 3]), or in self-similar distribution of interaction constants [4], or in cases when this invariance takes place in both cases [5, 6]. The first interest to this phenomenon is appeared in the renorm-group theory for description of critical phenomena [7]. But a constant interest to this phenomenon is appeared later because these oscillations have small amplitudes and they cannot be detected easily in evaluation of critical indexes. But the last studying of a wide class of aperiodic models, where fluctuations of interacting constants produce essential changes of the original model, evokes a constant interest to these oscillations again.

The discovery of quasi-crystals in 1984 year [8] evoked again a big interest to quasi-periodic structures that is confirmed by large number of theoretical and experimental papers. The general feature for such structures is the self-similar (fractal) structure of the energy spectrum (see, for example [9]). In general, the behavior of the fractal spectrum exhibits a complex behavior and in order to find and evaluate the thermodynamic peculiarities of these systems some simple models are used. In this sense, in paper [10] as the simplest model of the fractal spectrum the triadic Cantor set was chosen. It was shown the heat capacitance has an interesting behavior: it oscillates log-periodically near the mean value, which is equaled to the fractal dimension of the spectrum considered.

In papers [11] one of the authors of this paper (R.R.N) suggested the theory of dielectric relaxation based on the fractional kinetics. It has been shown that many self-similar dynamic processes taking place on microscopic scales are averaged on the mesoscopic scales; from the mathematical point of view this averaging procedure leads to power-law dependencies with real or complex-conjugated exponents (log-periodic dependencies). It means that in the range of mesoscopic scales the kinetic equations for the total polarization $P(t)$ contains differential operators with non-integer or even complex-conjugated values. This general approach helps to understand the reasons of some empirical expressions which were suggested for description of dielectric permittivity in frequency domain.

In this paper the authors suggest new general approach for detection and evaluation of log-periodic oscillations of different thermodynamic values for spin-models using as an example the 1D Ising model. The selection of this model is not random. This model admits the exact solution which, in turn, is expressed in terms of elementary functions. This peculiarity of the selected model gives a possibility to realize almost all analytical calculations and demonstrate all possibilities of the method suggested. The essence of the approach suggested is related to rather general averaging of thermodynamic functions over the self-similar ensemble of copies. The details of this approach are demonstrated on the ensemble of Ising chains with spin $S=½$ and interaction covering the nearest neighbors.

2. Formulation of the model

We consider a system of spins that is located in a medium and in each point the pressure is transmitted. Under an external tension/pressure applied in the medium the strongly nonequilibrium states can be created. Also, in this medium the process of relaxation can have hierarchy subordinated nature. The nature of this subordination is related to the loss of ergodicity. The creation of this hierarchy subordinated states leads finally to a fractal structure of distribution of thermodynamic potential in configuration space. The fractal properties of the system considered essentially change its thermodynamic and kinetic properties. These modifications are related to partition of configuration space on sets of aggregates (valleys) and each valley has own statistical ensemble. So, in this situation the calculation of average values are realized on two stages. At first, it is necessary to average over ensemble belonging to a 'pure' valley and then realize the second averaging over ensemble describing the distribution of all valleys. The kinetic aspect that presents in this system and conditioned on a weak restoration of ergodicity is determined by slow unification of initial valleys to clusters containing large-scale aggregates. As the result of this unification the relaxation process is divided on hierarchy levels. Each level is associated with statistical ensemble of spins and it is located in space region of the certain pressure. Taking into account that the value of the exchange integral depends, in turn, on the value of the hydrostatic pressure applied we obtain finally self-similar ensembles (clusters) of spins that differ from each other by the value of exchange integral and number of spins entering in it. These values we obey to self-similar conditions.

$$N_l = N_0 b^l, \quad J_l = J_0 \xi^l, \quad 0 < b, \xi < 1, 0 < l < \infty, \tag{1}$$

where N_0, J_0, b, ξ are parameters of the model. Then we realize the averaging procedure of specific free energy of the spin system and find the final expression for the total free energy

$$\overline{f(T)} = N^{-1} \sum_l N_l f(T, J_l). \tag{2}$$

The last expression helps to restore other thermodynamic values of the spin-system considered.

3. Log-periodic oscillations the specific heat in the self-similar Ising spin system

Having in mind only pure demonstration purposes we show how to calculate the specific heat of the model spin system defined above. In this model each hierarchy level is associated with ensemble of spins belonging to the Ising chain with $S=½$ and only the interaction between two nearest neighbors is considered. For this model system expression for the specific thermodynamic potential of the l-th ensemble of spins is determined by well-known expression

$$f(\beta,J_l) = -T\ln\left[2\operatorname{ch}(J_l/T)\right] = -(J_0/\beta)\ln\left[2\operatorname{ch}(\beta\xi^l)\right], \tag{3}$$

where $\beta = J_0/T$. Then the averaged thermodynamic potential по формуле (2) is determined by the following expression

$$\overline{f(\beta)} = N^{-1}\sum\nolimits_l N_l f(\beta,J_l) = -(N_0 J_0/N)\sum\nolimits_{l=0}^{\infty} b^l\beta^{-1}\ln\left[2\operatorname{ch}(\beta\xi^l)\right]. \tag{4}$$

where the value of N determines the total number of spins in the model considered. From the normalization condition $\sum_{l=0}^{\infty} N_l = N$ we find $N_0/N = 1-b$. We introduce the important characteristic of the self-similar system considered as the "fractal" dimension

$$d = \ln b/\ln\xi. \tag{5}$$

Then $b = \xi^d$ and d is chosen as independent parameter of the system. Taking into account these definitions expression (4) is rewritten in the form

$$\overline{f(\beta)} = -(1-\xi^d)J_0\sum\nolimits_{l=0}^{\infty}\xi^{d\cdot l}\beta^{-1}\ln\left[2\operatorname{ch}(\beta\xi^l)\right]. \tag{6}$$

The specific heat of the self-similar system of the Ising chains is determined by expression

$$\overline{C}(\beta) = -\beta^2 d^2\left(\beta\overline{f(\beta)}\right)/d\beta^2 = \left(1-\xi^d\right)\sum\nolimits_{l=0}^{\infty}\xi^{d\cdot l}(\beta\xi^l)^2\operatorname{ch}^{-2}(\beta\xi^l). \tag{7}$$

For calculation of the infinite sum (7) it is convenient to use the Mellin transform. Applying the Mellin transform to both parts of (7) and using the formulae

$$q(\beta) = \beta^2\operatorname{ch}^{-2}\beta \overset{\text{MT}}{=} \hat{Q}(s) = \int_0^{\infty}\beta^{s-1}\beta^2\operatorname{ch}^{-2}\beta d\beta = 2^{-s}(1-2^{-s})\Gamma(s+2)\zeta(s+1), \quad \gamma = \operatorname{Re}s > -2, \tag{8}$$

$q(\beta\xi^l) \overset{\text{MT}}{=} \xi^{-ls}\hat{Q}(s)$, where $\Gamma(x)$ is gamma-function and $\zeta(x)$ is zeta Riemann function, then for the Mellin image of the function $\overline{C}(\beta)$ one can obtain the following expression:

$$\hat{C}(s) = (1-\xi^d)\hat{Q}(s)\sum_{l=0}^{\infty}\xi^{(d-s)l} = \frac{(1-\xi^d)\hat{Q}(s)}{1-\xi^{(d-s)}}, \quad -2<\gamma=\operatorname{Re}s<d, \xi<1. \tag{9}$$

Realizing the inverse Mellin transform of (9) we get

$$\overline{C}(\beta) = \frac{1}{2\pi i}\int_{\gamma-i\infty}^{\gamma+i\infty}\hat{C}(s)\beta^{-s}ds = \varepsilon(1-\xi^{\mp d})\frac{1}{2\pi i}\int_{\gamma-i\infty}^{\gamma+i\infty}\frac{\hat{Q}(s)\beta^{-s}ds}{1-\xi^{(d-s)}}, \tag{10}$$

where the parameter γ is chosen in accordance with condition (9). The value of $\varepsilon=\pm1$ determines the chosen contour of integration. For calculation of the Mellin-Barns integral in the right-hand side of (10) the integration line we complete up to the closed counter forming correspondingly the left ($\varepsilon=+1$) or the right ($\varepsilon=-1$) semicircle and after that we use the Cauchy theorem

$$\overline{C}(\beta) = \varepsilon(1-\xi^d)\sum_k\operatorname*{Res}_{s_k}\left[\hat{Q}(s)\beta^{-s}\left(1-\xi^{(d-s)}\right)^{-1}\right]. \tag{11}$$

Here s_k are the poles belonging to the chosen contour and form two groups: s'_k are the poles of the function $\hat{Q}(s)$ (that is determined by expression (8)) and s''_k are zeros of the function $1-\xi^{(d-s)}$ which in turn are determined by expression

$$s''_k = d + 2\pi ik/\ln\xi, \quad k\in\mathbb{Z}. \tag{12}$$

The poles of the function $\hat{Q}(s)$ are determined by the poles of gamma-function figuring in (8) and these poles are well-known and equaled to $s'_k = -k-2 \ (k = 0,1,2,...)$.

For the case $\xi < 1$ the poles $s''_k = d + 2\pi ik / \ln\xi$ are located on the right-hand side from γ, so the temperature interval is divided onto regions: $T < T_m$ and $T > T_m$, where T_m corresponds to the temperature of the heat capacitance global maximum. For region $T < T_m$ the contribution to $\overline{C}(\beta)$ gives only poles s''_k and so the integration line in (10) we close by the right semicircle. In the result we have

$$\overline{C}(T) = -(1-\xi^d)\sum_k \operatorname*{Res}_{s''_k}\left[\hat{Q}(s)\beta^{-s}\left(1-\xi^{(d-s)}\right)^{-1}\right] = \ln^{-1}(1/\xi)\left(1-\xi^d\right)(T/J_0)^d w_d\left(\ln T/\ln(1/\xi)\right), \quad (13)$$

где $w_d(x)$ - periodic function having the unit period.

$$w_d(x) = \sum_{k=-\infty}^{\infty} \hat{Q}\left(d + 2\pi ik / \ln(1/\xi)\right)e^{2\pi ikx}. \quad (14)$$

In the region of temperatures $T > T_m$ the integration line is closed by the left semicircle and contribution to heat capacitance comes from the poles s'_k. Taking into account the known relationship $\operatorname*{Res}_{-k-2}\Gamma(s+2) = (-1)^k / k!$ we obtain

$$\overline{C}(\beta) = (1-\xi^d)\sum_k \operatorname*{Res}_{s'_k}\left[\frac{\hat{Q}(s)\beta^{-s}}{1-\xi^{(d-s)}}\right] = (1-\xi^d)\sum_{k=0}^{\infty}\frac{2^{k+2}(2^{k+2}-1)(-1)^{k+1}\zeta(-k-1)\beta^{k+2}}{k!\left(1-\xi^{(2+d+k)}\right)}. \quad (15)$$

Taking into account the values $\zeta(-2m) = 0$, $\zeta(-(2m-1)) = -B_{2m}/2m$, $m = 1,2,...$, where B_{2m} are Bernoulli numbers, expression (15) can be rewritten in the form

$$\overline{C}(\beta) = (1-\xi^d)\sum_{k=1}^{\infty}\frac{4^k(4^k-1)B_{2k}\beta^{2k}}{2k\cdot(2k-2)!\left(1-\xi^{(d+2k)}\right)}. \quad (16)$$

The series (16) converges at $T > T_m = 0{,}63694 J_0$. (The convergence radius is found for power law series as the limit of the ratio a_{k-1}/a_k at $k\rightarrow\infty$). From expression (13) it follows that in the interval $T < T_m$ we have power-law dependence T^d in behaviour of heat capacitance that is modified by log-periodic oscillations. Above this temperature T_m the capacitance is decreasing monotonically with increasing of temperature. On Fig. 1 (a) we give temperature dependencies of heat capacitance calculated in accordance with expression (7) at $\xi = 0{,}25$ for different values of fractal dimension d and Fig. 1 (b) demonstrate its behavior at d=0,25 for different values of parameter ξ. These two plots clearly shows the oscillating character of heat capacitance at low temperatures and the character of this oscillations is changed with increasing of dimension d and they decay clearly at increasing of the parameter ξ.

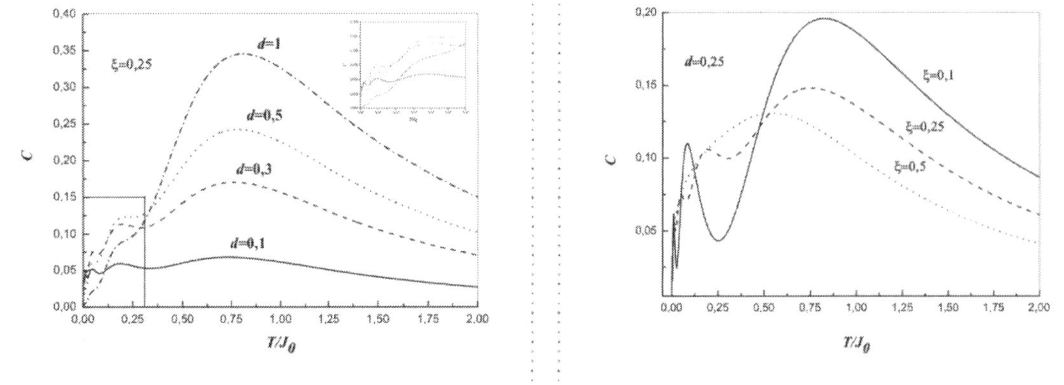

a) b)

Figure 1. The dependence of the specific heat with respect to the dimensionless temperature (T/J_0) at various values of the parameters ξ and d.

Figs. 2 show the dependence of the function $C(T)\cdot T^{-d}$ from temperature for some values of fractal dimension d (Fig. 2 (a)) and parameter ξ (Fig. 2 (b)). Fig. 2 (b) clearly shows the decay of these oscillations with increasing of the value of the scaling parameter ξ. The region of parameters ξ and d where the influence of log-periodic oscillations is essential one can evaluate as the ratio of the first decomposition coefficient to the zero one in behaviour of the log-periodic function $w_d(x)$ (14).

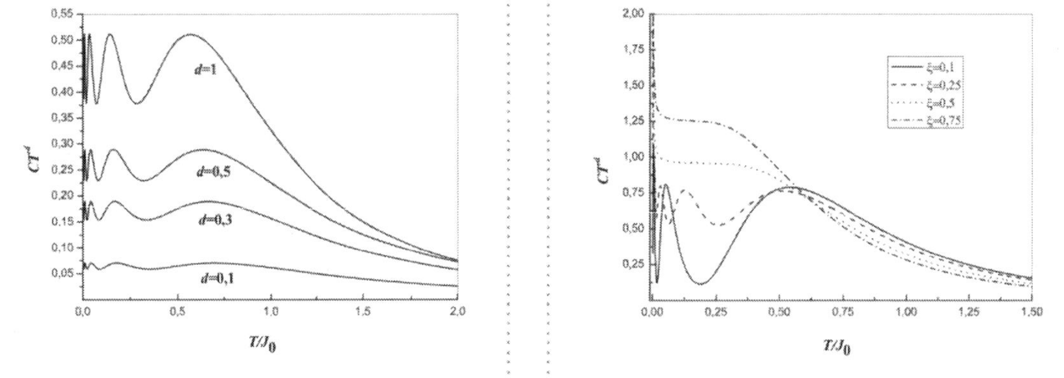

a) b)

Figure 2. The dependence $C(T)\cdot T^{-d}$ against the temperature for $\xi = 0,25$ at various values of the fractal dimension d (Fig. 2 (a)) and for d=1,5 at various values of ξ (Fig. 2 (b)).

4. Conclusions

In this paper we considered the thermodynamics of the spin system of the Ising-type having self-similar structure. The self-similar structure is presented by rather simple model: self- similar conditions are fulfilled for the constants of the exchange interaction and number of spins in a cluster having the same value of the exchange integral. It is clear that the suggested model of such type is not a specific one. It can be applied for a wide class of spin and electron many-body systems. The suggested model of the self-similar medium is characterized by only two parameters: scaling parameter ξ characterizing the hierarchy of the constants of interaction and the generalized fractal dimension *d*.

The equilibrium properties of the model suggested were based on the analysis of the specific heat depending on temperature, scaling parameter ξ and fractal dimension *d*. The averaging of the specific heat over self-similar ensemble allows to receive the log-periodic oscillations which form a mixed combination with power-law temperature dependence. Temperature region of the existence of the log-periodic oscillations is determined by the parameter ξ. In the case $\xi < 1$ which is considered in the paper the desired oscillations take place at low temperatures but for $\xi > 1$ these oscillations are appeared at high-temperatures. In the last case these oscillations have small amplitude because the log-periodic component is appeared together with monotone dependence. The character of the monotonic component is determined presumably by the value of the fractal dimension *d*, and contains alongside with power-law terms the logarithmic corrections also. So, one can state that log-periodic oscillations associate closely with the fractal (self-similar) structure of the system considered, which, in turn, accompanies a wide class of different processes. So, any accurate and general approach that enables to consider and model of hierarchy structure of disordered systems opens new possibilities in analysis of a wide class of physical phenomena that take place in the systems of such kind. The suggested approach combining the simplicity of the calculated scheme allows to receive the important characteristics of the self-similar system as the log-periodic oscillations of a wide class of thermodynamic values. It creates a general base for calculation and analytical evaluations of equilibrium values of various disordered structures based, in turn. on different self-similar (fractal) models.

5. Acknowledgements

The work was done in the frame of the research plan "Dielectric spectroscopy and kinetics of complex systems" (Number: 12-18 02 0210 021000018) that has been accepted by the KFU for 2012 year.

6. References

[1] Sornette D 1998 *Phys. Rep.* **297** 239.
[2] Kutnjak-Urbanc B, Zapperi S, Milosevic S and Stanley H E 1996 *Phys. Rev. E* **54** 272.
[3] Lessa J C and Andrade R F C 2000 *Phys. Rev E* **61**.
[4] Berche P E and Berche B 1997 *J. Phys. A* **30** 1347.
[5] Andrade R F C 1999 *Phys. Rev E* **59** 150.
[6] Andrade R F C 2000 *Phys. Rev E* **61** 7196.
[7] Niemeijer T and van Leeuwen J M J 1976 in *Phase Transitions and Critical Phenomena*, ed C Domb and M S Green, vol 6 (New York: Academic) p 425.
[8] Schechtman D, Blech I, Gratias D and Cahn J W 1984 *Phys. Rev. Lett.* **53** 1951.
[9] Fu X, Liu T, Zhou P and Sritrakool W 1997 *Phys. Rev. B* **55** 2882.
[10] Tsallis C, da Silva L R, Mendes R S, Vallejos R O and Mariz A 1997 *Phys. Rev. E* **56** R4922.
[11] Nigmatullin R R 2006 *Physica A* **363** 282; 2005 *Physica B* **358** 201.

I-theorem and self-organization in the van der Pol generator

R G Zaripov

Institute of Mechanics and Engineering, Kazan Scientific Center,
Russian Academy of Sciences,
Lobachevsky str. 2/31, Kazan, 420111, Russia

E-mail: zaripov@mail.knc.ru

Abstract. Based on the S- and I-theorems of evolution of q- entropy and q- discrimination information, the process of self-organization developing in the van der Pol generator with variation of the feedback parameter at induced transitions between stationary states in an open nonextensive system is studied.

To date, many properties of nonextensive (nonadditive) statistical systems that are described by non-Gaussian and non-Gibbs distributions in various fields of physics have been studied. A new area of research - nonextensive statistical mechanics and thermodynamics - is being developed. A detailed review of related works is given in [1-3].

For additive systems, self-organization in the generator was considered in [4], where the S-theorem on decrease in the renormalized Boltzmann–Gibbs–Shannon entropy (this entropy corresponds to the equal values of the average energy) was formulated for the first time. The *I*-theorem states that the renormalized Kullback discrimination information increases during self-organization of open systems [5, 6].

The van der Pol generator is described by the Langevin equations

$$\frac{dx}{dt}=v,\ \frac{dv}{dt}+(-a+bE)v+\omega_0^2 x=\sqrt{D}y(t), \tag{1}$$

where $a=a_f-\gamma$, a_f is the feedback coefficient (control parameter), γ and b are the coefficients of linear and nonlinear friction, $E=m(v^2+\omega_0^2x^2)/2$ is the oscillations energy, and D is the noise intensity. Moments $y(t)$ of the Langevin source are determined by the expressions $\langle y(t)\rangle=0$ and $\langle y(t)y(t')\rangle=2\delta(t-t')$. In additive statistical thermodynamics, induced transitions between states of an open system for γ, $|a|$, $b\langle E\rangle<<\omega_0$ are described by the Fokker–Planck equation for the distribution [4],

$$\frac{\partial p}{\partial t}=\frac{\partial}{\partial E}\left[(-a+bE)Ep+D(E)E\frac{\partial p}{\partial E}\right]. \tag{2}$$

Its solution for a stationary state with given noise $D(E)=D$ has the form of the "canonical" Gibbs distribution

$$p_0 = \exp\frac{F-H}{D},\ H = -aE + \frac{1}{2}bE^2,\ \int_0^{\infty} p_0 \mathrm{d}E = 1 \tag{3}$$

with effective "free energy" F, effective Hamilton function H and $k = 1$. The control parameter here is feedback coefficient a_f.

In the case of nonextensive systems, equation (2) can be generalized so that the dependence of the diffusion coefficient on the distribution is taken into account. This yields the quasi-parabolic equation [7]

$$\frac{\partial p}{\partial t} = \frac{\partial}{\partial E}\left[(-a+bE)Ep + D_q(E,p)E\frac{\partial p}{\partial E}\right], \tag{4}$$

which describes the Brownian motion. The properties of this equation are of particular interest. In the stationary state at $q = 1$, the diffusion coefficient is $D_1(E,p) = D$; at $q \neq 1$, determination of the diffusion coefficient yields a solution in the known form of non-Gibbs equilibrium distribution with $D = \mathrm{const}$. These conditions are satisfied by function $D_q(E,p) = Dp^{1-q}$. As a result, for the stationary state, we have the distribution

$$p_0 = \left[1-(1-q)D^{-1}\left(-aE + bE^2/2\right)\right]^{1/(1-q)} Z_q^{-1}, \tag{5}$$

which depends on control parameter a_f.

Equation (4) is nonlinear with respect to distribution, and its solution describes the Brownian motion, which will be used to reveal self-organization in the van der Pol generator at induced transitions between states.

In accordance with [4], we determine two values of the control parameter, apply approximations that lead to two distributions and consider induced transitions between the corresponding stationary states.

First, we set the initial condition $a_f = 0$ for the state of "physical chaos" [4] and assume that $b\langle E\rangle/\gamma \sim Db/\gamma^2 \ll 1$. Then, equation (5) yields the "equilibrium" distribution

$$\begin{gathered} p_{(1)} = \left[1-(1-q)D^{-1}\gamma E\right]^{1/(1-q)} Z_{q(1)}^{-1}, \\ Z_{q(1)} = \left[\frac{D}{\gamma(q-1)}\right]\Gamma\left(\frac{2-q}{q-1}\right)\Gamma^{-1}\left(\frac{1}{q-1}\right), \end{gathered} \tag{6}$$

where $\Gamma(\xi)$ is the gamma function and $1 < q < 2$.

Entropy Havrda-Charvat-Daroczy [8, 9] and the mean energy have the form

$$\begin{gathered} S_q\left(p_{(1)}\right) = \frac{1-\int p_{(1)}^q \mathrm{d}E}{q-1} = \frac{1}{q-1}\left\{1\ \left[\frac{D}{\gamma(q-1)}\right]\Gamma\left(\frac{1}{q-1}\right)\Gamma^{-1}\left(\frac{q}{q-1}\right)Z_{q(1)}^{-q}\right\}, \\ E_{q(1)} = \int E p_{(1)}^q \mathrm{d}E = \left[\frac{D}{\gamma(q-1)}\right]^2 \Gamma\left(\frac{2-q}{q-1}\right)\Gamma^{-1}\left(\frac{q}{q-1}\right)Z_{q(1)}^{-q}. \end{gathered} \tag{7}$$

Second, we consider a generation threshold with $a_f = \gamma$ and a distribution with $1 < q < 3$,

$$p_{(2)} = \left[1-(1-q)D^{-1}\left(bE^2/2\right)\right]^{1/(1-q)} Z_{q(2)}^{-1},$$
$$Z_{q(2)} = \sqrt{\frac{\pi D}{2b(q-1)}}\,\Gamma\left(\frac{3-q}{2(q-1)}\right)\Gamma^{-1}\left(\frac{1}{q-1}\right). \tag{8}$$

The entropy and mean energy is obtained in the same fashion,

$$S_q\left(p_{(2)}\right) = \frac{1}{q-1}\left\{1-\left[\frac{\pi D}{2b(q-1)}\right]^{1/2}\Gamma\left(\frac{q+1}{2(q-1)}\right)\Gamma^{-1}\left(\frac{q}{q-1}\right)Z_{q(2)}^{-q}\right\},$$
$$E_{q(2)} = \left[\frac{D}{b(q-1)}\right]\Gamma\left(\frac{1}{q-1}\right)\Gamma^{-1}\left(\frac{q}{q-1}\right)Z_{q(2)}^{-q}. \tag{9}$$

At high energies, equations (6) and (8) yield asymptotic power-type distributions $p_{(1)} \sim E^{-1/(q-1)}$ and $p_{(2)} \sim E^{-2/(q-1)}$, which are typical of fractal systems. If $q=1$, we obtain the respective well-known distributions $p_{(1)} = (\gamma/D)\exp(-\gamma E/D)$ and $p_{(2)} = \sqrt{2b/\pi D}\exp\left(-b^2E/2D\right)$, which are used in additive statistical thermodynamics [4]. The limiting value $q=2$ yields the powerfunction distributions $p_{(1)} = C_1E^{-1}$ and $p_{(2)} = C_2E^{-2}$, which are also valid for fractal systems at low energies.

To find the degree of order of microstates in the generator, consider the renormalized distribution

$$\tilde{p}_{(1)} = \left[1-(1-q)\tilde{D}^{-1}\left(a_f\right)\gamma E\right]^{1/(1-q)}\tilde{Z}_{q(1)}^{-1},$$
$$\tilde{Z}_{q(1)} = \left[\frac{\tilde{D}\left(a_f\right)}{\gamma(q-1)}\right]\Gamma\left(\frac{2-q}{q-1}\right)\Gamma^{-1}\left(\frac{1}{q-1}\right), \tag{10}$$

where $\tilde{D}\left(a_f\right) = D$ and $\tilde{Z}_{q(1)} = Z_{q(1)}$ at $a_f = 0$. The entropy and the mean energy take the form

$$S_q\left(\tilde{p}_{(1)}\right) = \frac{1}{q-1}\left\{1-\left[\frac{\tilde{D}\left(a_f\right)}{\gamma(q-1)}\right]\Gamma\left(\frac{1}{q-1}\right)\Gamma^{-1}\left(\frac{q}{q-1}\right)\tilde{Z}_{q(1)}^{-q}\right\},$$
$$\tilde{E}_{q(1)} = \left[\frac{\tilde{D}\left(a_f\right)}{\gamma(q-1)}\right]^2\Gamma\left(\frac{2-q}{q-1}\right)\Gamma^{-1}\left(\frac{q}{q-1}\right)\tilde{Z}_{q(1)}^{-q}. \tag{11}$$

Analyzing the transition between $p_{(2)}$ and $\tilde{p}_{(1)}$, we find Rathie-Kannappan discrimination information [10]

$$I_q\left(p_{(2)} : \tilde{p}_{(1)}\right) = \frac{1}{q-1}\left[1-\int p_{(2)}^q\tilde{p}_{(1)}^{1-q}\mathrm{d}E\right] =$$
$$= \tilde{Z}_q^{q-1}\left\{-\left[S_q\left(p_{(2)}\right)-S_q\left(\tilde{p}_{(1)}\right)\right]+\tilde{D}^{-1}\left(a_f\right)\left[E_{q(2)}-\tilde{E}_{q(1)}\right]\right\}. \tag{12}$$

Now we require that the mean energies be equal,

$$\int Ep_{(2)}^q\,\mathrm{d}E = \int E\tilde{p}_{(1)}^q\,\mathrm{d}E. \tag{13}$$

This equality follows from the additivity condition [2]

$$I_q\left(p_{(2)}:\tilde{p}_{(1)}\right)\tilde{Z}_{q(1)}^{1-q}=\left[I_q\left(p_{(2)}:p_{(1)}\right)-I_q\left(p_{(2)}:\tilde{p}_{(1)}\right)\right]Z_{q(1)}^{1-q}. \tag{14}$$

whatever the value of the control parameter. Then, we find a dependence between the renormalized and given values of noise,

$$\begin{gathered}\left[\frac{\tilde{D}(a_f)}{\gamma(q-1)}\right]^{2-q}\Gamma\left(\frac{2-q}{q-1}\right)\Gamma^{-q}\left(\frac{2-q}{q-1}\right)=\\ =\frac{D}{b(q-1)}\left[\frac{2b(q-1)}{\pi D}\right]^{q/2}\Gamma\left(\frac{1}{q-1}\right)\Gamma^{-q}\left(\frac{3-q}{2(q-1)}\right),\ \tilde{D}(a_f)>D.\end{gathered} \tag{15}$$

As a result, we prove the validity of the S- and I- theorems,

$$\begin{gathered}I_q\left(p_{(2)}:\tilde{p}_{(1)}\right)\tilde{Z}_{q(1)}^{1-q}=\left[I_q\left(p_{(2)}:p_{(1)}\right)-I_q\left(p_{(2)}:\tilde{p}_{(1)}\right)\right]Z_q^{1-q}=-\left[S_q\left(p_{(2)}\right)-S_q\left(\tilde{p}_{(1)}\right)\right]=\\ =\left[\frac{\pi D}{2b(q-1)}\right]^{1/2}\Gamma\left(\frac{q+1}{2(q-1)}\right)\Gamma^{-1}\left(\frac{q}{q-1}\right)Z_{q(2)}^{-q}-\\ -\left[\frac{\tilde{D}(a_f)}{\gamma(q-1)}\right]\Gamma\left(\frac{1}{q-1}\right)\Gamma^{-1}\left(\frac{q}{q-1}\right)\tilde{Z}_{q(1)}^{-q}>0.\end{gathered} \tag{16}$$

Thus, it follows from equations (16) that, as feedback parameter a_f increases from zero to γ, renormalized Havrda-Charvat-Daroczy entropy $S_q\left(\tilde{p}_{(1)}\right)<S_q\left(p_{(2)}\right)$ decreases (S- theorem) and simultaneously renormalized Rathie-Kannappan discrimination information $I_q\left(p_{(2)}:p_{(1)}\right)>I_q\left(p_{(2)}:\tilde{p}_{(1)}\right)$ increases (I-theorem). This is an indication of self-organization in the van der Pol generator. The same result follows from analysis of the third stationary state with developed generation and corresponding renormalization relative to the state with $a_f=0$. In this case, a change in entropy $D\left[S_q\left(p_{(1)}\right)-S_q\left(\tilde{p}_{(2)}\right)\right]=R_{\min}$ [5, 6] (where $k^{-1}D$ plays the role of the effective temperature) is due to work on ordering in the generator.

Calculations lead to the inequality $\mathrm{d}\left[S_q\left(\tilde{p}_{(1)}\right)\right]/\mathrm{d}a_f<0$, which proves that the state of the generator with the distribution $\tilde{p}_{(1)}$ is stable. Note that equality (14) and, accordingly, condition (13) need not be fulfilled in the general case. Then, inequality (16) does not hold and one cannot argue that a given process of generation is self-organizing.

Thus, it has been shown that an increase in the feedback parameter to the generation threshold in the Van der Pol generator for the case of nonextensive systems with $1 < q < 2$ causes self-organization with an increase in the extent of order and decrease in the extent of disorder. At $q = 1$, the results coincide with those known for additive systems [4]. In this work, only two stationary states and the transition between them have been considered. The results obtained here can be applied in analysis of fractal electrical oscillatory systems for the presence of self-organization.

Only one work is known [11] in which the van der Pol generator is also considered using Havrda-Charvat-Daroczy entropy and Bregman's measure [12]

$$D(p:p_0)=\int\left[f(p)-f(p_0)-(p-p_0)\frac{\partial f(p_0)}{\partial p_0}\right]\mathrm{d}X \qquad (17)$$

as the discrimination information. Functional (17) was first introduced in [12] as a convex function in problems of linear and convex programming. Function $f(p)=k(q-1)^{-1}(p^q-p)-p$ follows from the definition of entropy in the form $S_q=-\int f(p)\mathrm{d}X$; consequently, equation (17) yields the measure [11, 13]

$$\begin{aligned}D(p:p_0)&=-\left[S_q(p)-S_q(p_0)\right]-\int(p-p_0)\frac{\partial f(p_0)}{\partial p_0}\mathrm{d}X\\&=\frac{k}{q-1}\int p^q\mathrm{d}X+k\int p_0^q\mathrm{d}X\quad\frac{kq}{q-1}\int pp_0^{q-1}\mathrm{d}X,\end{aligned} \qquad (18)$$

used here. At q = 1, equation (18) gives the Kullback discrimination information borrowed from additive statistical mechanics.

We present some remarks. A disadvantage of Bregman's measure (17) is that it does not provide any composition law for independent systems. Note that the author of [11] applies distributions borrowed from [4] that characterize the stationary states of additive systems described by equation (2). In addition, he considers transitions between distributions $p_{(1)}$ and $p_{(2)}^q/\int p_{(2)}^q\mathrm{d}X$ from various families, which is inconsistent with the approach used in [14, 15]. That only such an averaging procedure over unnormalized distributions of p^q is valid [8, 9] also follows from the condition of thermodynamic equilibrium for nonextensive systems [16]. Also in work [11] value of a derivative $\mathrm{d}\left[S_q\left(\tilde{p}_{(1)}\right)\right]/\mathrm{d}a_f$ is not determined.

Since the equations use unnormalized distributions p^q and p_0^q from the same family, the new functional [7]

$$D_q(p:p_0)=\int\left[f(p)-f(p_0)-(p^q-p_0^q)\frac{\partial f(p_0)}{\partial(p_0^q)}\right]\mathrm{d}X \qquad (19)$$

is a natural generalization of measure (17) for nonextensive systems to determine the discrimination information. If the above function $f(p)$ is used as the Havrda-Charvat-Daroczy entropy, this functional takes the form

$$D_q(p:p_0)=-\left[S_q(p)-S_q(p_0)\right]-\int(p^q-p_0^q)\frac{\partial f(p_0)}{\partial(p_0^q)}\mathrm{d}X=\frac{1}{1-q}\left(1\quad\int p^q p_0^{1-q}\mathrm{d}X\right), \qquad (20)$$

which expectedly coincides with the expression for Rathie-Kannappan discrimination information [10]. This allows one to find the respective discrimination information in the case of other expressions for the nonextensive entropy and use it to detect the self-organization process in open systems.

References

[1] Tsallis C 2009 *Introduction to Nonextensive Statistical Mechanics. Approaching a Complex World* (New York: Springer) p 382

[2] Zaripov R G 2010 *Principles of Nonextensive Statistical Mechanics and Geometry of Measures of the Disorder and the Order* (Kazan: Kazan A.N. Tupolev State Technical University Press) p 404 (in Russian)

[3] Naudts Jan 2011 *Generalized Thermostatistics* (London: Springer) p 201
[4] Klimontovich Yu L 1983 *Soviet Technical Physics Letters* **9** 606
[5] Zaripov R G 1988 *Soviet Technical Physics* **33** 1366
[6] Zaripov R G 1999 *Discrimination information and disorder-order transitions* (Kazan: Kazan A.N. Tupolev State Technical University Press) p 155 (in Russian)
[7] Zaripov R G 2009 *Technical Physics* **54** 6
[8] Havrda J and Charvat F 1967 *Kybernetika* **3** 30
[9] Daroczy Z 1970 *Inform. Control.* **16** 36
[10] Rathie P N and Kannappan Pl 1972 *Inform. Control.* **20** 38
[11] Bagci G B 2007 *Arxiv preprint arXiv:* cond-mat/0705.2053
[12] Bregman L M 1967 *USSR Comput. Math. Math. Phys.* **7** 200
[13] Abe S and Bagci G B 2005 *Phys. Rev.* E **71** 016139
[14] Zaripov R G 2001 *Russian Phys. J.* **44** 1159
[15] Zaripov R G 2004 *Russian Phys. J.* **47** 647
[16] Zaripov R G 2006 *Technical Physics* **51** 1393

Journal of Physics: Conference Series **394** (2012) 012010 doi:10.1088/1742-6596/394/1/012010

Gauge theory approach to glass transition

Mikhail Vasin

Physical-Technical Institute, Ural Branch of Russian Academy of Sciences, 426000 Izhevsk, Russia

E-mail: dr_vasin@mail.ru

Abstract. This approach combines the gauge field theory of spin-glasses [1, 2, 3, 4] with the critical dynamics methods in order to describe the glass transition. The glass transition is regarded as a phase transition, which is interrupted because of premature critical slowing down, caused by the frustrations. The frustrations induce sources of the gauge field which are additional to the thermal activated spin vortexes. Interaction of this field with the order parameter field leads to the system's critical slowing at the glass transition temperature, and to the freezing of disordered structure.

1. Introduction

The formulation of the microscopic glass transition theory has remained one of the most intriguing but still unresolved problems of condensed matter physics for a long time. Many systems which manifest this phenomenon regardless of their nature enable us to conclude that this phenomenon does not depend on any microscopic details, but is determined by the symmetry properties of the systems, as in the case of phase transitions. On the other hand this universality allows to conclude that the glass transition is only the result of general dynamical properties of condensed matter. Therefore, the question "Is the glass transition a phase transition, or it is the dynamic effect related with limited diffusion dynamics?" remains to be actual. We have tried to conciliate these positions using the gauge theory technique and critical dynamics.

The key problem of the "phase transition"-like theories is the problem of "order parameter", $s(\bar{r})$, since the existence criterion of phase transition, $\langle s(0)s(\bar{r})\rangle_{|r|=\infty} \neq 0$, does not work in the disordered amorphous phase. The main problem of the "dynamical" glass transition theories is the description of dynamical transition from the ergodic state to the non-ergodic state of the system and physical interpretation of this transition. The first problem can be solved introducing additional degrees of freedom by means of gauge field, which describes the elastic stresses induced by frustrations. Below it will be shown that the glass transition corresponds to the critical point in the gauge field subsystem. The dynamical features of this transition, such as the Vogel-Fulcher-Tamman temperature dependence of the relaxation time and the plateau on the time dependence of the order parameter correlation function, can be revealed by means of functional methods of non-equilibrium dynamics.

2. Two conditions for glass transition

We are of opinion that necessary condition of glass transition is the presence of a thermodynamic phase transition in near region of temperatures. Certainly, it is a virtual phase transition, which is caused by a local ordering but can not occur because of some causes. In case of spin systems it could be the paramagnetic to ferromagnetic transition, which is locked by quenched disorder. In the system of atoms it is a virtual phase transition induced by a local ordering to the state with a non-Fedorov crystal group. In this case a full-fledged thermodynamic phase transition is not possible, because all space can not be mapped by the cells with non-Fedorov symmetry.

Another condition for the glass transition is the presence of frustrations in the system. In the first case the frustrations are quenched and sessile because of quenched disorder. In the second case the frustrations are geometrical. They differ from the quenched frustrations because they are movable topological defects (disclinations). In both cases the frustrations depress the phase transition dynamically, and don't allow system to reach of the long- range ordered state.

3. Model

In more detail the example of the three-dimensional Heisenberg spin-glass model, whose order parameter, $\mathbf{s}$, is considered to be a local magnetization vector with the SO(3) continuous symmetry group, was considered in [5, 6]. One of most elegant method of description of the frustrated systems is given by gauge theory [1, 2, 3, 4]. The static action of a system close to the second order phase transition in general case has the form of

$$S = \int \left[\frac{1}{2}(\partial_i \mathbf{s})^2 + U(\mathbf{s}) \right] d\mathbf{r}, \tag{1}$$

where $\mathbf{s}$ is the some order parameter field, and $U(\mathbf{s}) = \mu^2 \mathbf{s}^2 + v\mathbf{s}^4$. The structure of vitrescent systems is inhomogeneous in relative orientation of local ordering elements and in density. Congruent connections between the orientations are introduced by means of a gauge field, A^a_μ, when ordinary derivative, $\partial_i \mathbf{s}$, is replaced by the covariant derivative, $D_i \mathbf{s}$:

$$\partial_i s^a \to D_i s^a = \partial_i s^a + g\varepsilon^{abc} A^b_i s^c. \tag{2}$$

Besides, the system's action should contain the term $\sim (D_i A^a_\mu)^2$, since we suppose that A^a_μ is a continuous function. Therefore, one gets the gauge-symmetric action:

$$S = \int \left[\frac{1}{2}(D_i \mathbf{s})^2 + U(\mathbf{s}) + \frac{1}{4} F^a_{\mu\nu} F^a_{\mu\nu} \right] d\mathbf{r}, \quad \text{where} \quad F^a_{\mu\nu} = \partial_\mu A^a_\nu - \partial_\nu A^a_\mu + g\varepsilon^{abc} A^b_\mu A^c_\nu. \tag{3}$$

Expansion of the local magnetization field, $\mathbf{s}$, near one of the vacuum states, for instance $\langle \mathbf{s} \rangle_0 = (0,\, 0,\, i\mu/\sqrt{2v})$, in small $\phi = s - i\mu/\sqrt{2v}$ deviations, and use of the gauge transformation properties allow to rewrite the action (3) in the form of the functional of two massive vector bosons, A^κ_μ ($\kappa = \{1,\, 2\}$), with the mass $M_0 = ig\mu/\sqrt{2v}$, one massless vector boson, A^3_μ, and one scalar field, ϕ.

The method of introduction of disorder in the theory is very important and plays a key part. One of the important features of the above model (3) is the existence of soliton solutions, which correspond to the vortexes system. These thermally activated vortexes disordering the spin structure. They are the movable sources of the gauge field, and their system above the glass transition temperature can be considered as a system which remains in thermal equilibrium [3]. But in $T = T_c$ the gauge field becomes massless and the vortexes vanish. Therefore, in order that a system possess glass properties at $T < T_c$ we should inject to this system a disorder which gives rise to frustration of its structure at $T < T_c$. It seems evident that gauge field should be treated as a dynamical one, and full gauge invariance or homogeneity maintained as much as possible. In this case only the sources of gauge fields need be quenched in glass [4]. At $T > T_c$ these sources are additional to the thermally activated vortexes but as opposed to them do not vanish in the low-temperature state. They can be introduced into the model by means of the sources field, $J_{a\mu}$:

$$S = \int \left[\frac{1}{2}(\partial_\mu \phi)^2 + 2\mu^2\phi^2 + \frac{g^2\mu^2}{4v} A^\kappa_\mu A^\kappa_\mu + \frac{1}{4} F^a_{\mu\nu} F^a_{\mu\nu} + v\phi^4 + \frac{g^2}{2}\phi^2 A^a_\mu A^a_\mu + J^a_\mu A^a_\mu \right] d\mathbf{r}. \tag{4}$$

Since above the T_c the source subsystem is believed to be in thermal equilibrium [3]. Averaging over J_μ^a leads to redefinition of the partition function:

$$Z = \int \left[\int \exp\left(-S - \frac{1}{4} \int I_0^{-1} J_\mu^{a2} d\mathbf{r} \right) DJ_\mu^a \right] D\phi DA_\mu^a, \tag{5}$$

where $\int \ldots Dx$ is the continual integral. It leads to additional contribution to the A_μ^a "mass", which takes the form of

$$M^2 = -I_0 + \mu^2 g^2/4v. \tag{6}$$

Thus, the frustrations lead to the renormalization of the gauge field mass.

The renormalization of the gauge field mass affects the critical behavior of the system, since it shifts $M^2 = 0$ singularity to the temperature range above the virtual phase transition point, T_c. If we assume that $\mu^2 = \alpha k_B(T - T_c)$, where α is some constant, then from (6) we have the critical divergence of the A_μ^a field correlation radius at $T_g = T_c + 4I_0 v/\alpha k_B g^2$ in the disordered solid phase. In [6] it was shown that this leads to critical slowing-downs of the fluctuations. Thus, the disorder-induced frustrations inhibit the growth of the ϕ field correlation length, and the system freezes in a disordered state.

4. Analysis of the static spin-glass model

In T_g the gauge field becomes massless, whereas the local magnetization field still has got the "mass", μ. Analysis of the model close to this point allows to assert that T_g is the nothing else than the glass transition temperature. First of all it is evident from (4) that the linear susceptibility, $\chi_L = \partial\langle\phi\rangle/\partial h \sim \mu^{-2} = g^2/4I_0 v$ (h is an external source of the field ϕ), is finite at $T = T_g$. Similar unsophisticated estimation gives that the correlation length, $r_{cor} \sim \sqrt{g^2/4I_0 v}$, is finite too. However, nonlinear susceptibility, $\chi_N = \partial^3\langle\phi\rangle/\partial h^3 = \langle\phi^4\rangle$ [7] (fig. 1 a), diverges near T_g, because of the divergence of the A-field loop contribution (fig. 2). This picture is typical

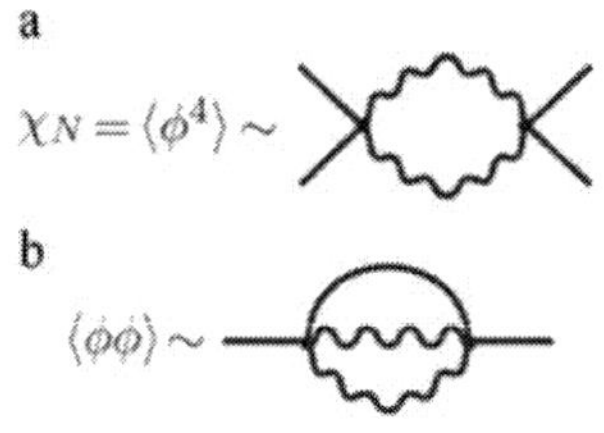

Figure 1. The straight line corresponds to the Green function of the "massive" field, $\langle\phi\phi\rangle$, the wavy line corresponds to the gauge field Green function, $\langle AA\rangle$.

for the glass transition.

Using this theory allows us to determine the temperature dependence of the system heat capacity at $T \to T_g^+$: $c_p = dU/dT \propto (T - T_g)^{-1/2}$. One can see (fig. 2), that the obtained temperature dependence of the heat capacity near T_g diverges and is not characteristic for glass transition. One can show that this is a result of only the static character of the theory. Below the analysis of the critical dynamics of this system close to T_g allows to make sure in it, and to show that T_g is the glass transition temperature.

5. Analysis of the dynamic spin-glass model

In order to check the another glass transition properties of the model, one can examine the non-equilibrium dynamics of the system near to T_g by means of functional methods of non-equilibrium dynamics [8] near the critical point. It leads to the representation of the partition

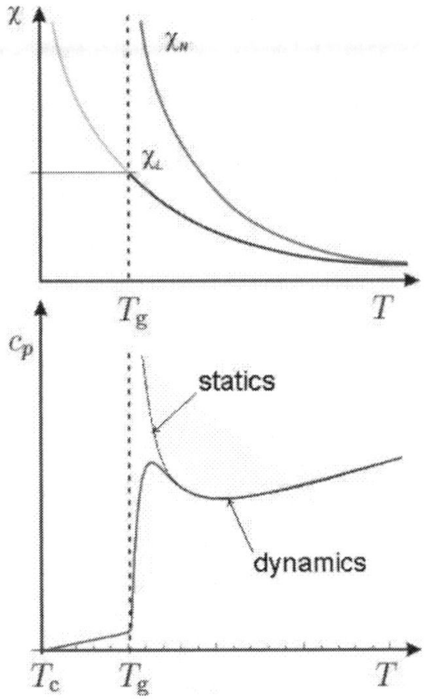

Figure 2. The qualitative presentation of the temperature function of the nonlinear susceptibility, $\chi_N \propto 1/(T-T_g)^{\gamma}$, and linear susceptibility, $\chi_L \propto 1/(T-T_c)$. The χ_L curve is not described by this theory at $T < T_g$, it is finite in this region. The qualitative picture of the temperature dependence of the heat capacity near T_g, which is obtained from the theory.

function of the system in the form of

$$Z = \int \exp(-S^*) D\vec{\phi} D\vec{A}^a_\mu, \tag{7}$$

with

$$\begin{gathered} S^* = \frac{1}{2}\int \Big[\vec{\phi}(t,\mathbf{r})\hat{G}^{-1}(t-t',\mathbf{r}-\mathbf{r}')\vec{\phi}(t',\mathbf{r}') \\ +\vec{A}^a_\mu(t,\mathbf{r})\hat{\Delta}^{-1}_{\mu\nu}(t-t',\mathbf{r}-\mathbf{r}')\vec{A}^a_\nu(t',\mathbf{r}')\Big] d\mathbf{r}d\mathbf{r}'dtdt' \\ +\int \Big[g\varepsilon^{abc}(\partial_\mu \bar{A}^a_\nu)A^b_\mu A^c_\nu + g\varepsilon^{abc}(\partial_\mu A^a_\nu)\bar{A}^b_\mu A^c_\nu \\ +g\varepsilon^{abc}(\partial_\mu A^a_\nu)A^b_\mu \bar{A}^c_\nu + g^2\varepsilon^{abc}\varepsilon^{aij}\bar{A}^b_\mu A^c_\nu A^i_\mu A^j_\nu + g^2\bar{A}^a_\mu A^a_\mu \phi^2 + g^2(A^a_\mu)^2\bar{\phi}\phi + v4\,\bar{\phi}\phi^3\Big] d\mathbf{r}dt, \end{gathered} \tag{8}$$

where $\vec{\phi} = \{\bar{\phi}, \phi\}$, and $\vec{A}^a_\mu = \{\bar{A}^a_\mu, A^a_\mu\}$ are vectors, the components of which are named "quantum" and "classical" respectively in the Keldysh representation [8], G^{-1} and $\Delta^{-1}_{\mu\nu}$ are matrices, inverse to the Green functions matrices having the following form:

$$\hat{G} = \begin{pmatrix} G^K & G^A \\ G^R & 0 \end{pmatrix}, \quad \hat{\Delta}_{\mu\nu} = \begin{pmatrix} \Delta^K_{\mu\nu} & \Delta^A_{\mu\nu} \\ \Delta^R_{\mu\nu} & 0 \end{pmatrix}, \tag{9}$$

where

$$G^{R(A)}(k,\omega) = \frac{1}{k^2+\mu^2 \pm i\Gamma_\phi \omega}, \qquad G^K(k,\omega) = \frac{2\Gamma_\phi}{(k^2+\mu^2)^2+\Gamma^2_\phi\omega^2}, \tag{10}$$

Γ_ϕ is the kinetic coefficient of the local magnetization,

$$\Delta^{R(A)}_{\mu\nu}(k,\,\omega) = \frac{\delta_{\mu\nu}}{k^2+M^2 \pm i\Gamma_A\omega}, \qquad \Delta^{K}_{\mu\nu}(k,\,\omega) = \frac{2\Gamma_A\delta_{\mu\nu}}{(k^2+M^2)^2+\Gamma_A^2\omega^2}, \tag{11}$$

Γ_A is the kinetic coefficient of the gauge field.

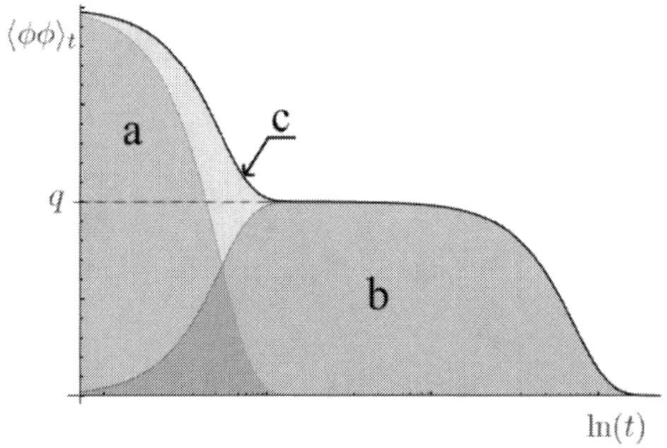

Figure 3. The dependence $\langle\phi\phi\rangle_t$ on $\ln(t)$: a) is the contribution of $\langle\phi\phi\rangle_t^0$, which is the Debye-relaxation; b) is the contribution of the second term which is given by the cooperative effects; c) is the sum of first and second terms.

The renormalization group analysis of the dynamical model shows that the theory becomes not renormalizable on small time scales, $t \ll \Gamma_\phi g^2/4I_0 v$ [6]. On these scales the fluctuation-dissipation theorem is broken, and one can interpret this as appearing of the dynamical heterogeneity where $\Gamma_\phi g^2/4I_0 v$ is the characteristic scale of time homogeneity. Subject to this heterogeneity one can estimate the temperature dependence of the relaxation time [6]:

$$\tau_{rel} = \Gamma_\phi \propto \exp\left(\frac{2vg^2T_g}{\alpha\pi^2(T-T_g)}\right), \tag{12}$$

which corresponds to the well known Vogel-Fulcher-Tammann law. Besides, in the one-loop approximation the ϕ-field correlation function can be represented as the sum of the unperturbed and cooperative parts:

$$\langle\phi\phi\rangle_t \simeq G^K(t) + \frac{g^4\mu^2\ln(L/a_0)}{\Gamma_\phi\pi^2}\int\limits_0^t G^K(t_1)e^{-4I_0v|t-t_1|/\Gamma_\phi g^2}dt_1. \tag{13}$$

This function has the form which is characteristic for glass systems (fig.3) and gives the Edwards-Anderson parameter value in static limit when $T = T_g$.

One can obtain the qualitative form of the temperature dependence of the heat capacity close to the glass transition:

$$c_p(T) \propto \frac{\partial}{\partial T}\left[e^{-CT_g/(T-T_g)}\right] = \frac{e^{-CT_g/(T-T_g)}}{(T-T_g)^2}CT_g, \tag{14}$$

where $C = 2vg^2/\alpha\pi^2$. The calculation of heat capacity gives that it has the sharp growth near the glass transition temperature which then drops in T_g, that is characteristic for vitrescent systems. With using the suggested theory it is not difficult to give a simple qualitative explanation of this dependence: At the approaching the system's temperature to T_g, the fluctuations are increase, and consequently, the heat capacity increases too like in the case of second order phase transition; however, the quenched frustrations block this increase over the scales, which exceed some finite scale, which is determined by these frustrations. As a result the heat capacity does not infinitely diverge in T_g, but gets a finite maximum in $T_{\text{max}} > T_g$, and then falls (fig. 2).

6. Conclusions

According to the above theory one can offer the following physical picture of the glass transition processes: when the temperature approximates T_g, the correlation length of the gauge field diverges, while the correlation length of the order parameter is finite. The infinite correlation of the gauge field means that the relative rotations are correlated at an infinitely large distance. Then the growth of the order parameter field fluctuations becomes impossible, since the spins can not already turn independently of each other. This picture can be described in framework of the quasi-equilibrium static theory as a phase transition in the gauge field subsystem, but very important dynamic features of this transition can be explained only in the dynamic approach.

The statement about that the correlation length of the gauge field grows faster than the correlation length of the order parameter can be explained in the following way: without the frustrations the correlation lengths of both the gauge field and the order parameter field grow equally, because they are components of the same field. In the dual representation the system contains thermally activated vortexes, whose concentration tends to zero when $T \to T_c^+$. The priority growth rate of the correlation length of the gauge field becomes possible when the system contains frustration. The frustration induces vortexes additional to the thermally activated vortexes, as a result in the equilibrium state the vortex concentration is fixed non-zero. The reason for this is that the fluctuations of the gauge field are sure to develop around these sources. In the case of a nonzero concentration of these static sources the fluctuations around them can interflow, which leads to the faster growth of the effective correlation length of the gauge field. With the given source concentration and the appropriate correlation length of the order parameter (or the gauge field without frustrations) it leads to the formation of the percolation cluster which is associated with these sources. Thus, the effective correlation length of the gauge field becomes infinitely large, and the relative rotations at infinitely remote points become correlated and the relaxation time of the system infinitely grows up, but the order parameter correlation length remains to be relatively small. Therefore, the system freezes in a disordered state.

This work was supported partly by the RFBR grants No. 10-02-00882-a and No. 10-02-00700-a.

References

[1] Volovik G E, Dzyaloshinskii I E 1978 *Sov. Phys. JETP* **48** 555
[2] Hertz J A 1978 *Phys. Rev.* B **18** 4875
[3] Rivier N, Duffy D M 1982 *J. Physique* **43** 295
[4] Rivier N 1985 *Revista Brasileira de Flsica* **15** 311
[5] Nusinov Z 2004 *Phys. Rev.* B **69** 014208-1
[6] Vasin M G 2011 *J. Stat. Mech.* P05009
[7] Binder K, Young A P 1986 *Reviews of Modern Physics* **58** 801
[8] Kamenev A 2005 *Nanophysics: Coherence and Transport, edited by H. Bouchiat* (Elsevier, Amsterdam)

Anomalous heat capacity for nematic MBBA near clearing point

D A Lukashenko[1] and B.M. Khasanov[2]
Theoretical Physics Department, Institute of Physics, Kazan university, 420008, Kremlevskaya, 18; Kazan, Russia

[1]E-mail: [1] Shadow-LD@yandex.ru,
[2]E-mail: [2] bulat.khasanov@ksu.ru

Abstract. The pretransition properties of the heat capacity on the both sides of the nematic – isotropic phase transition for the liquid crystal MBBA (n-methoxybenzylidene-n –butylaniline) are discussed on the basis of the Landau-de Gennes theory. Agreement between theory and experiment allows to obtain the critical exponents gamma and beta.

1. Introduction

Experimental and theoretical investigation of the statistical properties of heat capacity in nematic liquid crystals, especially near the phase transition to isotropic phase, is important both for the physics of phase transition and for nematic liquid crystal themselves. The analysis of the temperature dependence of the specific heat showed that the fluctuations of the OP are not small near phase transition point. Experiments on heat capacity are usually interpreted from the point of view of fluctuations of the modulus order parameter (OP), which are the principal cause of anomalously temperature dependence near phase transition point. But in addition to these modulus OP fluctuations there are the fluctuations of the orientation and biaxial fluctuations that disturb the uniaxiality of the OP.

Critical phenomena in phase transitions in liquid crystals, accompanied by the vanishing of long-rang orientational order of the long axes of the molecules, have been sufficiently studied. The existing experimental data [1] indicate that, as the nematic-isotropic transition point is approached the specific heat increases. This characteristic anomalies in the temperature are connected with the spatially-inhomogeneous fluctuations of the complex order parameter of nematic.

In this work we have attempted to carry out quantitative analysis of the possibility of the application of the Landau-de Gennes theory for the description of critical anomalies both above and below the nematic-isotropic (NI) phase transition.

2. Landau-de Gennes theory and fluctuations

In the phenomenological Landau-de Gennes theory [2] the free energy associated with the long-wavelength part of the tensor OP $Q_{\alpha\beta} = Q\left(n_\alpha n_\beta - 1/3 \cdot \delta_{\alpha\beta}\right)$fluctuations for a uniaxial nematic in the neighborhood of the phase transition ($\vec{n}$ is a unit vector, director) in the one constant approximation, is given by

$$F=V\left(\frac{1}{2}AQ^2-\frac{1}{3}BQ^3+\frac{1}{4}CQ^4\right)+\frac{1}{2}\int d\vec{r}[(A-2BQ+3CQ^2)P_2^0P_2^0+L\nabla P_2^0\nabla P_2^0 \quad , (1)$$
$$+2(A-BQ+CQ^2)P_2^1P_2^{-1}+2L\nabla P_2^1\nabla P_2^{-1}+2(A+2BQ+CQ^2)P_2^2P_2^{-2}+2L\nabla P_2^2\nabla P_2^{-2}]$$

where $A=a(T-T^*)$ and B, C are temperature independent constants, Q is a module of OP, and T^* - the lowest possible temperature for isotropic phase, L is an elastic constant. The first part of F is the Landau expansion in terms of the scalar OP. $Q=0$ corresponds to the isotropic phase, while in the nematic phase:

$$Q_N=B\cdot\left(1+\tau^{1/2}\right)/2C,\quad \tau=\left(T^{**}-T\right)\left(T^{**}-T^*\right)^{-1}. \tag{2}$$

Here $T^{**}=T^*+B^2/4ac$ is the upper limit of stability for the nematic phase. From the first part of expression (1) we can easily find the temperature and latent heat of the transition:

$$T_c=T^*+\frac{2}{9}\frac{B^2}{aC},\qquad H=\frac{1}{2}\frac{aB^2}{C^2}T_c \tag{2a}$$

as well as the susceptibility and heat capacity above and below the transition point:

$$\chi^{-1}=\begin{cases}a(T-T^*), & T>T_c\\ 2a(T^{**}-T^*)(\tau+\sqrt{\tau}), & T<T_c\end{cases},\quad C_p=\begin{cases}0, & T>T_c\\ a^2(1+\tau^{-1/2})\cdot(2c)^{-1}< & T\quad T_c\end{cases} \tag{3}$$

The susceptibility in the isotropic phase shows the linear divergence. At the same time the susceptibility in the nematic phase has a linear plus a square-root divergence, i.e. a deviation from the Curie-Weiss law in the mean field approximation and γ exponent is less than one [3].

Let us consider nematic as a continuous homogeneous medium. Characteristic lengths of considerable change of $\langle P_2^m\rangle$ must be greater than the distance between molecules, i. e. $|\vec{k}|<k_{\max}$.

Now we use the equipartition theorem to find the thermal averages of the fluctuations of the longitudinal, transverse and biaxial modes. In the isotropic phase one can find

$$\langle P_2^0P_2^0\rangle=\langle P_2^1P_2^{-1}\rangle=\langle P_2^2P_2^{-2}\rangle=k_BT\left(\chi^{-1}+Lk^2\right)^{-1} \tag{4}$$

While in the nematic phase we obtain

$$\langle P_2^0P_2^0\rangle=\frac{k_BT}{\chi^{-1}+Lk^2},\quad \langle P_2^1P_2^{-1}\rangle=\frac{k_BT}{Lk^2},\quad \langle P_2^2P_2^{-2}\rangle=\frac{k_BT}{3BQ+Lk^2} \tag{5}$$

Here χ^{-1} is determined by the expression (3), and

$$3BQ=6a(T^{**}-T^*)(1\quad\sqrt{\tau}) \tag{6}$$

In the isotropic phase the fluctuations of all the five modes are degenerate. They increase in the long-wave limit ($k\to 0$) as $T\to T^*$ according to the Curie-Weiss law.

In the nematic phase the fluctuations spectrum splits. $\langle P_2^0P_2^0\rangle$ is the OP longitudinal fluctuations. $\langle P_2^1P_2^{-1}\rangle$ is the transverse fluctuations (director), which are not critical and represent Goldstone modes in the symmetry-broken phase. $\left\langle P_2^2 P_2^{-2} \right\rangle$ determine biaxial fluctuations, where the biaxial susceptibility $(3BQ)^{-1}$, contrary to the longitudinal χ^{-1}, tends to a finite value at $T = T^{**}$ $(\tau = 0)$.

In the isotropic phase the excess specific heat capacity per unit volume due to OP fluctuations is given by

$$\Delta C_p = \frac{5k_B T^2}{8\pi^2}\left(\frac{\partial \chi^{-1}}{\partial T}\right)^2 \left(\chi^{-1} L^3\right)^{-1/2}\left(arctg(x) \quad \frac{x}{1+x^2}\right), \tag{7}$$

where $x = k_m \xi$, $\xi = \xi_0 \left(\frac{T^*}{T - T^*}\right)^{1/2}$ and $\xi_0 = \left(\frac{L}{aT^*}\right)^{1/2}$ is the bare OP correlating length.

Here k_m is the upper limit of integration with respect to the wave vector of fluctuation modes. The continuous model we are used assumes averaging over a small but macroscopic volume. Therefore all our expressions are correct up to wave vector k_m.

To compare the expression (7) with the experimental data, it is convenient to rewrite it in the form

$$\begin{gathered} C_p = \Delta C_{p_0} + \frac{BT^2}{\sqrt{T - T^*}} arctg \frac{A}{\sqrt{T - T^*}} \quad \frac{ABT^2}{T - T^* + A^2}, \\ A = \frac{k_m L^{1/2}}{a^{1/2}} = k_m \xi_0 \sqrt{T^*}, \; B = \frac{5}{8\pi^2} \frac{k_B a^{3/2}}{L^{3/2}} \frac{M}{\rho} \end{gathered} \tag{8}$$

Because direct experimental observation of T^* is not possible, it have to be treated as adjustable parameter (together with A and B) in the fits. Furthermore, no data points in the range $T - T^* < 1$ was covered in the fits. We consider the Gaussian approximation only that is correct not to close to the transition point. The best agreement with experiment was obtained for the parameter values $A = 22.12$, $T^* = 317.97$ K, $\Delta C_{p_0} = 499.11\ J\cdot(mol\cdot K)^{-1}$, $B = 1.24\cdot 10^{-4}$. The fitting of experimental data [4] result is presented at figure 1. The fitting parameters A and B are connected with each other through the equation (8), which immediately allows to get $(k_m)^{-1} = 5.5$ Å, $\xi_0 = 6.82$ Å. Our value of ξ_0 is close to the value obtained in [5].For density and the molecular weight of MBBA we used $\rho = 1.049\ g\cdot cm^{-3}$, $M = 267.37\ g\cdot mol^{-1}$.

The treatment of the fluctuation part of the specific heat at the nematic phase is more complex. The entropy density associated with fluctuations is obtained by differentiating (1) with respect to the temperature. The subsequent ensemble average gives the entropy change due to fluctuations. The final expression for the fluctuation induced heat capacity has the form

$$\begin{gathered} \Delta C_p = \frac{T}{2}\frac{\partial^2 \chi^{-1}}{\partial T^2}\sum_k \left\langle \left|P_2^0\right|^2 \right\rangle + T\frac{\partial^2 (3BQ)}{\partial T^2}\sum_k \left\langle P_2^2 P_2^{-2} \right\rangle \\ -\frac{T}{2}\left(\frac{\partial \chi^{-1}}{\partial T}\right)^2 \sum_k \left\langle \left|P_2^0\right|^2 \right\rangle^2 - T\left(\frac{\partial^2 (3BQ)}{\partial T^2}\right)^2 \sum_k \left\langle P_2^2 P_2^{-2} \right\rangle^2 \end{gathered} \tag{9}$$

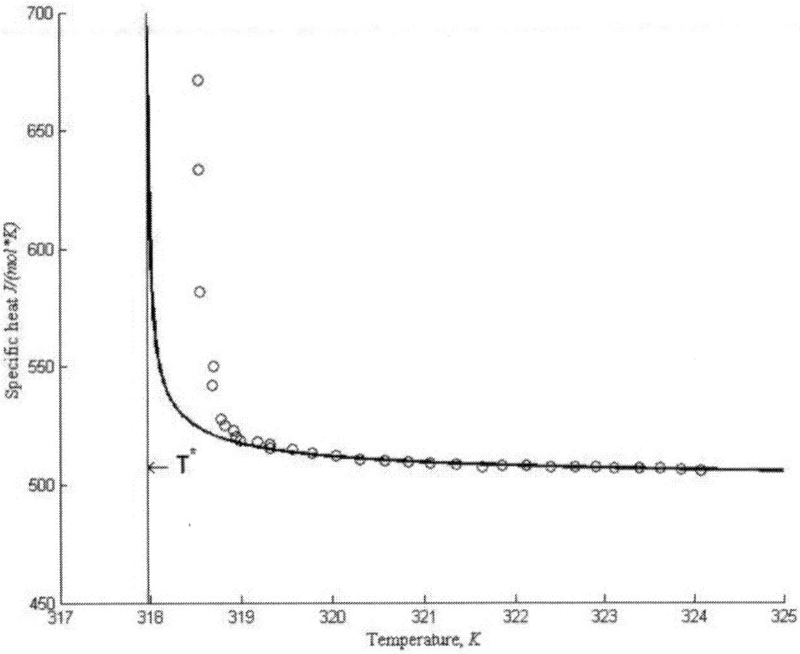

Figure 1. Specific heat above the transition temperature. Solid line show theoretical data and circles show the experimental data of Anisimov [4]

Note that in (9) there are present components which are proportional to the second derivative with respect to reverse susceptibilities. It is caused by the fact that the longitudinal susceptibility in nematics does not obey Curie-Weiss law, and the reverse biaxial susceptibility is directly proportional to OP. The formula (9) can be used to describe the experimental data [4].

The results of the experimental research of the heat capacity, OP and susceptibility demonstrate that the use of expressions (2), (3) and (6) have the difficulties of describing the data in the nematic phase.

Changing to some other temperature scale $T_1 > T^{**}$ we can approximate the expression (2) in the corresponding temperature range to get the following formula:

$$Q_N = \frac{B}{2C}\frac{T_1}{T^{**} - T^{*}}t^{\beta}, \tag{10}$$

where $t = (T_1 - T)\cdot T_1^{-1}$. β and T_1 are fitting parameters. The equation (10) has the form of widely used the Haller relation [6].The reverse of the longitudinal and biaxial susceptibilities in this case have the following forms $\chi^{-1} = \chi_0^{-1}t^{\gamma}$, $\chi_0 = (2aT)^{-1}$, $3BQ = 3\chi_0^{-1}t^{\beta}$.

Using these definitions, we find the expression for heat capacity in the nematic phase:

$$C_p = \Delta C_{p_0} + C_{landau} + \Delta C_p = \Delta C_{p_0} + 2\kappa^2 H\beta \frac{T}{T_c \cdot T_1} t^{2\beta-1} +$$

$$+\frac{k_B T^2}{2}\frac{1}{2\pi^2}\frac{2^{3/2}}{\xi_0^3 T_1^2} t^{\frac{3\gamma-4}{2}} \left\{\left[arctg(\overline{x_0}) - \frac{\overline{x_0}}{1+\overline{x_0}^2}\right]\frac{\gamma^2}{2} - \gamma(\gamma-1)\left[\overline{x_0} - arctg(\overline{x_0})\right]\right\} \quad , \quad (11)$$

$$+\frac{k_B T^2}{2}\frac{1}{2\pi^2}\frac{6^{3/2}}{\xi_0^3 T_1^2} t^{\frac{3\gamma-4}{2}} \left\{\left[arctg(\overline{x_2}) - \frac{\overline{x_2}}{1+\overline{x_2}^2}\right]\beta^2 - 2\beta(\beta-1)\left[\overline{x_2} - arctg(\overline{x_2})\right]\right\}$$

where $\overline{x_0} = \frac{k_m \xi_0}{\sqrt{2}} t^{-\gamma/2}$, $\overline{x_2} = \frac{k_m \xi_0}{\sqrt{6}} t^{-\beta/2}$, $\kappa = \left(\frac{T_1}{T_1 - T_c}\right)^{\beta}$, $\xi_0 = \sqrt{\frac{L_N}{2aT_1}}$

Here H is the transition latent heat and for the nematic MBBA its experimental value is $H = 0.78\ J \cdot g^{-1}$ [1]. L_N - is the elastic constant in nematic phase which is not equal to one in isotropic phase. The second term in equation (11) is the heat capacity obtained in the Landau theory. In order to determine quantitative agreement of (11) with experimental data we treat variable parameters T_1, $\xi_0, \Delta C_{p_0}$ and make a fitting in temperature range $T < T^{**} - 1^{\circ}$, because of the approximation used. After that we get following parameter values that give the best fit: $T_1 = 320.97K$, $\xi_0 = 9.08\overset{0}{\mathrm{A}}$, $\Delta C_{p_0} = 489.5\ J \cdot (mol \cdot K)^{-1}$ (see the figure 2), as well as the critical exponents $\beta = 0.2$ [7] and $\gamma = 0.68$ [8].

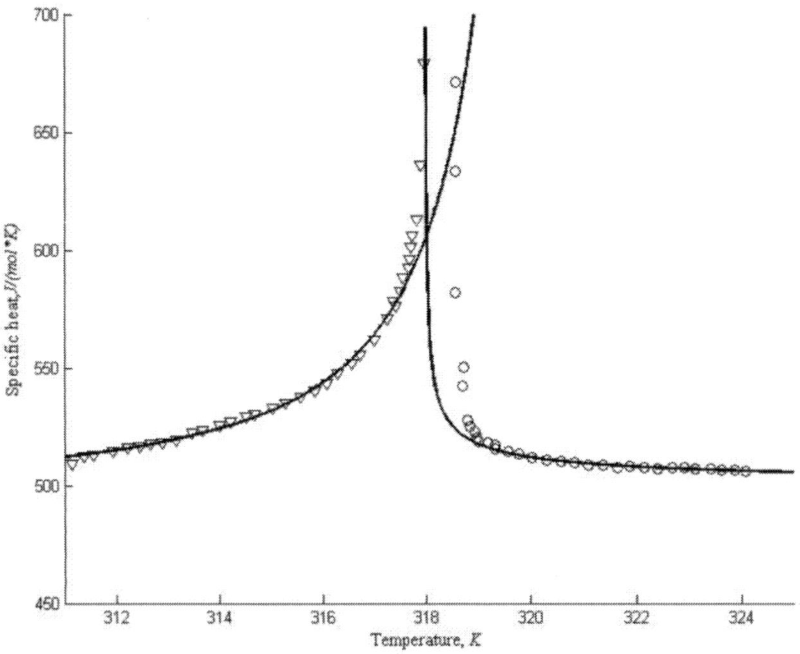

Figure 2. Specific heat above and below the transition temperature. Solid lines show theoretical data and circles and triangles show the experimental data [4].

The agreement of theoretical formula (11) with experiment shows us that the use of empirical approximation procedure of Haller for OP and similar procedure for the longitudinal susceptibility in the Landau-de Gennes model gives a satisfactory description of heat capacity experiments in nematic phase. The critical exponent β of order parameter is in good agreement with the experimental data for MBBA [9]. The regular part of the heat capacity ΔC_{p_0} in nematic phase is a bit less than the corresponding value in the area higher than the transition point, and it coincides with experiment.

3. Conclusions

In this paper we have presented the analysis of the specific heat anomaly near the NI transition. Experiments show a complex behavior in which Gaussian and non-Gaussian fluctuations cannot be neglected [2]. The absence of the divergence at the NI transition point allows us to expect of using the Gaussian approximation or weakly interacted fluctuation limit. The existence of the cubic invariant in (1) is probably the main reason for the failure of the mean-field description in the vicinity of the NI transition. From the analysis due to Landau-de Gennes model we obtained for the critical exponents $\gamma = 0.68$, $\beta = 0.2$, which are in qualitative agreement with those obtained for MBBA from different sources. The temperature dependence of the heat capacity in the nematic phase is connected with the change in the degree of nematic order and with the increase of the longitudinal and biaxial fluctuations. In the isotropic phase the fluctuation contribution is the only one.

It is interesting to note that from the renormalization group study [10] it follows that the critical anomalies can be connected either with the presence of the fixed point, which described the transition near the isolated singular point, or with the two fixed points which are due to the specific character of the fluctuations in the nematic. The last of them are a saddle and a stable focus. In the neighbourhood of a first-order transition line of the phase diagram of the renormalization group equations the transition is controlled by saddle fixed point, in the vicinity of which $\gamma = 0.58$, $\beta = 0.23$.

4. References

[1] Anisimov M A 1991 *Critical phenomena in liquids and liquid crystals* (Gordon and Breach Science Publishers)
[2] De Gennes P G and Prost J 1993 *The Physics of Liquid Crystals 2nd Edition* (Oxford: Clarendon Press)
[3] Lei Lin 1982 *J. de Phys.* **43** 251
[4] Anisimov M A *et al* 1977 *Sov.Phys.-JETP* **72** 1983
[5] Stinson T W, Litster J D 1973 *Phys. Rev. Lett.* **30** 688
[6] Haller I 1975 *Prog. Solid State Chem.* **10** 103
[7] Kobinata S *et al* 1981 *Mol. Cryst. Liq. Cryst.* **66** 67
[8] Aver'yanov E M 1990 *Sov.Phys.-JETP* **97** 855-62
[9] Simoes M and Simeao D S 2006 *Phys. Rev. E* **74** 051701
[10] Korshenevskii A L, Shalaev B N 1979 *Sov.Phys.-JETP* **49** 1094

Journal of Physics: Conference Series **394** (2012) 012012 doi:10.1088/1742-6596/394/1/012012

Microscopic Structural and Dynamical Properties of Amorphous Metallic Alloy $Ni_{33}Zr_{67}$ at the Temperature T=300K

R M Khusnutdinoff, A V Mokshin and I I Khadeev
Kazan Federal University, Kremlevskaya Street 18, 420008 Kazan, Russia

E-mail: r.khusnutdinoff@gmail.com

Abstract. We study the structural properties and the collective microscopic dynamics of atoms in the amorphous metallic alloy $Ni_{33}Zr_{67}$ at the temperature *T=300K* by molecular dynamics simulations. The calculated equilibrium structural and dynamical characteristics are compared with the experimental data on neutron diffraction and on inelastic X-ray scattering. We present the interpretation of observed structural relaxation of the microscopic density fluctuations of particles for amorphous metallic alloy in the framework of the recurrent relation approach. The results of theoretical calculations of the intensity of scattering $I(k,\omega)$ for amorphous $Ni_{33}Zr_{67}$ are in a good agreement with the results of computer simulation as well as with the experimental data on inelastic X-ray scattering.

1. Introduction

Design and production of new materials are important fields in the continuous development of science and technology. Bulk metallic glasses (BMG) belong to the advanced materials that attract the special interest of scientists and technologists because of their unique physicochemical and mechanical properties [1-3]. For example, amorphous alloys containing transition metal (Fe, Co, Ni) as a main component have high tensile strength, which exceeds by more than twice of their crystalline counterparts. Some of these amorphous alloys also have a high corrosion resistance, the excellent magnetic and electrical properties. The unique properties of the BMG are related with the local structural disorder in the atomic components of the mixture [4].

Amorphous metal alloys Ni_xZr_{100-x} are characterized by the following interesting features [5]: (i) the structure of the alloys has strong short-range topological order, which significantly affects on the dispersion curve in the amorphous state; (ii) the alloys have a large coherent neutron scattering cross section, which is an important feature at the comparison of the simulation results and the theoretical calculations with the experimental data on neutron diffraction. In addition, some metal alloys (such as the known, $Zr_{41.2}Ti_{13.8}Cu_{12.5}Ni_{10.0}Be_{22.5}$ – Vitreloy 1) form the bulk metallic glass phase at the critical cooling rates of ~1 K/sec [6]. These features have stimulated many experimental and theoretical studies. So, for example, the influence of the concentration dependence of the Ni_xZr_{100-x} alloys (x = 5, 10, 16.7, 33.3, 0.5, 66.7, 83.3, 90.0, 95.0) on the processes of crystallization of the bulk metallic glass was investigated by using the computer simulation methods in Ref [7]. The authors demonstrated that the highest degree of crystallization of the alloy is related with high nickel concentration in the system (in particular, $Ni_{66.7}Zr_{33.3}$-alloy), where crystallization in the alloys is related with increasing of the number of quasicrystalline icosahedral clusters.

Further, the study of the dynamic properties of the $Ni_{33}Zr_{67}$-alloy in amorphous and crystalline phase was performed in Ref [8] by using the inelastic neutron scattering experiment. By comparing the spectra of the dynamic structure factor, $S(k,\omega)$, in amorphous and crystalline phases, the authors have concluded that the origin of the collective excitations in $Ni_{33}Zr_{67}$ metallic glass is closely associated with the optical excitations which observed in the crystalline phase. In addition, the three optical modes, which exist in the crystalline $Ni_{33}Zr_{67}$-alloy, can also be observed on the short time scales in the amorphous phase. However, the recent experimental studies have questioned this conclusion [9].

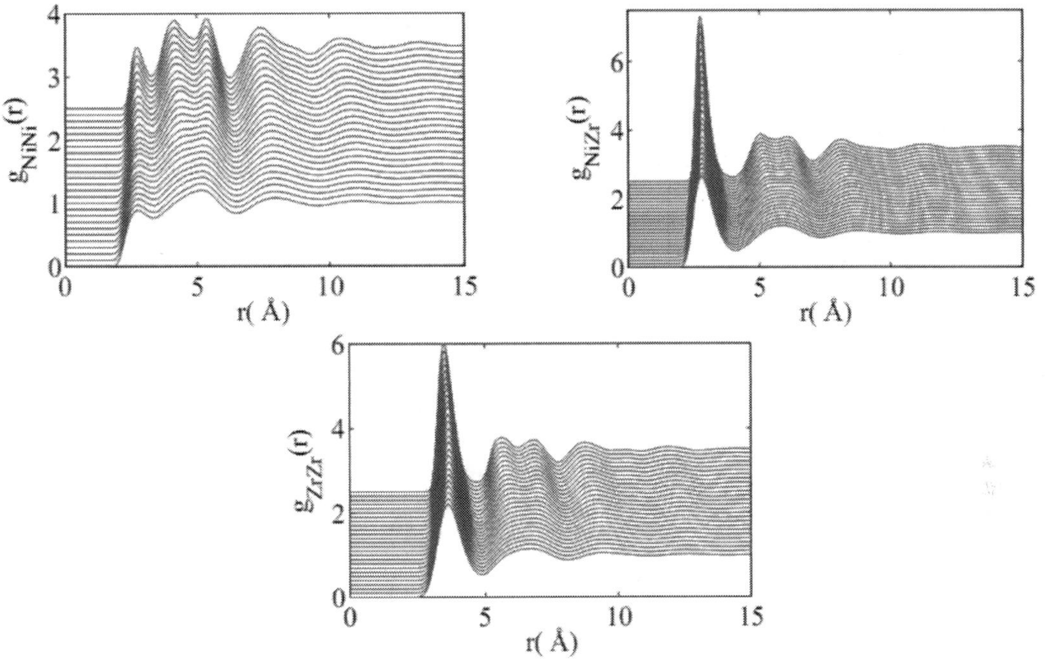

Figure 1. Partial radial distribution functions of particles for $Ni_{33}Zr_{67}$ metallic alloy at the temperatures *T* from *300* (top curve) to *3000K* (bottom curve).

In this paper we investigate the microscopical structural and dynamical properties of amorphous $Ni_{33}Zr_{67}$ metallic alloy at the temperature $T=300K$ to elucidate the mechanism of propagation of collective excitations.

2. Details of simulation

The molecular dynamics (MD) simulation of the amorphous metallic alloy $Ni_{33}Zr_{67}$ was performed at the temperature $T=300K$. The investigated system was consisted of $N=N_{Ni}+N_{Zr}=10976$ particles, where $N_{Ni}=3535$ and $N_{Zr}=7441$; the numerical density was $n=0.0427$ Å^{-3}. The interaction of atoms was carried out with the potential of interparticle interaction [10]

$$U_{\alpha\beta}(r) = A_{\alpha,\beta}\left[\frac{1}{\left(\varsigma_{\alpha,\beta}r - a_{\alpha,\beta}\right)^{n}} - 1\right]\exp\left(\frac{1}{\varsigma_{\alpha,\beta}r - b_{\alpha,\beta}}\right) \qquad \left(r < \frac{a_{\alpha,\beta}}{\varsigma_{\alpha,\beta}}\right). \qquad (1)$$

Here, $\alpha, \beta \in \{Ni, Zr\}$ and $A_{\alpha,\beta}$, $\varsigma_{\alpha,\beta}$, $a_{\alpha,\beta}$, $b_{\alpha,\beta}$ are the potential parameters, taken from Ref. [10].

All simulations are carried out in the isobaric-isothermal ensemble (NpT). To maintain the system in the thermal equilibrium, the Berendsen's thermostat and barostat were applied. The amorphous alloy at the temperature $T=300K$ was obtained by a rapid cooling from the liquid state at the temperature $T=3000K$ with the rate $dT/dt=10^{13}$ *K/sec*. The integration of the equations of motion of the particles was performed using the velocity Verlet algorithm with the timestep $\tau=10^{-15}$ *sec*.

3. Simulation results and discussion

The structural features of metallic alloy were analyzed by means of the partial radial distribution function:

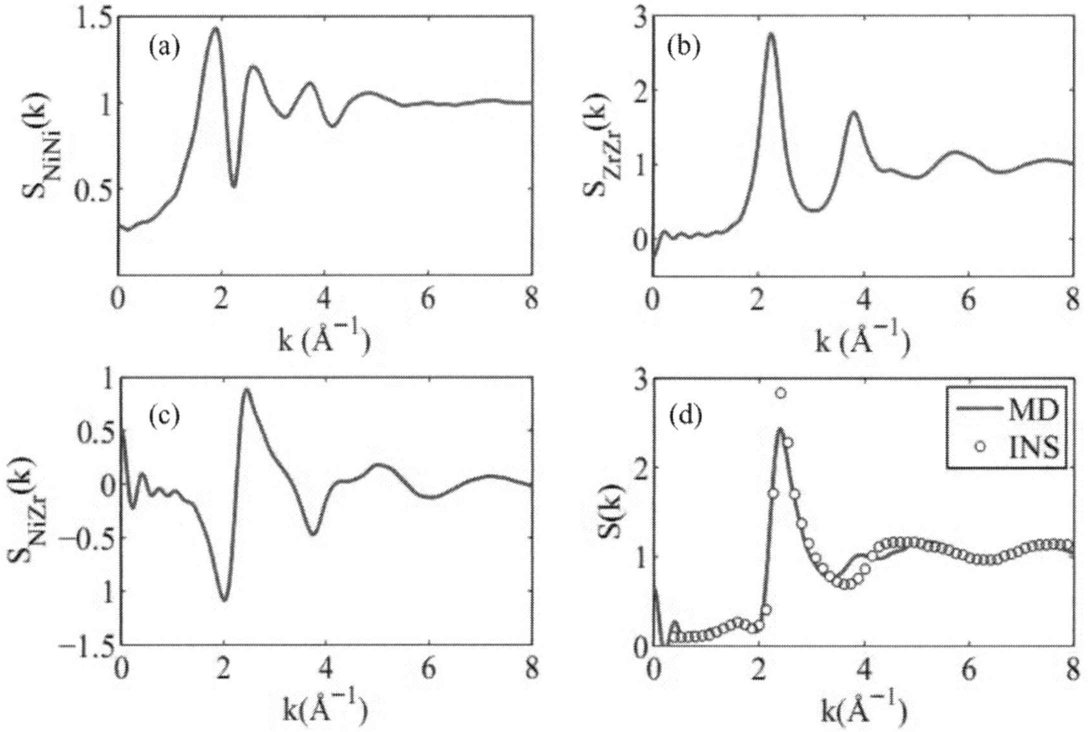

Figure 2. Static structure factor of $Ni_{33}Zr_{67}$-alloy at the temperature *T=300K*.

$$g_{\alpha\beta}(r) = \frac{L^3}{N_\alpha N_\beta}\left\langle \sum_{j=1}^{N_\alpha} \frac{n_{j\beta}(r)}{4\pi r^2 \Delta r} \right\rangle, \qquad \alpha,\beta \in \{Ni, Zr\} \qquad (2)$$

and the partial static structure factor $S(k)$ [11]:

$$S_{\alpha\beta}(k) = \delta_{\alpha,\beta} + 4\pi n \int_0^\infty r^2 \left[g_{\alpha,\beta}(r) - 1\right] \frac{\sin(kr)}{kr} dr. \qquad (3)$$

Here, *k* is the wave-number. The total static structure factor was found by means of the spatial Fourier transformation of the total radial distribution function [12]:

$$g(r) = \frac{1}{(c_1 b_1 + c_2 b_2)^2} \sum_{\alpha,\beta}^{2} \sqrt{c_\alpha b_\beta} f_\alpha f_\beta g_{\alpha,\beta}(r), \qquad \alpha,\beta \in \{Ni, Zr\} \qquad (4)$$

where, *c* is the concentration of particles, *f* is the scattering length (in the case of neutron scattering).

Fig. 1 shows the temperature dependence of the particle distribution functions, $g_{\alpha\beta}(r)$, calculated on the basis of molecular dynamics simulation. The figure demonstrates that the $Ni_{33}Zr_{67}$-system is completely transformed into an amorphous state with a cooling: the distribution function is characterized a pronounced first peak and a splitting of the second one [13].

Fig. 2 (a, b, c) shows the partial static structure factor obtained on the basis of computer simulation while the comparison of the MD results for *S(k)* with the experimental data on neutron diffraction [14] is presented in Fig 2(d). It can be seen from the figure that the chosen model of interatomic interaction potential [10] reproduces in detail the structure of amorphous alloy.

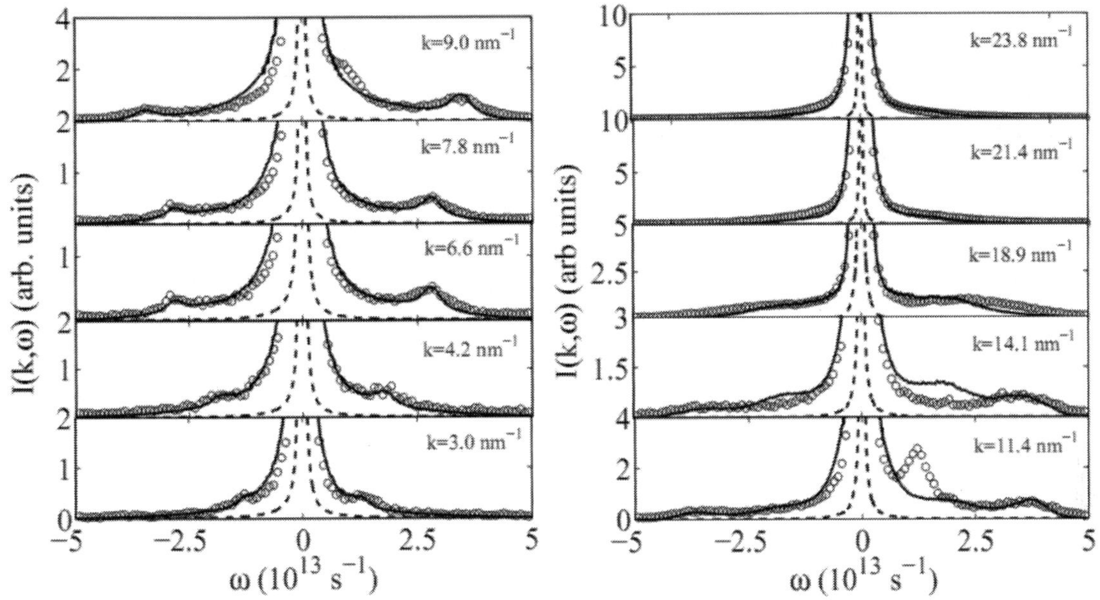

Figure 3. Intensity of scattering for the metallic alloy $Ni_{33}Zr_{67}$ at the temperature $T=300K$. The solid line represents the simulation results, taking into account the detailed balance condition (6) and the experimental resolution function (7); (o o o) – the experimental data on inelastic X-ray scattering [9]. The dotted line shows the experimental resolution function $R(k,\omega)$.

The microscopic dynamics of the amorphous metal alloy was analyzed by means of the dynamic structure factor $S(k,\omega)$. To compare the simulation results with the experimental data, we have calculated the intensity of the inelastic X-ray scattering, $I(k,\omega)$, which is related with the dynamic structure factor by the following relations [15]:

$$I(k,\omega) = E(k)\int R(k,\omega-\omega')S_q(k,\omega')d\omega', \tag{5}$$

$$S_q(k,\omega) = \frac{\hbar\beta\omega}{1-e^{-\hbar\beta\omega}}S(k,\omega). \tag{6}$$

Here $S_q(k,\omega)$ is the quantum dynamic structure factor, $\beta = 1/(k_B T)$ is the inverse temperature, $E(k)$ is a normalization factor, and $R(k,\omega)$ is the experimental resolution function

$$R(k,\omega) = \frac{1}{\sqrt{\pi}\omega_0(k)}\exp\left(-\omega^2/\omega_0(k)^2\right) \tag{7}$$

which satisfies the normalization condition

$$\int_{-\infty}^{\infty} R(k,\omega)d\omega = 1. \tag{8}$$

Comparison of the simulation results for the scattering intensity $I(k,\omega)$ with the experimental data on inelastic X-ray scattering is given in Fig. 3. As can be seen from the figure, the simulation results reproduce accurately the high frequency features of the X-ray scattering intensity in a wide range of wave numbers: they predict the positions, heights, and decrease of the side peaks. However, the simulation results do not reveal the experimentally-observed low-frequency excitations in $I(k,\omega)$ at the values of $k=9\div 11.4\ nm^{-1}$.

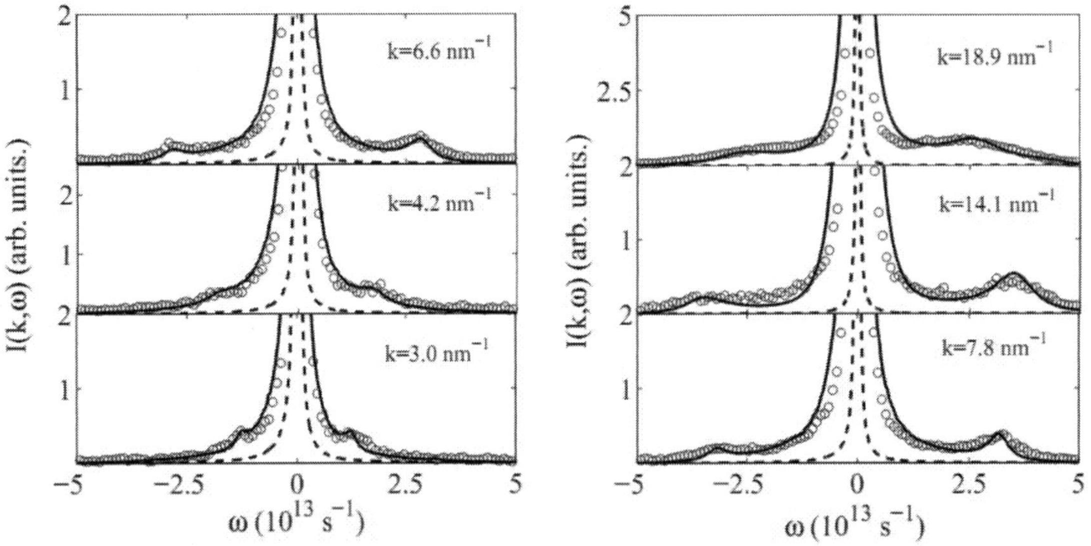

Figure 4. Intensity of scattering for metallic alloy $Ni_{33}Zr_{67}$ at the temperature $T=300K$. The solid line represents the results of theoretical calculations, taking into account the detailed balance condition (6) and the experimental resolution function (7); (o o o) – the experimental data on inelastic X-ray scattering [9]. The dotted line shows the experimental resolution function $R(k,\omega)$.

Accordingly to the theoretical approach developed [16,17] within Lee's recurrent relation formalism [18,19] the dynamic structure factor can be taken in the form [20]:

$$S(k,\omega)=\frac{S(k)}{2\pi}\Delta_1\Delta_2\Delta_3\sqrt{4\Delta_4-\omega^2}\left\{\Delta_2^2\Delta_3^2+\omega^2\left[-2\Delta_1\Delta_3^2+\right.\right.$$
$$+\Delta_1^2\Delta_4-\Delta_1^2\Delta_3+2\Delta_1\Delta_2\Delta_4-\Delta_1\Delta_2\Delta_3+\Delta_2^2\Delta_4\left.\right]+$$
$$+\omega^4\left[\Delta_3^2-2\Delta_1\Delta_4+2\Delta_1\Delta_3-2\Delta_2\Delta_4+\Delta_2\Delta_3\right]+$$
$$\left.+\omega^6\left[\Delta_4-\Delta_3\right]\right\}^{-1}. \tag{9}$$

Here, $S(k)$ is the static structure factor; Δ_1, Δ_2, Δ_3 and Δ_4 are the frequency relaxation parameters:

$$\begin{cases}\Delta_1=\frac{k_BT}{m}\frac{k^2}{S(k)},\\ \Delta_2=\frac{k_BT}{m}k^2\left(3-\frac{1}{S(k)}\right)+\frac{n}{m}\int\nabla_l^2U(r)\left[1-\exp\left(i\vec{k}\vec{r}\right)\right]g(r)d^3\vec{r},\\ \ldots\end{cases} \tag{10}$$

First two parameters were calculated from the molecular dynamics simulation data and the parameters Δ_3 and Δ_4 were taken as fitting values.

Fig. 4 shows the results of theoretical calculations for the scattering intensity $I(k,\omega)$ [Eqs. (5) and (9)] with the condition of the detailed balance (6) and the experimental resolution function (7) in comparison with the experimental data on inelastic X-ray scattering [9]. It is seen that Eq. (9) yields a good agreement with the experimental data. The results of theoretical calculations reproduce the high-frequency features of the experimental spectra on inelastic X-ray scattering for all values of wave numbers: the correct predictions for the position as well as for the height and damping of the lateral peaks.

4. Conclusion

This paper presents the results of the atomic/molecular dynamics simulations of the $Ni_{33}Zr_{67}$ bulk metallic glass at the temperature $T=300K$. The results indicate on the possibility to perform the correct description of the structural and dynamic properties of the amorphous metallic alloys containing the polyvalent (transition) metals in the framework of pair-wise effective potentials. The simulation results of structural and dynamic characteristics are in a good agreement with the experimental data on neutron diffraction and inelastic X-ray scattering data. The theoretical model developed recently to describe the structural relaxation in liquid metals reproduces the features of the $I(k,\omega)$ spectra for $Ni_{33}Zr_{67}$ metallic glass associated, in particular, with the appearance and disappearance of the high-frequency peaks.

5. Acknowledgments

The authors acknowledge to Dr. T. Scopigno (University of Rome "La Sapienza", Italy) and Prof. J.-B. Suck (University of Technology Chemnitz, Germany) for providing the experimental data and discussion of results. This work was supported by the grant of Academy of Sciences of the Republic of Tatarstan (project № 31-02/2012 (G)) and the grant of Ministry of Education SEC competition, the event 1.4, 2012.

References

[1] Clement W, Willens R H, Duwez P 1960 *Nature* **187** 869
[2] Greer A L 1995 *Science* **276** 1947
[3] Johnson W L 1999 *MRS Bulletin* **24** 42
[4] Khusnutdinoff R M, Mokshin A V 2010 *Bull. Russ. Acad. Sci. Phys.* **74** 640
[5] Omoto T, Arai M, Inamura Y et al 1998 *J. Non-Cryst. Solids* **232-234** 613
[6] Johnson W L 2002 *JOM* **54** 40
[7] Yang Q W, Zhangm T 2007 *J. Phys.: Condens. Matter.* **19** 086212
[8] Omoto T, Arai M, Suck J-B et al 2002 *J. Non-Cryst. Solids* **312-314** 599
[9] Scopigno T, Suck J-B et al 2006 *Phys. Rev. Lett.* **96** 135501
[10] Hausleitner C, Hafner J 1990 *Phys. Rev. B* **42** 5863
[11] Allen M P and Tildesley D J 1987 *Computer Simulation of Liquids* (Oxford: Clarendon Press)
[12] Andonov P, Chieux P 1985 *Journal de Physique* **46** C8-81
[13] Mokshin A V, Yulmetyev R M, Khusnutdinoff R M, Hänggi P 2006 *Phys. Solid State* **48** 1760
[14] Lee D, Lee A et al 1982 *Journal de Physique* **43** C9-19
[15] Khusnutdinoff R M, Mokshin A V, Yulmetyev R M 2009 *JETP* **108** 417
[16] Mokshin A V, Yulmetyev R M, Khusnutdinoff R M, Hänggi P 2007 *J. Phys.: Condens. Matter* **19** 046209
[17] Yulmetyev R M, Mokshin A V, Hänggi P, Shurygin V Yu 2002 *JETP Lett.* **76** 147
[18] Lee M H 1983 *Phys. Rev. Lett.* **49** 1072
[19] Lee M H 1983 *Phys. Rev. B* **26** 2547
[20] V, Yulmetyev R M, Khusnutdinov R M and Hänggi P 2006 *JETP* **103** 841

Description of the anomalous dielectric relaxation in disordered systems in the frame of the Mori-Zwanzig formalism

A A Khamzin, R R Nigmatullin and I I Popov

Kazan (Volga Region) Federal University, 420008, Kremlevskaya 18, Kazan, Russia.

e-mail: airat.khamzin@rambler.ru

Abstract. In the frame of Mori-Zwanzig formalism the empirical Navriliak-Negami expression for the complex dielectric permittivity has been derived. The derivation is based on construction of kinetic equations for relaxation and memory functions accordingly. They contain integro-differential operators of non-integer order.

1. Introduction

One of the most striking features of the dielectric relaxation phenomenon in different disordered materials such as glass-forming liquids and amorphous polymers is the failure of the Debye theory [1]. This simple theory of dielectric relaxation cannot describe properly the low-frequency spectrum, where the real relaxation behavior can deviate essentially from the conventional exponential Debye's pattern and actually it is characterized by a broad distribution of relaxation times. Such behavior is reflected in the title *anomalous dielectric relaxation* and in the first time was suggested empirically in the pioneer paper [2] of the Cole's brothers in 1941 year for description of dielectric relaxation in polar liquids. These and subsequent investigators have proposed [3, 4] various empirical formulas describing the departure from the Debye behavior. In specific terms, the normal Debye relaxation process is characterized by a normalized complex dielectric permittivity (NCDP) function

$$\hat{\varepsilon}(i\omega) = \frac{\varepsilon(i\omega) - \varepsilon_\infty}{\varepsilon_0 - \varepsilon_\infty} = \frac{1}{1 + i\omega\tau}, \tag{1}$$

where $\varepsilon(i\omega)$ determines the complex dielectric permittivity (CDP), ε_0 and ε_∞ are low-frequency and high-frequency limits of the CDP $\varepsilon(i\omega)$, correspondingly, τ is a characteristic relaxation time defined in the present context as the Debye relaxation time. Equation (1) adequately describes the low frequency behavior of the observed complex susceptibility of many simple polar liquids.

A significant amount of experimental data on disordered systems supports the following empirical expressions for dielectric loss spectra, namely, the Cole-Cole equation [5]

$$\hat{\varepsilon}(i\omega) = \left(1 + (i\omega\tau)^\alpha\right)^{-1}, \quad 0 < \alpha \le 1, \tag{2}$$

the Cole-Davidson equation

$$\hat{\varepsilon}(i\omega) = (1 + i\omega\tau)^{-\beta}, \quad 0 < \beta \leq 1, \tag{3}$$

and the Havriliak-Negami equation

$$\hat{\varepsilon}(i\omega) = \left(1 + (i\omega\tau)^{\alpha}\right)^{-\beta}, \quad 0 < \alpha, \beta \leq 1, \tag{4}$$

which is a combination of the Cole-Cole and Cole-Davidson equations.

As for physical mechanism determining the grounds of the Cole-Cole expression we see it in the hierarchical subordination of the dielectric relaxation process in many disordered systems [6,7], which, in turn, is related with a loss of ergodicity of the system considered. The hierarchical subordination is expressed in formation of clusters of relaxing dipoles having a self-similar structure. In paper [8] we formulated a model of anomalous dielectric relaxation based on hierarchical subordination of dynamic degrees of freedom. In the frame of this model it is possible to derive the Cole-Cole expression for the CDP and give the physical interpretation of the stretching power-law exponent α. In this paper, based on the results obtained in [8] we give a semi-phenomenological derivation for the Cole-Davidson and Havriliak-Negami expression for the CDP in the frame of the Mori-Zwanzig formalism.

2. Kinetic equations of the anomalous dielectric relaxation. General grounds

In this section we want to demonstrate the general approach of derivation of the desired kinetic equations for relaxation function. Their solutions lead (in frequency region) to the empirical expressions for the CDP reminded above. The relaxation function $\phi(t)$ is defined as

$$\phi(t) = \begin{cases} P(t)/P(0), & t \geq 0, \\ 0, & t < 0, \end{cases} \tag{5}$$

where $P(t)$ determines an experimental relaxation function (the electric polarization), $P(0) = \varepsilon_0 - \varepsilon_\infty$. The relaxation function $\phi(t)$ is associated with the NCDP $\hat{\varepsilon}(i\omega)$ by known relationship [5]

$$\hat{\varepsilon}(i\omega) = \hat{L}\left[-\dot{\phi}(t); i\omega\right] = 1 - i\omega\hat{L}\left[\phi(t); i\omega\right], \tag{6}$$

where $\hat{L}\left[\phi(t); i\omega\right]$ is the Laplace transform of the relaxation function $\phi(t)$. Let us present the empirical expression for the NCDP $\hat{\varepsilon}(i\omega)$ in the following form

$$\hat{\varepsilon}(i\omega) = \left(1 + \hat{R}^{-1}(i\omega)\right)^{-1}, \tag{7}$$

where $\hat{R}(i\omega) = \left(\hat{\varepsilon}^{-1}(i\omega) - 1\right)^{-1}$ we determine as the Laplace image of some function $R(t)$: $\hat{R}(i\omega) = \hat{L}[R(t); i\omega]$. Then for the functions NCDP mentioned above the functions $\hat{R}(i\omega)$ и $R(t)$ has the following forms:

Debye: $$\hat{R}_D(i\omega) = (i\omega\tau)^{-1}, \quad R_D(t) = \tau^{-1}; \tag{8}$$

Cole-Cole: $$\hat{R}_{CC}(i\omega) = (i\omega\tau)^{-\alpha}, \quad R_{CC}(t) = (\Gamma(\alpha))^{-1}\tau^{-\alpha}t^{\alpha-1}; \tag{9}$$

Cole-Davidson: $$\hat{R}_{CD}(i\omega) = \sum_{k=0}^{\infty}(1 + i\omega\tau)^{-\beta(k+1)}, \quad R_{CD}(t) = e^{-t/\tau}\tau^{-\beta}t^{\beta-1}E_{\beta,\beta}\left[\tau^{-\beta}t^{\beta}\right], \tag{10}$$

where $E_{\rho,\mu}[z] = \sum_{n=0}^{\infty} z^n / \Gamma(\rho n + \mu)$ determines two-parametric Mittag-Leffler function.

Havriliak-Negami: $$\hat{R}_{HN}(i\omega) = \frac{1}{(1+(i\omega\tau)^{\alpha})^{\beta}-1} = \sum_{k=0}^{\infty}(1+(i\omega\tau)^{\alpha})^{-\beta(k+1)},$$

$$R_{HN}(t) = \sum_{k=0}^{\infty}\tau^{-\alpha\beta(k+1)}t^{\alpha\beta(k+1)-1}E^{\beta(k+1)}_{\alpha,\alpha\beta(k+1)}[-\tau^{-\alpha}t^{\alpha}] = \tau^{-\alpha\beta}\exp\left(-\alpha^{-1}\tau^{-\alpha}\,{}_0D_t^{1-\alpha}t\right)t^{\alpha\beta-1}E_{\alpha\beta,\alpha\beta}[\tau^{-\alpha\beta}t^{\alpha\beta}], \quad (11)$$

where $E^{\gamma}_{\rho,\mu}[z] = \sum_{n=0}^{\infty}(\gamma)_n z^n / n!\Gamma(\rho n+\mu)$ determines the generalized Mittag-Leffler function, ${}_0D_t^{1-\alpha} = (d/dt)\,{}_0D_t^{-\alpha}$ determines the fractional derivative in the Riemann-Liouville form and expression

$${}_0D_t^{-\alpha}\phi(t) = \Gamma^{-1}(\alpha)\int_0^t (t-t')^{\alpha-1}\phi(t')dt', \quad (12)$$

determines the Riemann-Liouville fractional integral [9].

The substitution of expression (7) in equation (6) leads to the following operator equation for the desired relaxation function

$$i\omega\hat{L}[\phi(t);i\omega] = -i\omega\hat{R}(i\omega)\hat{L}[\phi(t);i\omega]. \quad (13)$$

This equation being transformed to time-domain restores the desired kinetic equation in the most general form

$$\frac{d}{dt}\phi(t) = -\frac{d}{dt}R(t)*\phi(t), \quad (14)$$

where $R(t)*\phi(t) = \int_0^t R(t-t')\phi(t')dt'$ determines the convolution of $R(t)$ and $\phi(t)$. The substitution of expressions for $R(t)$ (8)-(11) into equation (14) immediately leads us to the desired kinetic equations that corresponds to empirical expressions mentioned above, *viz.*, Debye, Cole-Cole, Cole-Davidson and Havriliak-Negami, correspondingly

$$\frac{d\phi(t)}{dt} + \tau^{-1}\phi(t) = 0, \quad (15)$$

$$\frac{d\phi(t)}{dt} + \tau^{-\alpha}\,{}_0D_t^{1-\alpha}\phi(t) = 0, \quad (16)$$

$$\frac{d\phi}{dt} + \tau^{-\beta}\frac{d}{dt}e^{-t/\tau}\mathbb{E}^{1}_{\beta,\beta,\tau^{-\beta};0+}e^{t/\tau}\phi(t) = 0, \quad (17)$$

$$\frac{d\phi}{dt} + \sum_{k=0}^{\infty}\tau^{-\alpha\beta(k+1)}\frac{d}{dt}\mathbb{E}^{\beta(k+1)}_{\alpha,\alpha\beta(k+1),-\tau^{-\alpha};0+}\phi(t) = 0. \quad (18)$$

Here we use the determination of the following integral operator [10]

$$\mathbb{E}^{\gamma}_{\rho,\mu,\omega;+0}f(t) = \int_0^t (t-t')^{\mu-1}E^{\gamma}_{\rho,\mu}[\omega(t-t')^{\rho}]f(t')dt' = \sum_{m=0}^{\infty}\left((\gamma)_m\omega^m/m!\right){}_0D_t^{-(\rho m+\mu)}. \quad (19)$$

So, we proved that for the NCDP expressions corresponding to anomalous dielectric relaxation (including also the Debye's relaxation as a partial case) one can write the kinetic equations for the corresponding relaxation functions. The distinct feature of these kinetic equations is the presence of differential operators of non-integer order.

3. The Mori-Zwanzig formalism and memory function phenomenon

In the previous section we formulated an alternative approach to solution of the problem of the anomalous dielectric relaxation. It is based on derivation of kinetic equations for relaxation function that differs from the conventional Debye's relaxation function. We demonstrated the general approach of derivation of these kinetic equations that leads to the well-known empirical expressions for the CDP. Then in order to make next step it is necessary to construct a theory which can give a physical justification of these empirical expressions suggested earlier. Among known semi-phenomenological approaches (as far as we know) related to derivation of kinetic equations and associated with anomalous dielectric relaxation that similar to our approach we want to mark the follows. In paper [11] the kinetic equation corresponding to the Havriliak-Negami expression for the NCDP is derived by means of generalization of the noninertial Fokker-Plank equation for Brownian motion. In paper [12] the authors suggest an approach to analysis of the non-Debye's dielectric relaxation based on the Mori-Zwanzig formalism. The last approach from our point of view is the most acceptable for the purpose of derivation the kinetic equations of anomalous dielectric relaxation. This conclusion is based on the following reasons. Firstly, the projection operator method introduced by Mori-Zwanzig for construction of kinetic equations for temporal correlation functions is based on the first principles and actually can be considered as exact. Secondly, this method naturally introduces the memory function conception that is convenient in interpretation of physical results.

Equation for the temporal correlation function $\psi(t)$ obtained in the frame of Mori-Zwanzig formalism can be written in the form [13]

$$\frac{d\psi(t)}{dt} = -\int_0^t K(t-t')\psi(t')dt', \tag{20}$$

where $K(t)$ determines the memory function (MF).

In the frame of the linear response approximation, the fluctuations of polarization caused by thermal motion are the same as for the macroscopic reconstruction induced by the electric field [14]. Thus, one can equate the relaxation function and the macroscopic dipole correlation function (DCF) $\psi(t)$ as follows

$$\phi(t) \cong \psi(t) = \frac{<\mathbf{M}(t)\mathbf{M}(0)>}{<\mathbf{M}(0)\mathbf{M}(0)>}, \tag{21}$$

where $\mathbf{M}(t)$ is the macroscopic fluctuating dipole moment of the sample volume unit, which is equaled to the vector sum of all entering in it molecular dipoles. So, the relaxation function $\phi(t)$ satisfies to the same equation with memory (20). If one introduces the integral memory function (IMF) $M(t) = \int_0^t K(t')dt'$ then the equation for the relaxation function (20) accepts the following form

$$\frac{d\phi(t)}{dt} = -\frac{d}{dt}\int_0^t M(t-t')\phi(t')dt' \equiv -\frac{d}{dt}M(t) * \phi(t), \tag{22}$$

which mathematically coincides with equation (14). So the function $R(t)$ in (14) can be interpreted as the IMF that is $R(t) \equiv M(t)$. So, comparing equation (22) with equations (15)-(18) we obtain the following expressions for the IMF that correspond to empirical expressions discussed earlier

Debye: $$M_D(t) = \tau^{-1}, \tag{23}$$

Cole-Cole: $$M_{CC}(t;\alpha) = \tau^{-\alpha}t^{\alpha-1} / \Gamma(\alpha), \tag{24}$$

Cole-Devidson: $$M_{CD}(t;\beta) = e^{-t/\tau}\sum_{k=0}^{\infty} M_{CC}(t;\beta(k+1)) = e^{-t/\tau}\tau^{-\beta}t^{\beta-1}E_{\beta,\beta}\left[\tau^{-\beta}t^{\beta}\right], \tag{25}$$

Havriliak-Negami: $$M_{HN}(t;\alpha,\beta)=\exp\left(-\alpha^{-1}\tau^{-\alpha}\,{}_0D_t^{1-\alpha}t\right)\tau^{-\alpha\beta}t^{\alpha\beta-1}E_{\alpha\beta,\alpha\beta}[\tau^{-\alpha\beta}t^{\alpha\beta}]. \quad (26)$$

Now it is necessary to develop a method for derivation of expressions for the IMF. We note that the Debye's expression for the IMF (23) corresponds to the absence of memory (markovian process) and for this case we have $K_D(t)=\tau^{-1}\delta(t)$. Non-Debye's expressions for IMF correspond the cases of the presence of a memory in the system considered.

4. Derivation of the Havriliak-Negami expression for the IMF

For derivation of the IMF (25) and (26) that correspond to Cole-Davidson and havriliak-Negami expressions it is necessary to construct the corresponding integro-differential (kinetic) equations. The Mori-Zwanzig formalism implies to calculate the memory function $K(t)$ from the following engaging system of integro-differential equations [15]

$$\frac{dK_n(t)}{dt}=-\int_0^t K_{n+1}(t-t')K_n(t')dt', \quad n=1,\,2,3,..., \quad (27)$$

where a set of $K_n(t)$ determines the so-called memory functions of the *n*th order and $K_1(t)\equiv K(t)$. For further conveniences we write these equations for the corresponding IMF

$$M_n(t)=tK_n(0)-\int_0^t M_{n+1}(t-t')M_n(t')dt', \quad n=1,\,2,3,..., \quad (28)$$

where $M_n(t)=\int_0^t K_n(t')dt'$. Let us analyze the equation for $M_1(t)$

$$M_1(t)=tK_1(0)-\int_0^t M_2(t-t')M_1(t')dt'. \quad (29)$$

If $K_1(0)=0$, and $M_2(t)=M_{CC}(t;\alpha)=\Gamma^{-1}(\alpha)\tau_\alpha^{-\alpha}t^{\alpha-1}$ then the last equation accepts the form

$$M_1(t)=-\tau_\alpha^{-\alpha}\,{}_0D_t^{-\alpha}M_1(t) \;\Rightarrow\; \frac{dM_1(t)}{dt}=-\tau_\alpha^{-\alpha}\,{}_0D_t^{1-\alpha}M_1(t). \quad (30)$$

Let us suppose that $M_1(t)$ contains two parts: $M_1(t)=M_1^{(1)}(t)+M_1^{(2)}(t)$, where $M_1^{(2)}(t)$ is determined by equation (29) at $K_1(0)=0$. At these conditions we obtain the following basic equation for $M_1(t)$:

$$M_1(t)=M_1^{(1)}(t)-\int_0^t M_2(t-t')M_1(t')dt'. \quad (31)$$

Let us suppose that this relaxation process considered in subsystems is superimposed on the second relaxation process that takes place between systems. The temporal evolution of the second process we determine by the second temporal argument t' in the corresponding IMF as $M_1(t,t')$. Let us suppose that the second relaxation process obeys to kinetic equation (30)

$$dM_1(t,t')/dt'=-\tau_\alpha^{-\alpha}\,{}_0D_t^{1-\alpha}M_1(t,t'). \quad (32)$$

The solution of this equation can be written as

$$M_1(t,t')=\exp\left(-\tau_\alpha^{-\alpha}\,{}_0D_t^{1-\alpha}t'\right)M_1(t). \quad (33)$$

To the Havriliak-Negami expression for the IMF the following relaxation mechanism described above is realized: $M_1^{(1)}(t)=M_{CC}(t;\alpha\beta)$, $M_2(t)=-M_{CC}(t;\alpha\beta)$. Then as it follows from (31) the corresponding equations for the IMF describing the relaxation inside the systems accepts the form

$$M_1(t) = \tau^{-\alpha\beta} t^{\alpha\beta-1} / \Gamma(\alpha\beta) + \tau^{-\alpha\beta} {}_0D_t^{-\alpha\beta} M_1(t) . \tag{34}$$

The solution of this equation is known [10]

$$M_1(t) = \tau^{-\alpha\beta} t^{\alpha\beta-1} E_{\alpha\beta,\alpha\beta}[\tau^{-\alpha\beta} t^{\alpha\beta}] . \tag{35}$$

The IMF for the combined relaxation process is determined from (33) at $t' \to t$ and accepts the form

$$M_1(t,t) = M_{HN}(t;\alpha,\beta) = \exp\left(-\alpha^{-1}\tau^{-\alpha} {}_0D_t^{1-\alpha} t\right) \tau^{-\alpha\beta} t^{\alpha\beta-1} E_{\alpha\beta,\alpha\beta}[\tau^{-\alpha\beta} t^{\alpha\beta}], \tag{36}$$

where $\alpha = \tau_\alpha^\alpha / \tau^\alpha$. The Cole-Davidson expression for IMF is realized at $\alpha = 1$. In this case the relaxation process between systems (as it follows from (32)) is described by Debye equation.

5. Conclusions

In this paper the authors suggest a method of description of the anomalous dielectric relaxation based on construction of kinetic equations containing non-integer differentiation/integration operators. For the derivation of the desired kinetic equations we use the Mori-Zwanzig formalism for correlation functions. This formalism determines also the important characteristics as memory function. Finally the problem of justification of empirical expressions for CDP is reduced to the derivation of the corresponding to them the desired IMF. For justification of the Havriliak-Negami empirical expression the following relaxation model is suggested. In this model relaxation in the system considered unifies the relaxation in dipole clusters and relaxation of dipole clusters themselves. In accordance with suggested relaxation model the scheme of the calculation of the desired IMF is suggested. It implies the construction of the corresponding kinetic equations for calculation of the desired IMF including each component of the relaxation process. The kinetic equations for the IMF are derived also in the frame of the Mori-Zwanzig formalism and based on utilization of the second order memory functions.

Acknowledgements

The work was done in the frame of the research plan "Dielectric spectroscopy and kinetics of complex systems" (Number: 12-18 02 0210 021000018) that has been accepted by the KFU for 2012 year.

References

[1] Debye P 1954 *Polar Molecules* (New York: Dover).
[2] Cole K S and Cole R H 1941 *J. Chem. Phys.* **9** 341.
[3] Davidson D W and Cole R H 1951 *J. Chem. Phys.* **19** 1484.
[4] Havriliak S and Negami S 1966 *J. Polym. Sci.*, Part A-1 **14**, 99; 1967 *Polymer* **8** 161.
[5] Scaife B K P 1998 *Principles of Dielectrics* 2nd ed. (London: Oxford University Press).
[6] Olemskoi A I, Flat A Ya 1993 *Phys. Usp.* **36** 1087.
[7] Novikov V V, Privalko V P 2001 *Phys. Rev. E* 64 031504.
[8] Khamzin A A, Nigmatullin R R and Popov I I 2012 *Theor. Math. Phys.* In press.
[9] Samko S G, Kilbas A A and Marychev O I 1993 *Fractional, Integrals and Derivatives. Theory and Applications* (Amsterdam: Gordonand Breach).
[10] Kilbas A A, Saigo M and Saxena R K 2004 *Integral Transforms and Special Functions* **15** 31.
[11] Kalmykov Y P, Coffey W T and Titov S V 2004 *Phys. Rev. E* **70** 041103.
[12] Douglas J F and Hubbard J B 1991 *Macromolecules* **24** 3163.
[13] Boon J P and Yip S 1980 *Molecular Hydrodynamics* (New York: Dover).
[14] Williams G 1972 *Chem. Rev.* **72** 55.
[15] Mori H 1965 *Prog. Theor. Phys.* **33** 423; 1965 **34** 399.

Spin excitations and electron paramagnetic resonance in two-layer antiferromagnetic system $Y_{1-x}Yb_xBa_2Cu_3O_{6+y}$

A A Vishina[1], A Maisuradze[2], A Shengelaya[3], B I Kochelaev[1] and H. Keller[2]

[1]Institute of Physics, Kazan Federal University, Kremlevskaya 18, Kazan 420008, Russia
[2]Physik-Institut der Universität Zürich, Winterhurerstrasse 190, ZH-8057, Switzerland
[3]Department of Physics, Tbilisi State University, Chavchavadze av. 3, GE-0128, Georgia

E-mail: alenavishina@gmail.com

Abstract. An experimental and theoretical study of the EPR spectrum of $Y_{0.98}Yb_{0.02}Ba_2Cu_3O_{6.1}$ is reported. A strong anisotropy of the EPR linewidth is revealed for the parallel and perpendicular orientations of the external magnetic field relative to the tetragonal symmetry axis. This anisotropy is related to the indirect spin-spin interaction between the ytterbium ions via antiferromagnetic spin-waves. The explicit expressions for the coupling of ytterbium ions with the spin waves and the corresponding indirect spin-spin interactions are derived. An estimate of the exchange coupling between the ytterbium and copper ions is found from the comparison of the theory with the experiment.

1. Introduction

Antiferromagnetic properties and spin fluctuations of quasi-two-dimensional copper oxides remain a subject of intensive investigations due to their transformation into high-temperature superconductors by doping of CuO_2 planes with electronic holes. Soon after the discovery of high-T_c superconductivity in $La_{2-x}Ba_xCuO_4$ [1] the attention was attracted to the YBCO family. Detailed inelastic neutron scattering experiments in the bilayer compound $YBa_2Cu_3O_{6+y}$ reveal a very large superexchange integral between nearest-neighbor Cu sites within the CuO_2 layers ($J \approx 2000\ K$ [2], $J \approx 1400\ K$ [3]), a rather strong coupling between the nearest layers J_1 ($\delta = J_1 / J \geq 4 \cdot 10^{-2}$ [2], $\delta \geq 7 \cdot 10^{-2}$ [3]), and an extremely small coupling to the next-nearest-layer with $J_2 / J \approx 2 \cdot 10^{-5} - 3 \cdot 10^{-4}$ [2,3]. These results were obtained on the basis of the standard spin-wave theory, which for the bilayer system predicts two acoustical and two optical modes of spin waves. The observed energy gap for the modes involving spin deviations out of the CuO_2 planes was modeled by taking into account a week planar anisotropy $\alpha_{xy} = \Delta J / J \approx 2 \cdot 10^{-4} - 7 \cdot 10^{-4}$[2,3]. The Neel temperature $T_N = 415$ K at $y < 0.15$ could be explained by taking into account the quantum effect and a small anisotropy of the exchange integral taking the value the Cu-Cu coupling $J \approx 1700\ K$ [4]. Recently the analytic expression for T_N was obtained by the Green's function method in the Tyablikov approximation as a function of both exchange integrals between the CuO_2 planes for $J_1, J_2 = J$ and $\alpha_{xy} = 0$ [5]. NMR and EPR were found to be very powerful techniques for probing the local magnetic susceptibility and studying the spin kinetics of high-T_c oxides [6]. The EPR method is usually based on the EPR signal from the magnetic probes using ions having local *d*- or *f*-electrons. In particular, the use of Mn^{2+} as an EPR probe in the CuO_2 plane allowed revealing a very fast spin relaxation rate of the Cu-ions due to their spin-phonon interaction, which prevents the observation of the EPR signal directly from the Cu-ions [7]. A detailed EPR study of the Yb^{3+} spin relaxation as a function of temperature in $YBa_2Cu_3O_{6+y}$ ($y = 0.1,\ 0.4,\ 0.5,\ 0.6,\ 0.98$) shows that at high temperatures the main contribution to the relaxation rate is given by the Raman two-phonon process due to the modulation of the crystal electric field by optical modes of the oxygen ions in the CuO_2 planes [8]. In the superconducting samples ($y > 0.4$) an electronic contribution of the Korringa-type relaxation was found. However, the nature of the temperature independent relaxation rate remains unclear.

In this work we report our theoretical and experimental study of spin excitations and relaxation due to the interaction of Yb^{3+} ions with antiferromagnetic (AF) spin waves in $YBa_2Cu_3O_{6+y}$. At low oxygen doping ($y < 0.15$) the CuO_2 planes are not doped with electronic holes. We find that indirect spin-spin interactions of Yb^{3+} ions via AF magnons (Suhl-Nakamura interaction) are very long-range and give an important contribution to the EPR linewidth and position of the resonance line. From the comparison of our calculations with the experiment we estimate the exchange coupling between the Yb^{3+} and Cu^{2+} ions to be $|A| \approx 120\ K$.

The paper is organized as follows. In Sec. 2 we find the energy gaps of AF spin waves in $YBa_2Cu_3O_{6+y}$ due to an external magnetic field and dipole-dipole interactions. The derivation of the Yb-Cu interaction and an indirect interaction between Yb^{3+} ions via magnons is given in Sec. 3. The experimental results on the EPR spectra and a comparison with experiment follow in Sec. 4.

2. The energy gaps in the spin wave dispersion due to the external magnetic field and dipole-dipole interactions

2.1. The external magnetic field

In the $YBa_2Cu_3O_{6+x}$ the CuO_2 plains are grouped in bilayers with a small exchange interaction between different bilayers. In this study we neglect this coupling. The exchange spin Hamiltonian can be written in the form:

$$\frac{H_{ex}}{J} = \frac{1}{2}\sum_{\mathbf{nm}}\left[\mathbf{S}^a_{\mathbf{n}1}\mathbf{S}^b_{\mathbf{m}1} + \mathbf{S}^b_{\mathbf{n}2}\mathbf{S}^a_{\mathbf{m}2} - \alpha_{xy}\left(S^{za}_{\mathbf{n}1}S^{zb}_{\mathbf{m}1} + S^{zb}_{\mathbf{n}2}S^{za}_{\mathbf{m}2}\right)\right] + \delta\sum_{\mathbf{n}}\mathbf{S}^a_{\mathbf{n}1}\mathbf{S}^b_{\mathbf{n}2} + \delta\sum_{\mathbf{m}}S^b_{\mathbf{m}1}S^a_{\mathbf{m}2}. \qquad (1)$$

Here a and b indicate two sub-lattices in every plane. In the layer 1 the sites **n** and **m** are related to the a and b sub-lattices correspondingly, while vise versa in the layer 2. Hereafter we take into account the nearest neighbors only. J is the exchange constant within the plane, δ and α_{xy} correspond to the exchange coupling between the two planes and an anisotropic exchange term both measured in units J, respectively. We choose the direction of the spontaneous magnetizations along the x-axis, since due to the axial symmetry it can be chosen arbitrary.

We shall now take into account the external magnetic field applied to the system within the plane and directed along the y axis:

$$H_{ZCu} = B_{Cu}\sum_{\mathbf{n}}\left(S^{ya}_{\mathbf{n}1} + S^{yb}_{\mathbf{n}2}\right) + B_{Cu}\sum_{\mathbf{m}}\left(S^{yb}_{\mathbf{m}1} + S^{ya}_{\mathbf{m}2}\right); \quad B_{Cu} = g_{Cu}\mu_B H_0. \qquad (2)$$

Here g_{Cu} is the g-factor for the Cu ion, μ_B is the Bohr magneton, and H_0 – the external magnetic field.

We assume the field rotates the magnetizations of both sub-lattices slightly toward the y direction by the angle φ. It is convenient then to turn the axes so that the new x-axis is directed along the corresponding magnetizations of the sub-lattices:

$$\begin{aligned} S_x^a &= S'_x \cos\varphi - S'_y \sin\varphi, \\ S_y^a &= S'_x \sin\varphi + S'_y \cos\varphi, \\ S_x^b &= S'_x \cos\varphi + S'_y \sin\varphi, \\ S_y^b &= -S'_x \sin\varphi + S'_y \cos\varphi. \end{aligned} \tag{3}$$

Using the standard Holstein-Primakoff formalism we make the following transformation for the layer:

$$\begin{aligned} S_{\mathbf{n}1}^{xa} &= \frac{1}{2} - a_{\mathbf{n}1}^{+} a_{\mathbf{n}1}, \quad S_{\mathbf{n}1}^{+} = \left(1 - a_{\mathbf{n}1}^{+} a_{\mathbf{n}1}\right)^{1/2} a_{\mathbf{n}1}, \quad S_{\mathbf{n}1}^{-} = a_{\mathbf{n}1}^{+} \left(1 - a_{\mathbf{n}1}^{+} a_{\mathbf{n}1}\right)^{1/2}, \\ S_{\mathbf{m}1}^{xb} &= -\frac{1}{2} + b_{\mathbf{m}1}^{+} b_{\mathbf{m}1}, \quad S_{\mathbf{m}1}^{+} = b_{\mathbf{m}1}^{+} \left(1 - b_{\mathbf{m}1}^{+} b_{\mathbf{m}1}\right)^{1/2}, \quad S_{\mathbf{m}1}^{-} = \left(1 - b_{\mathbf{m}1}^{+} b_{\mathbf{m}1}\right)^{1/2} b_{\mathbf{m}1}, \end{aligned} \tag{4}$$

with $S^{\pm} = -S^{z} \pm iS^{x}$. For the layer $j = 2$ the transformation can be obtained by interchanging indices **n** and **m**. Here a^{+}, a, b^{+}, b are the boson creation and annihilation operators for the two sub-lattices. Then we perform the Fourier transformation to the reciprocal lattice:

$$\begin{aligned} a_q &= N^{-1/2} \sum_{\mathbf{n}} e^{i\mathbf{q}\mathbf{r}_{\mathbf{n}}} a_{\mathbf{n}}, \\ b_q &= N^{-1/2} \sum_{\mathbf{n}} e^{i\mathbf{q}\mathbf{r}_{\mathbf{n}}} b_{\mathbf{n}}, \end{aligned} \tag{5}$$

where N is the number of unit cells in the bilayer. From the energy minima condition for the spin system ground state in the external magnetic field we obtain the value of the rotation angle

$$\sin\varphi = -\frac{B_{Cu}}{J(4+\delta)} = -b_{Cu}. \tag{6}$$

As a result of these steps the total spin Hamiltonian $H_{tot} = H_{ex} + H_{ZCu}$ takes the form:

$$\begin{aligned} \frac{H_{tot}}{2J} &= \left(1 + \frac{\delta}{4}\right) \sum_q \left(a_{q1}^{+} a_{q1} + a_{q2}^{+} a_{q2} + b_{q1}^{+} b_{q1} + b_{q2}^{+} b_{q2}\right) + \\ &+ \left(1 - b_{Cu}^2 - \frac{\alpha_{xy}}{2}\right) \sum_q \gamma_q \left(a_{q1} b_{-q1} + a_{q1}^{+} b_{-q1}^{+} + a_{q2} b_{-q2} + a_{q2}^{+} b_{-q2}^{+}\right) + \\ &+ \left(b_{Cu}^2 - \frac{\alpha_{xy}}{2}\right) \sum_q \gamma_q \left(a_{q1} b_{q1}^{+} + a_{q1}^{+} b_{q1} + a_{q2} b_{q2}^{+} + a_{q2}^{+} b_{q2}\right) + \\ &+ \left(1 - b_{Cu}^2\right) \frac{\delta}{4} \sum_q \left(a_{q1} b_{-q2} + a_{q1}^{+} b_{-q2}^{+} + a_{2q} b_{-q1} + a_{q2}^{+} b_{-q1}^{+}\right) + \\ &+ \frac{\delta}{4} b_{Cu}^2 \sum_q \left(a_{q1} b_{q2}^{+} + a_{q1}^{+} b_{q2} + a_{q2} b_{q1}^{+} + a_{q2}^{+} b_{q1}\right). \end{aligned} \tag{7}$$

Here

$$\gamma_q = 1/2\left[\cos\left(q_x a\right) + \cos\left(q_y a\right)\right] \tag{8}$$

To diagonalize this Hamiltonian we perform the Bogoliubov transformation to new creation and annihilation boson operators α_q^+, β_q^+, η_q^+, κ_q^+, α_q β_q, η_q, κ_q, which must satisfy the commutation relations of the following type:

$$\left[\alpha_q, H_{tot}\right] = E_{\alpha q}\alpha_q \tag{9}$$

Using these equations we obtain the energy eigenvalues and explicit expressions for the eigenoperators for two acoustical and two optical modes of spin waves.

For the acoustical modes we have:

$$\begin{aligned} \left(\frac{E_{\alpha q}}{2J}\right)^2 &= \left[1-\gamma_q + 2b_{Cu}^2\left(\gamma_q + \frac{\delta}{4}\right)\right]\left[1+\gamma_q + \frac{\delta}{2} - \alpha_{xy}\gamma_q\right], \\ \left(\frac{E_{\beta q}}{2J}\right)^2 &= \left[1-\gamma_q + \alpha_{xy}\gamma_q\right]\left[1+\gamma_q + \frac{\delta}{2} - 2b_{Cu}^2\left(\gamma_q + \frac{\delta}{4}\right)\right]. \end{aligned} \tag{10}$$

One can see that at the Brillouin zone center $q = (0,0)$ the gapless mode $E_{\alpha q}$ has now a gap at the value close to the Zeeman energy $E_{\alpha 0} \approx 4b_{Cu}J \approx g_{Cu}\mu_B H_0$, while the other mode $E_{\beta q}$ maintains almost the same gap due to anisotropy of the exchange interaction. The eigenoperators of these modes are:

$$\begin{aligned} \alpha_q &= \frac{1}{2}\left\{u_{\alpha q}\left(a_{1q} + a_{2q} + b_{1q} + b_{2q}\right) - v_{\alpha q}\left(a_{1-q}^+ + a_{2-q}^+ + b_{1-q}^+ + b_{2-q}^+\right)\right\}, \\ \beta_q &= \frac{1}{2}\left\{u_{\beta q}\left(a_{1q} + a_{2q} - b_{1q} - b_{2q}\right) - v_{\beta q}\left(a_{1-q}^+ + a_{2-q}^+ - b_{1-q}^+ - b_{2-q}^+\right)\right\}. \end{aligned} \tag{11}$$

The coefficients of the transformation in (11) are given by:

$$\begin{aligned} u_{\alpha,\beta}^2 &= \frac{J}{E_{\alpha,\beta}}\left[1 + \frac{\delta}{4} \mp \frac{\alpha_{xy}\gamma_q}{2} \pm b_{Cu}^2\left(\gamma_q + \frac{\delta}{4}\right) + \frac{E_{\alpha,\beta}}{2J}\right], \\ v_{\alpha,\beta}^2 &= \frac{J}{E_{\alpha,\beta}}\left[1 + \frac{\delta}{4} \mp \frac{\alpha_{xy}\gamma_q}{2} \pm b_{Cu}^2\left(\gamma_q + \frac{\delta}{4}\right) \quad \frac{E_{\alpha,\beta}}{2J}\right]. \end{aligned} \tag{12}$$

For the optical modes the eigenvalues of energy are:

$$\begin{aligned} \left(\frac{E_{\eta q}}{2J}\right)^2 &= \left[1+\gamma_q - \alpha_{xy}\gamma_q\right]\left[1-\gamma_q + \frac{\delta}{2} + 2b_{Cu}^2\left(\gamma_q - \frac{\delta}{4}\right)\right], \\ \left(\frac{E_{\kappa q}}{2J}\right)^2 &= \left[1+\gamma_q - 2b_{Cu}^2\left(\gamma_q - \frac{\delta}{4}\right)\right]\left[1-\gamma_q + \frac{\delta}{2} + \alpha_{xy}\gamma_q\right]. \end{aligned} \tag{13}$$

At the zone boundary for $\mathbf{q} = (\pi/a, \pi/a)$ the parameter $\gamma_{\pi,\pi} = -1$, and the mode $E_{q\kappa}$ now has a gap of about the Zeeman energy, while the other mode has about the same gap as it had before. The eigenoperators for the optical modes are the following:

$$\begin{aligned} \eta_q &= \frac{1}{2}\left\{u_{q\eta}\left(a_{q1} - a_{q2} + b_{q1} - b_{q2}\right) - v_{q\eta}\left(a_{-q1}^+ - a_{-q2}^+ + b_{-q1}^+ - b_{-q2}^+\right)\right\}, \\ \kappa_q &= \frac{1}{2}\left\{u_{q\kappa}\left(a_{q1} - a_{q2} - b_{q1} + b_{2q}\right) - v_{q\kappa}\left(a_{-q1}^+ - a_{-q2}^+ - b_{-q1}^+ + b_{-q2}^+\right)\right\}. \end{aligned} \tag{14}$$

The coefficients of the transformation in (13) are given by

$$u_{\eta,\kappa}^2 = \frac{J}{E_{\eta,\kappa}}\left[1+\frac{\delta}{4} \mp \frac{\alpha_{xy}\gamma_q}{2} \pm b_{Cu}^2\left(\gamma_q - \frac{\delta}{4}\right) + \frac{E_{\eta,\kappa}}{2J}\right],$$
$$v_{\eta,\kappa}^2 = \frac{J}{E_{\eta,\kappa}}\left[1+\frac{\delta}{4} \mp \frac{\alpha_{xy}\gamma_q}{2} \pm b_{Cu}^2\left(\gamma_q - \frac{\delta}{4}\right) - \frac{E_{\eta,\kappa}}{2J}\right]. \tag{15}$$

The obtained expressions are valid for the external magnetic field below the critical value, at which the two sub-lattices collapse into one with magnetic moments directed along the magnetic field.

2.2. *Dipole-dipole interactions*

The internal field due to magnetic dipole-dipole interactions between the Cu ions may be more important than the external magnetic field below the critical value. The spin Hamiltonian of the dipole-dipole interactions has the form:

$$H_{dd} = g^2\mu_B^2 \sum_{ij} \frac{(\mathbf{S}_i\mathbf{S}_j)R_{ij}^2 - 3(\mathbf{R}_{ij}\mathbf{S}_i)(\mathbf{R}_{ij}\mathbf{S}_j)}{R_{ij}^5} \tag{16}$$

We take into account the second term only, since the first term is isotropic and much smaller than the isotropic exchange interaction. We have found that the nearest neighbors (nn) do not contribute to the energy gap, so we have to take into account the next nearest neighbors (nnn) also. To simplify the expressions we set the external magnetic field to $H_0 = 0$.

The following calculations are similar to the ones performed above: we make the Holstein-Primakoff and the Fourier transformations for $H_{dd} = H_{dd}^{nn} + H_{dd}^{nnn}$. The minima condition for the ground state is realized for the magnetization directed along the diagonal of the square lattice of the layers. The result of these calculations for the total Hamiltonian $H_{tot} = H_{ex} + H_{dd}$ is the following:

$$\begin{aligned}
\frac{H_{tot}}{2J} &= \left(1+\frac{\delta}{4}-\frac{d_1}{2}+2d_3+\frac{d_4}{2}\right)\sum_q\left(a_{q1}^+a_{q1}+a_{q2}^+a_{q2}+b_{q1}^+b_{q1}+b_{q2}^+b_{q2}\right)+\\
&+\sum_q\left(\gamma_q-\frac{\alpha_{xy}\gamma_q}{2}-\frac{d_1}{4}\gamma_{qy}\right)\left(a_{q1}b_{-q1}+a_{q1}^+b_{-q1}^+ + a_{q2}b_{-q2}+a_{q2}^+b_{-q2}^+\right)+\\
&+\sum_q\left(-\frac{\alpha_{xy}\gamma_q}{2}+\frac{d_1}{4}\gamma_{qy}\right)\left(a_{q1}b_{q1}^+ + a_{q1}^+b_{q1}+a_{q2}b_{q2}^+ + a_{q2}^+b_{q2}\right)+\\
&+\left(\frac{\delta}{4}-\frac{d_2}{8}\right)\sum_q\left(a_{q1}b_{-q2} \quad a_{q1}^+b_{-q2}^+ + a_{q2}b_{-q1}+a_{q2}^+b_{-q1}^+\right)+\\
&-\frac{d_2}{8}\sum_q\left(a_{q1}b_{q2}^+ + a_{q1}^+b_{q2}+a_{q2}b_{q1}^+ + a_{q2}^+b_{q1}\right)+\\
&+\frac{d_4}{8}\sum_q\gamma_{yq}\left(a_{q1}a_{-q2}+a_{q1}^+a_{-q2}^+ + b_{q1}b_{-q2}+b_{q1}^+b_{-q2}^+\right) \quad -\\
&-\frac{d_4}{8}\sum_q\gamma_{yq}\left(a_{q1}a_{q2}^+ + a_{q1}^+a_{q2}+b_{q1}b_{q2}^+ + b_{q1}^+b_{q2}\right),
\end{aligned} \tag{17}$$

Here $d_1 = 3g^2\mu_B^2/Ja^3$, $d_2 = 3g^2\mu_B^2/Jc^3$, $d_3 = 3g^2\mu_B^2/4\sqrt{2}Ja^3$, $d_4 = 3g^2\mu_B^2a^2/JR^5$ with $R=\sqrt{a^2+c^2}$, $\gamma_{yq} = \cos(q_y a)$; c is the distance between the layers. Performing then the Bogoliubov-transformation we obtain the eigenvalues for the spin wave energy.

The acoustical modes:

$$\left(\frac{E_{\alpha q}}{2J}\right)^2 = \left(1-\gamma_q+\Delta_\alpha\right)\left(1+\gamma_q+\frac{\delta}{2}-\alpha_{xy}\gamma_q\right),$$

$$\left(\frac{E_{\beta q}}{2J}\right)^2 = \left(1-\gamma_q+\alpha_{xy}\gamma_q+\Delta_\beta\right)\left(1+\gamma_q+\frac{\delta}{2}\right); \tag{18}$$

$$\Delta_\alpha = 2d_3, \quad \Delta_\beta = -\frac{d_1}{2}+\frac{d_2}{4}+2d_3+\frac{d_4}{2}.$$

Here were neglected the dipole-dipole terms, which do not contribute to the energy gap at the zone-center $q=0$. Since all the dipole-dipole terms $d_1,d_2,d_3,d_4 << \delta$ (exchange coupling between the layers), we have omitted them in the corresponding sum. One can see that the first mode has now the energy gap $E_{\alpha 0} \approx 4J\sqrt{d_3} >> g_{Cu}\mu_B H_0$

The optical modes:

$$\left(\frac{E_{q\eta}}{2J}\right)^2 = \left(1+\gamma_q-\alpha_{xy}\gamma_q+\Delta_\eta\right)\left(1-\gamma_q+\frac{\delta}{2}\right),$$

$$\left(\frac{E_{q\kappa}}{2J}\right)^2 = \left(1+\gamma_q+\Delta_\kappa\right)\left(1-\gamma_q+\frac{\delta}{2}+\alpha_{xy}\gamma_q\right); \tag{19}$$

$$\Delta_\eta = -\frac{d_1}{2}+\frac{d_2}{4}+2d_3+\frac{d_4}{2}, \quad \Delta_\kappa = 2d_3.$$

Here were neglected the dipole-dipole terms, which do not contribute to the energy gap at the zone-boundary $\mathbf{q}=(\pi/a,\pi/a)$. The dipole-dipole terms in the sum with $\delta/2$ were also omitted. It is evident that both energy gaps at the zone boundary are also much larger than the Zeeman energy.

3. Interactions of the Yb^{3+}-ions with the AF spin-waves and the Suhl-Nakamura interaction

3.1. The Hamiltonian of Yb-spin-wave interactions

An exchange interaction between the particular *j*-th Yb^{3+}-ion of the crystal lattice and the nearest eight Cu-ions laying in the two parallel CuO_2 planes can be written in the form:

$$H^j_{YbCu} = A\mathbf{Y}_j\left[\sum_{\mathbf{n}}\left(\mathbf{S}^a_{\mathbf{n}1}+\mathbf{S}^b_{\mathbf{n}2}\right)+\sum_{\mathbf{m}}\left(\mathbf{S}^b_{\mathbf{m}1}+\mathbf{S}^a_{\mathbf{m}2}\right)\right] \tag{20}$$

Here A is the coupling constant, $\mathbf{Y}_j$ is the spin of the *j*-th Yb^{3+}-ion. For the Cu-spins again the axes-rotation (3), the Holstein-Primakoff transformation (4), and the Fourier transformation (both for the magnon and ytterbium operators) are performed. Taking into account the Zeeman interaction we obtain after the Bogoliubov transformation (9) the following result for the Hamiltonian $H_{YbCu} = H^{(0)}_{YbCu} + H^{(1)}_{YbCu} + H^{(2)}_{YbCu}$ for the magnetic field oriented along the *y*-axis:

$$
\begin{aligned}
&H_{YbCu}^{(0)} = 4Ab_{Cu}Y_0^y N^{1/2} - Ab_{Cu}Y_0^y N^{-1/2}\sum_q \left\{v_{q\alpha}^2 + v_{q\beta}^2 + v_{q\eta}^2 + v_{q\kappa}^2\right\}, \\
&H_{YbCu}^{(1)} = 2A\sum_q \left\{F_q^{ac}\left[Y_q^z\left(u_{q\alpha}+v_{q\alpha}\right)\left(\alpha_q+\alpha_{-q}^+\right)+ib_{Cu}Y_q^x\left(u_{q\alpha}-v_{q\alpha}\right)\left(\alpha_q-\alpha_{-q}^+\right)\right]+\right. \\
&+F_q^{op}\left[Y_q^z\left(u_{q\kappa}+v_{q\kappa}\right)\left(\kappa_q+\kappa_{-q}^+\right)+ib_{Cu}Y_q^x\left(u_{q\kappa}-v_{q\kappa}\right)\left(\kappa_q-\kappa_{-q}^+\right)\right]+ \\
&\left.+iY_q^y\left(1-b_{Cu}^2\right)^{1/2}\left[F_q^{ac}\left(u_{q\beta}-v_{q\beta}\right)\left(\beta_q-\beta_{-q}^+\right)+F_q^{op}\left(u_{q\eta}-v_{q\eta}\right)\left(\eta_q-\eta_{-q}^+\right)\right]\right\}.
\end{aligned}
\tag{21}
$$

Here $Y_q^{x,y,z}$ is the Fourier-transform of the site operators similar to (5); F_q^{ac} and F_q^{op} are form factors for the acoustical and the optical modes:

$$
F_q^{ac} = \cos\frac{q_x a}{2}\cos\frac{q_y a}{2}, \quad F_q^{op} = \sin\frac{q_x a}{2}\sin\frac{q_y a}{2}. \tag{22}
$$

In the case of the magnetic field oriented along the z-axis we obtain:

$$
\begin{aligned}
&H_{YbCu}^{(0)} = 4Ab_{Cu}Y_0^z N^{1/2} - Ab_{Cu}Y_0^z N^{-1/2}\sum_q \left\{v_{q\alpha}^2 + v_{q\beta}^2 + v_{q\eta}^2 + v_{q\kappa}^2\right\}, \\
&H_{YbCu}^{(1)} = 2A\sum_q \left\{F_q^{ac}\left[-iY_q^y\left(u_{q\beta}-v_{q\beta}\right)\left(\beta_q-\beta_{-q}^+\right)-b_{Cu}Y_q^x\left(u_{q\beta}+v_{q\beta}\right)\left(\beta_q+\beta_{-q}^+\right)\right]+\right. \\
&+F_q^{op}\left[-iY_q^y\left(u_{q\eta}-v_{q\eta}\right)\left(\eta_q-\eta_{-q}^+\right)-b_{Cu}Y_q^x\left(u_{q\eta}+v_{q\eta}\right)\left(\eta_q+\eta_{-q}^+\right)\right]+ \\
&\left.+Y_q^z\left(1-b_{Cu}^2\right)^{1/2}\left[F_q^{ac}\left(u_{q\alpha}+v_{q\alpha}\right)\left(\alpha_q+\alpha_{-q}^+\right)+F_q^{op}\left(u_{q\kappa}+v_{q\kappa}\right)\left(\kappa_q+\kappa_{-q}^+\right)\right]\right\}.
\end{aligned}
$$

The term $H_{YbCu}^{(2)}$ quadratic in the boson operators is responsible for two-magnon processes, which are not considered here. The total Hamiltonian now takes the form:

$$
H = H_{ZYb} + \sum_q\left(E_{q\alpha}\alpha_q^+\alpha_q + E_{q\beta}\beta_q^+\beta_q + E_{q\eta}\eta_q^+\eta_q + E_{q\kappa}\kappa_q^+\kappa_q\right) + H_{YbCu}; \tag{23}
$$

Here H_{ZYb} is the Zeeman energy of the ytterbium ions which has to be renormalized by the H_{YbCu} interaction.

3.2. The Suhl-Nakamura interaction

An exchange between the ytterbium ions by the AF magnons creates an indirect spin-spin coupling known as the Suhl-Nakamura (SN) interaction. An effective spin-spin Hamiltonian can be obtained by the unitary transformation with the following elimination of the magnon operators. Using the Hamiltonian (23) we make the following transformation:

$$
\tilde{H} = e^S H e^{-S} = H_0 + H_{YbCu} + \left[S, H_0 + H_{YbCu}\right] + \frac{1}{2}\left[S,\left[S, H_0 + H_{YbCu}\right]\right] + \ldots \tag{24}
$$

Here S is some yet unknown operator linear in the Yb-Cu coupling. To eliminate the linear terms we put

$$
H_{YbCu} + \left[S, H_0\right] = 0; \quad \langle m|S|n\rangle = \frac{\langle m|H_{YbCu}|n\rangle}{E_m^0 - E_n^0} \tag{25}
$$

Here we have used the eigen states of the unperturbed Hamiltonian. The SN interaction can be obtained from the (21, 24, 25) in the second order coupling of Yb-Cu:

$$H_{SN} = -2A^2 \sum_q \left\{ \left(F_q^{ac}\right)^2 \left[\frac{\left(u_{q\alpha} + v_{q\alpha}\right)^2}{E_{q\alpha}} Y_q^z Y_{-q}^z + \frac{\left(u_{q\beta} - v_{q\beta}\right)^2}{E_{q\beta}} Y_q^y Y_{-q}^y \right] + \right.$$
$$\left. + \left(F_q^{op}\right)^2 \left[\frac{\left(u_{q\eta} - v_{q\eta}\right)^2}{E_{q\eta}} Y_q^y Y_{-q}^y + \frac{\left(u_{q\kappa} + v_{q\kappa}\right)^2}{E_{q\kappa}} Y_q^z Y_{-q}^z \right] \right\} \qquad (26)$$

Taking into account the explicit expressions for the Bogoliubov coefficients (12,15) we obtain the final result

$$H_{SN} = -\frac{2A^2}{J} \sum_q f_q \left(Y_q^y Y_{-q}^y + Y_q^z Y_{-q}^z \right); \quad f_q = \frac{\left(F_q^{ac}\right)^2}{1 + \frac{\delta}{2} + \gamma_q} + \frac{\left(F_q^{op}\right)^2}{1 + \frac{\delta}{2} - \gamma_q}. \qquad (27)$$

In the denominators the terms which are small compared to δ were omitted. In the coordinate representation the dependence of the SN interaction on the distance between the ytterbium ions R can be roughly approximated as $f(R) \propto \mathrm{J}_1(R)/R$, where $\mathrm{J}_1(R)$ is the Bessel function. It means that this interaction is more long-range than the dipole-dipole one.

4. The Electron Paramagnetic Resonance of Yb^{3+}-ions in antiferromagnetic YBCO compound

4.1. Experimental results for the ytterbium EPR spectra

The EPR spectra were measured in the sample $Y_{0.98}Yb_{0.02}Ba_2Cu_3O_{6.1}$. We expect that at this level of oxygen doping ($y = 0.1$) electronic holes are not yet present in the CuO_2 planes. Figure 1 shows typical Yb^{3+} EPR spectra taken at $T = 40$ K with the external magnetic field along and perpendicular to the crystal c-axis and with an alternating field within the xy-plane. It is well known that the ground-state multiplet $^2F_{7/2}$ of the Yb^{3+} ions $(4f^{13})$ is expected to be split by the crystal electric field of tetragonal symmetry into four Kramers doublets [9]. Inelastic neutron-scattering measurements showed that in $YbBa_2Cu_3O_7$ the first-excited doublet lies 1000 K above the ground state doublet [10]. Our measurements reveal a very strong anisotropy of the EPR linewidth in the antiferromagnetic sample $Y_{0.98}Yb_{0.02}Ba_2Cu_3O_{6.1}$. The g-factors $g_\perp = 3.54$, $g_\parallel = 3.23$ show an anisotropy similar to the one in the previous measurements for oxygen doping $y = 0.4$ (where the AF state is suppressed, while the symmetry remains tetragonal, and $g_\perp^0 = 3.49$, $g_\parallel^0 = 3.13$) with a shift of the resonance line to lower magnetic fields [8]. We suggest that these features of the Yb^{3+} EPR signal may be explained by the Yb-Cu exchange interaction and the corresponding coupling of Yb^{3+} ions with the AF spin-waves.

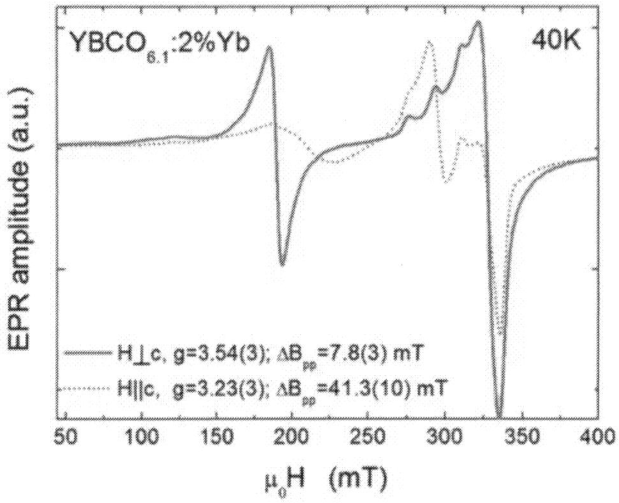

Figure 1. EPR spectra of Yb^{3+} in grain-oriented $Y_{0.98}Yb_{0.02}Ba_2Cu_3O_{6.1}$ at $T = 40$ K. The solid line corresponds to the external magnetic field perpendicular to the crystal c-axis, while the dotted line corresponds to the external magnetic field along the c-axis.

4.2. Analysis of the EPR linewidth

One can expect that at relatively low temperatures the broadening of the EPR signal is caused by the spin-spin interactions: the usual magnetic dipole-dipole interactions and the SN interactions between the Yb^{3+} ions. Their contribution can be calculated by the standard method of moments for the EPR line [9]. A contribution of the dipole-dipole interactions for the Yb^{3+} ion concentration $x = 0.02$ was estimated by calculations of the second and fourth moments using the experimental g-factors. For the external magnetic field perpendicular to the crystal c-axis the result is $\Delta B_{dd}^{\perp} = 8$ mT, while for the parallel orientation it is slightly larger $\left(\Delta B_{dd}^{\mathrm{P}} = 8.2\ \mathrm{mT}\right)$. The experimental values are significantly different: the calculated dipole-dipole contribution for the perpendicular orientation practically coincides with the experimental value $\Delta B_{\mathrm{exp}}^{\perp} = 7.8$ mT, while for the parallel orientation the experimental linewidth is five times larger $\left(\Delta B_{\mathrm{exp}}^{\mathrm{P}} = 41.3\ \mathrm{mT}\right)$. Such behavior can be related to the anisotropy of the SN interaction. For the external field below the critical value the AF magnetization is almost perpendicular to external field. It means that if $\mathbf{H}_0$ is directed along y-axis, the alternating field is directed along the x-axis and a contribution to the second moment is defined by a commutator of the total spin Y_0^x with H_{SN}. However, one can see from Eq. (27) that the operator structure of H_{SN} gives $\left[Y_0^x, H_{SN}\right] = 0$, which means that for this orientation there is no contribution from the SN-interaction to the EPR linewidth.

The situation is different for the external magnetic field oriented along the c-axis. In this case the alternating field can be directed arbitrarily in the xy-plane. In a high-temperature approximation the second moment can be calculated by the formula:

$$M_2 = -\frac{\mathrm{Sp}\left(\left[Y_0^-, H_{SN}\right]\left[Y_0^+, H_{SN}\right]\right)}{\mathrm{Sp}\left(Y_0^- Y_0^+\right)}, \quad Y_0^{\pm} = Y_0^x \pm i Y_0^y \tag{28}$$

The result is the following:

$$M_2 = \frac{A^4}{J^2}\frac{a^2}{\pi^2}\int_0^{\pi/a} dq_x \int_0^{\pi/a} dq_y f_q^2 = 0.14\frac{A^4}{J^2} \qquad (29)$$

Similar calculations for the fourth moment give

$$M_4 = \frac{8A^8}{J^4}\frac{a^2}{\pi^2}\int_0^{\pi/a} dq_x \int_0^{\pi/a} dq_y f_q^4 = 0.051\frac{A^8}{J^4} \qquad (30)$$

For the diluted ytterbium spin-system the peak-to-peak EPR linewidth for the parallel orientation can be calculated by the formula [9]:

$$\Delta B_{SN}^{\parallel} = \frac{\pi}{\sqrt{3}} x \left(\frac{M_2^3}{M_4}\right)^{1/2} \approx 0.42x\left(\frac{A^2}{J}\right) \qquad (31)$$

This result allows to estimate the exchange coupling A between the ytterbium and copper ions if we suppose that the main contribution for the parallel orientation comes from the SN interaction. Taking the experimental value $\Delta B_{\exp}^{\mathrm{P}} = 41.3$ mT and extracting the dipole-dipole contribution of 8.2 mT, we relate the rest to the SN interaction. Using $J = 1700$ K [4] and the Yb concentration $x = 0.02$ we find $|A| \approx 120$ K. The sign remains unknown.

4.3. Analysis of the EPR g-factors

According to the Eqs. (21, 6) the renormalized Zeeman energy can be written in the form;

$$\tilde{B}_{Yb} = \mu_B H_0 \left[g_{Yb}^0 - g_{Cu}\frac{A}{J}\left(1 - \frac{1}{4}v_{tot}^2\right)\right]; \quad v_{tot}^2 = \frac{1}{N}\sum_q \left\{ v_{q\alpha}^2 + v_{q\beta}^2 + v_{q\eta}^2 + v_{q\kappa}^2 \right\} \approx 0.614 \qquad (32)$$

We can now estimate the renormalized g-factors of the ytterbium ions due to their coupling with the AF spin waves. Following the above discussion we take the initial values of the Yb^{3+} g-factors for the experimental value of the paramagnetic state $g_{Yb}^0 = g_{\mathrm{P}}^0 = 3.13$ for the parallel orientation and $g_{Yb}^0 = g_{\perp}^0 = 3.49$ for the perpendicular orientation [8]. The typical g-factors of the copper ion in the tetragonal field with the orbital ground state of the type $(x^2 - y^2)$ are $g_{Cu}^{\mathrm{P}} = 2.4$ and $g_{Cu}^{\perp} \approx 2.1$ for the parallel and perpendicular orientations of the external magnetic field, respectively [9]. If we adobe a negative sign of the exchange coupling found above $A = -120$ K and take again $J = 1700$ K, we obtain for the AF state the theoretical values $g_{\mathrm{P}}^{\mathrm{th}} = 3.27$ and $g_{\perp}^{\mathrm{th}} = 3.61$, which are quite close to the experimental values $g_{\mathrm{P}} = 3.23$ and $g_{\perp} = 3.54$. Some corrections to the g-factors can appear due to the SN-interaction and the terms $H_{YbCu}^{(2)}$ quadratic in the boson operators. Our preliminary evaluations show that they are very small, especially for our system with low concentration of ytterbium ions.

In conclusion, the present study of spin excitations and EPR in the bilayer antiferromagnetic system reveals a strong anisotropy of the EPR linewidth due to the anisotropic SN interaction between the ytterbium ions via antiferromagnetic spin waves. A comparison of the theoretical and experimental results allows to estimate the exchange coupling between the Yb^{3+} and Cu^{2+} ions. The energy gaps in the spin-wave spectra due to the external magnetic field and internal magnetic dipole-dipole interactions were found as well.

Acknowledgements

This work was supported by the Swiss National Science Foundation, the SCOPES grant No. IZ73Z0_128242, and partially by the Ministry of Education and Science of the Russian Federation.

References

[1] Bednorz J G and Müller K A 1986 *Z. Phys. B - Condensed Matter* **64** 189
[2] Vettier C, Burlet P, Henry J Y, Jurgens M J, Lapertot G, Regnault L R and Rossat-Mignod J 1989 *Phys. Scr.* **T29** 110
[3] Shamoto S, Sato M, Tranquada J M, Sternlieb B J and Shirane G 1993 *Phys. Rev. B* **48** 13817
[4] Rossat-Mignod J, Regnault L P, Vettier C, Burlet P, Henry J Y and Lapertot G 1991 *Physica B* **169** 58
[5] Val'kov V V and Fedoseev A D 2011 *Theor. Math. Phys.* **168** 1216
[6] For a review on EPR in cuprates see a chapter by Kochelaev B I and Teitel'baum G B in the book "*Superconductivity in Complex Systems*" 2005 ed K A Müller and A Bussmann-Holder (Springer-Verlag, Berlin Heidelberg) pp 203-266
[7] Kochelaev B I, Kan L, Elschner B, and Elschner S 1994 *Phys. Rev. B* **49** 13106
[8] Maisuradze A, Shengelaya A, Kochelaev B I, Pomjakushina E, Conder K, Keller H, and Müller K A 2009 *Phys. Rev. B* **79** 054519
[9] Abragam A and Bleany B 1970 "*Electron Paramagnetic Resonance of Transition Ions*" (Clarendon Press, Oxford)
[10] Guillaume M, Allenspach P, Mesot J, Staub U, Furrer A, Osborn R, Tailor A D, Stucki F and Unternaher P 1992 *Solid State Commun.* **81** 999

The phase diagrams of the ferromagnet-superconductor trilayers in external magnetic field

Maxim Avdeev, Yurii Proshin
Kazan Federal University, 420008, 18 Kremlyovskaya St., Kazan, Russia.
E-mail: avdeyev_mv@mail.ru

Abstract. We investigate the critical properties of asymmetrical trilayers F_1F_2S and F_1SF_2 structures in an external parallel magnetic field. The different mutual orientations of the F layers magnetizations are examined. At this condition the triplet component of the superconducting condensate is arisen. Assuming that all F and S layers are dirty, we solve the boundary value problem for the Usadel function. Then we use Gor'kov's self-consistency equation and calculate the critical temperature for both trilayers as function of the F layers thicknesses in external magnetic field H. We predict the surprising appearance of re-entrant superconductivity with increasing magnetic field.

1. Introduction

At present, there is a greet interest to study of the layered structures, consisting of alternating layers of ferromagnetic (F) and superconducting (S) metals (see reviews [1, 2, 3] and references therein). It is common knowledge that superconductivity and ferromagnetism are two competing physical phenomena and their coexistence in homogeneous materials is difficult to observe. However, this coexistence is possible in artificial structures FS due to the proximity effect [4].

In the case of good contact S and F metals the superconducting correlations can penetrate from the S layer into the F metal on depth of order coherence length $\sim \sqrt{a_F l_f}$ ($a_F = v_F/2I$ is spin stiffness length, v_F is Fermi velocity, I is exchange field and l_f is free path length in the F metal). Note, that penetration depth in the ferromagnet is much smaller than that in the normal metal, since condition $I \gg T_c$ is satisfied in conventional ferromagnets.

As it was predicted Larkin, Ovchinnikov [5] and Fulde, Ferrell [6], the order parameter $\Delta(\mathbf{r})$, penetrating into F metal, is oscillating function of coordinate $\mathbf{r}$ and in the simplest case $\Delta(\mathbf{r}) = \exp i\mathbf{k}_f\mathbf{r}$. Such a nonuniform state with a nonzero momentum $k_f^2 \approx 2iI/D_f$ (D_f is diffusion coefficient) is called LOFF state.

The interplay between superconducting and ferromagnets layers leads in particular to nonmonotonic dependence of the critical temperature T_c on the F layers thickness d_f (see [7, 2, 3]) even for the two- (F/S) and trilayered (F/S/F) systems.

In recent years more theoretical and experimental works were devoted to studying the properties of trilayer FS systems with noncollinear magnetizations direction in the F layers [8, 9, 10, 11]. This leads to the generation of spin-triplet correlations, decaying in the F metal at much greater distances in comparison with the spin-singlet component of the superconducting condensate [12, 13]. This fact for the thin-film structures leads to strong dependence of the critical temperature T_c on the angle ϕ between the directions of the magnetizations of layers F [11].

The unique properties of layered F/S structures make them attractive for technological applications. Thus, in late 90s, a model of a spin switch based on a three-layered F/F/S [14] and F/S/F [15, 16] systems was proposed. External magnetic field changes the relative direction of magnetizations **M** of the F layers. As a result, the system can turn from the superconducting to the resistive state.

In the present paper we consider thin-films asymmetrical trilayers $F_1/F_2/S$ and $F_1/S/F_2$ in an external parallel magnetic field.

2. Method

The critical temperature T_c at the second-order transition is obtained from the self – consistency equation for the gap

$$\Delta(\mathbf{r})\ln t = \pi T_c \sum_{\omega>0}^{\infty} \mathrm{Sp}\left(\hat{F}(\mathbf{r},\omega) - \frac{\Delta(\mathbf{r})}{\omega}\right), \tag{1}$$

where $t = T_c/T_{c0}$ is the reduced temperature (T_{c0} is the transition temperature of isolated S layer), $\omega = \pi T_c(2n+1)$ is Matsubara frequency.

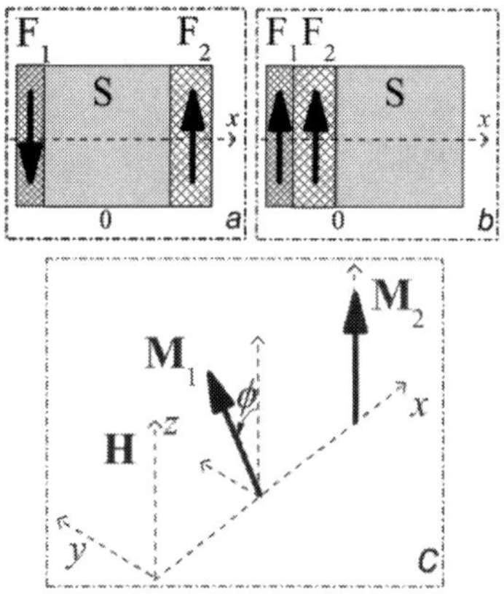

Figure 1. (a) F_1SF_2 system shown at $\phi = \pi$ (Antiparallel state);
(b) F_1F_2S system shown at $\phi = 0$ (Parallel state).
The magnetizations have vector form $\mathbf{M}_1(0, 0, M)$, $\mathbf{M}_2(0, M\sin\phi, M\cos\phi)$ (see panel (c)).

Anomalous Green function $\hat{F}$ near the second-order transition into the superconducting state, for S and F layers, satisfies the Usadel equation ($\omega > 0$) written on the external magnetic field presence [17, 18]

$$\left[|\omega| - \frac{1}{2}D_s\left(\nabla - \frac{2\pi i}{\Phi_0}\mathbf{A}\right)^2\right] F^s_{\alpha\beta}(\mathbf{r},\omega) = \delta_{\alpha\beta}\Delta(\mathbf{r}),$$

$$\left[|\omega| - iI\sigma^z_{\alpha\beta} - \frac{1}{2}D_{\alpha\beta}\left(\nabla - \frac{2\pi i}{\Phi_0}\mathbf{A}\right)^2\right] F^f_{\alpha\beta}(\mathbf{r},\omega) = 0, \tag{2}$$

$$D_{\alpha\beta} = \frac{D_f}{1 - i2I\tau_f\sigma^z_{\alpha\beta}}.$$

Here Φ_0 is the quantum magnetic flux, $D_{s,f}$ is the diffusion coefficient in S and F layers, respectively and $\tau_f = l_f/v_F$ is the elastic scattering time. Eq. (2) should be solved with boundary conditions on the internal SF interfaces of the Kupriyanov–Lukichev type [19] modified with consideration of the external magnetic field H (see, for example [20])

$$\frac{4D_s}{\sigma_s v_F^s}\left(\nabla - \frac{2\pi i}{\Phi_0}\mathbf{A}\right)\mathbf{n}F^s_{\alpha\beta} = \frac{4D_{\alpha\beta}}{\sigma_f v_F^f}\left(\nabla - \frac{2\pi i}{\Phi_0}\mathbf{A}\right)\mathbf{n}F^f_{\alpha\beta} = F^s_{\alpha\beta} - F^f_{\alpha\beta}, \tag{3}$$

and on the free boundaries

$$\left(\nabla - \frac{2\pi i}{\Phi_0}\mathbf{A}\right)\mathbf{n}F^{s,f}_{\alpha\beta} = 0, \tag{4}$$

where σ_s and σ_f are the parameters of the transparency of the junction on the S and F sides, respectively, $\mathbf{n}$ is the unit vector of the normal to the contact surface. The boundary conditions (3) take into account the condition of detailed balance $\sigma_s v_F^s N_s = \sigma_f v_F^f N_f$ where N_s and N_f are the densities of state on the Fermi surface in the S and F metals, respectively. Note that, if the magnetization $\mathbf{M}$ in F layer has the form $\mathbf{M}(0, M\sin\phi, M\cos\phi)$ then the solution of the Eq. (2) is $\hat{F}' = \hat{U}^\dagger \hat{F}\hat{U}$. Here $\hat{U}$ is unitary matrix, which has the form

$$\hat{U} = \begin{pmatrix} \cos\frac{\phi}{2} & i\sin\frac{\phi}{2} \\ i\sin\frac{\phi}{2} & \cos\frac{\phi}{2} \end{pmatrix}.$$

The external magnetic field is parallel to the plane of contact. Because the thicknesses d_s, $d_f \ll \lambda_H$ (λ_H is magnetic penetration depth), we may average the coefficients depending on magnetic field in Eq. (2) over the thicknesses of the S and F layers, i.e.

$$A^2(x) \approx \langle A^2 \rangle = \frac{1}{d_s}\int H^2(x - x_c)^2 dx \approx \frac{1}{12}d_s^2 H^2.$$

Similarly, we assume that $\Delta_s(x) \approx \langle\Delta\rangle$ and $\Delta_f(x) = 0$ in S and F layers respectively.

In Figure 1 the geometry of the problem is displayed for $F_1/F_2/S$ (panel (a)) and $F_1/S/F_2$ (panel (b)) trilayers. The external magnetic field is parallel to the plane of contact.

3. Results and discussion

At first, we discuss an influence of asymmetry on the critical temperature value. For convenience, we have introduced the parameter $n_{sf} = v_F^s N_s / v_F^f N_f$. In Figure 2 the difference $\Delta t = t^{AP} - t^P$ is displayed as function of the reduced thicknesses d_{f1}/a_{f1} and d_{f2}/a_{f2} for the $F_1/S/F_2$ (panel (a)) and $F_1/F_2/S$ (panel (b)) trilayers without external magnetic field ($H = 0$). Here t^{AP} and t^P denote the reduced temperature for the antiparallel and parallel orientation of the magnetizations, respectively. Note, that the function $\Delta t(d_{f1}, d_{f2})$ maximum corresponds to the asymmetric case, i.e. $d_{f1} \neq d_{f2}$ for both cases. It is interesting fact, because the stable operation of the spin valve device based on three-layered structure FSF [15, 16, 21] (see also references cited in [11]) needs in the high Δt values. Note also, Δt for the $F_1/F_2/S$ trilayer (Figure 2(b)) can be both positive or negative as opposed to above considered $F_1/S/F_2$ case (Figure 2(a)) [8, 11].

In Figure 3 we plot the phase diagrams for asymmetrical $F_1/S/F_2$ structures in the presence of external magnetic field, here $h = H/H_c$ (H_c is critical field for isolated S metal). In the top panel (a) the dependence of the different Δt on the reduced thickness F_2 layer is displayed. Without field this dependence has smooth maximum at the value $d_{f2}/a_{f2} \approx 0.5$ and then it reaches a plateau. With increasing field ($h = 0.2$) two peaks appear. Note, that

$\Delta t(h = 0.2) > \Delta t(h = 0)$. In the panel (b) the $t(d_{f2}/a_{f2})$ dependences are displayed and in the panel (c) the $t(h)$ curves for antiparallel state are shown. The magnetic field monotonically depresses the critical temperature, but it may also lead to qualitative change of dependence $t(d_f)$ (see Figure 3(b)). The effect of magnetic field is especially strong in vicinity of the re-entrant superconductivity appearance. In other word, the magnetic field makes possible to observe re-entrant superconductivity.

In Figure 4 we display the phase diagrams for $F_1/F_2/S$ system in the external magnetic field. In the panels (a, b) the dependences of the reduced critical temperature $t^{P,(AP)}$ on the reduced thickness F_1 layer d_{f1}/a_{f1} for $\phi = 0^0$ (a) and $\phi = 180^0$ (b) are shown. In figure 4(c) we plot the angular dependence of the critical temperature. We see that the temperature dependence $t(\phi)$ is not monotonic and it has a minimum at $\phi \approx 90^0$. It is interesting to note that the surprising appearance of re-entrant superconductivity with increasing magnetic field (see Figure 4(a, b)).

In conclusion, we shortly discuss the model of the spin switch, based on the $F_1/F_2/S$ trilayers (Figure 4). For technical realizations of a spin valve, the direction of the magnetization $\mathbf{M}_1$ is fixed, while the magnetization of the adjacent F_2 layer can be rotated by the external magnetic field. If the initial state of the system to choose at $t_{init}(\phi = 180^0, h = 0) = 0.28$ on the phase diagram (see Figure 4(c)) then the system is in the superconducting state ($t_{init} < t$). However, under the influence of a weak magnetic field, we can change the direction of magnetization $\mathbf{M}_2$ ($\phi = 90^0$). Thus, the system moves to the resistive state, since the condition $t(\phi = 90^0, h \geq 0) > t$ is satisfied. Note, that, it is important to choose the "working point" for a stable work of a spin valve in an optimum manner: on the one hand, it should be located in the phase diagram with a maximally large temperature t, but on the other hand, the difference Δt should remain larger than certain critical value Δt_c.

References

[1] Fominov Y V, Chtchelkatchev N M and Golubov A A 2002 *Phys. Rev.* B **66**(1) 014507
[2] Izumov Yu A, Proshin Yu N and Khusainov M G 2002 *Usp. Fiz. Nauk* **172** 114
[3] Buzdin A 2005 *Rev. Mod. Phys.* **77** 935
[4] de Gennes P G 1964 *Rev. Mod. Phys.* **36** 225
[5] Larkin A I and Ovchinnikov Yu N 1964 ZhETF, **47** (3), 1136
[6] Fulde P and Ferrell R 1964 *Phys. Rev.* **135** A550
[7] Khusainov M and Proshin Y 1997 *Phys. Rev.* B **56** R14283
[8] Fominov Y V, Golubov A A and Kupriyanov M Y 2003 *JETP Letters* **77** 510
[9] Bergeret F S, Volkov A F and Efetov K B 2005 *Rev. Mod. Phys.* **77** 1321
[10] Houzet M and Buzdin A I 2007 *Phys. Rev.* B **76** 060504
[11] Fominov Y V, Golubov A A, Karminskaya T Y, Kupriyanov M Y, Deminov R G and Tagirov L R 2010 *JETP Letters* **91** 329
[12] Bergeret F S, Volkov A F and Efetov K B 2001 *Phys. Rev. Lett.* **86** 4096
[13] Champel T and Eschrig M 2005 *Phys. Rev.* B **72** 054523
[14] Oh S, Youm D and Beasley M 1997 *Appl. Phys. Lett.* **71** 2376
[15] Tagirov L R 1999 *Phys. Rev. Lett.* **83** 2058
[16] Buzdin A I, Vedyayev A V and Ryzhanova N V 1999 *EPL (Europhysics Letters)* **48**(6) 686
[17] Radović Z, Dobrosavljević-Grujić L, Buzdin A I and Clem J R 1988 *Phys. Rev.* B **38** 2388
[18] Krunavakarn B and Yoksan S 2006 *Physica C: Superconductivity* **440** 25
[19] Kupriyanov M Yu and Lukichev V F 1988 *Zh. Eksp. Teor. Fiz.* **94** 139
[20] Avdeev M, Khusainov M, Proshin Y and Tsarevskii S 2010 *Superconductor Science and Technology* **23** 105005
[21] Gu J Y, You C Y, Jiang J S, Pearson J, Bazaliy Ya B and Bader S D 2002 *Phys. Rev. Lett.* **89**(26) 267001

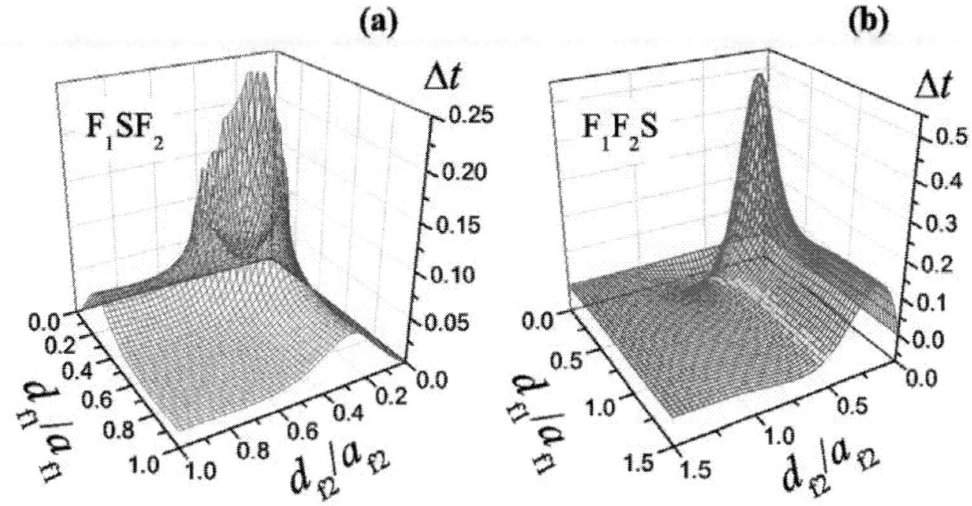

Figure 2. The $\Delta t(d_{f1}, d_{f2})$ dependence for the F_1SF_2 (a) and F_1F_2S (b) trilayers without external field ($H = 0$). The common parameters of calculation are $l_s/\xi_{s0} = 0.2$, $I/\pi T_{cs} = 10$. In panel (a) $d_s/\xi_{s0} = 0.7$, $n_{sf} = 1.5$, $l_f/a_f = 0.25$, $\sigma_s = 0.8$. In panel (b) $d_s/\xi_{s0} = 0.6$, $n_{sf} = 1$, $l_f/a_f = 0.3$, $\sigma_s = 2$, $\sigma_f = 10$.

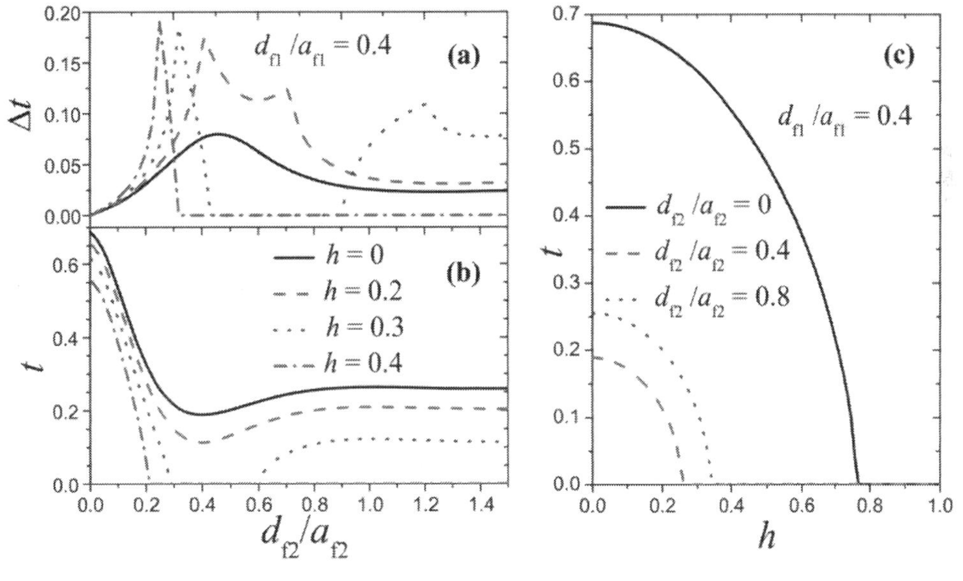

Figure 3. The phases diagrams displayed for asymmetrical $F_1/S/F_2$ system in external magnetic field $h = H/H_c$ (H_c is critical field of the isolated S film). In panel (a) shown the different $\Delta t = t^{AP} - t^P$ dependence on the reduced thickness d_{f2}/a_{f2} and fixed value $d_{f1}/a_{f1} = 0.4$. In panel (b) shown the dependence $t(d_{f2})$ for the antiparallel state. In panel (c) displayed dependence $t(h)$ for some values d_{f1}/a_{f1} and d_{f2}/a_{f2}. Other parameters are the same as in Figure 2(a).

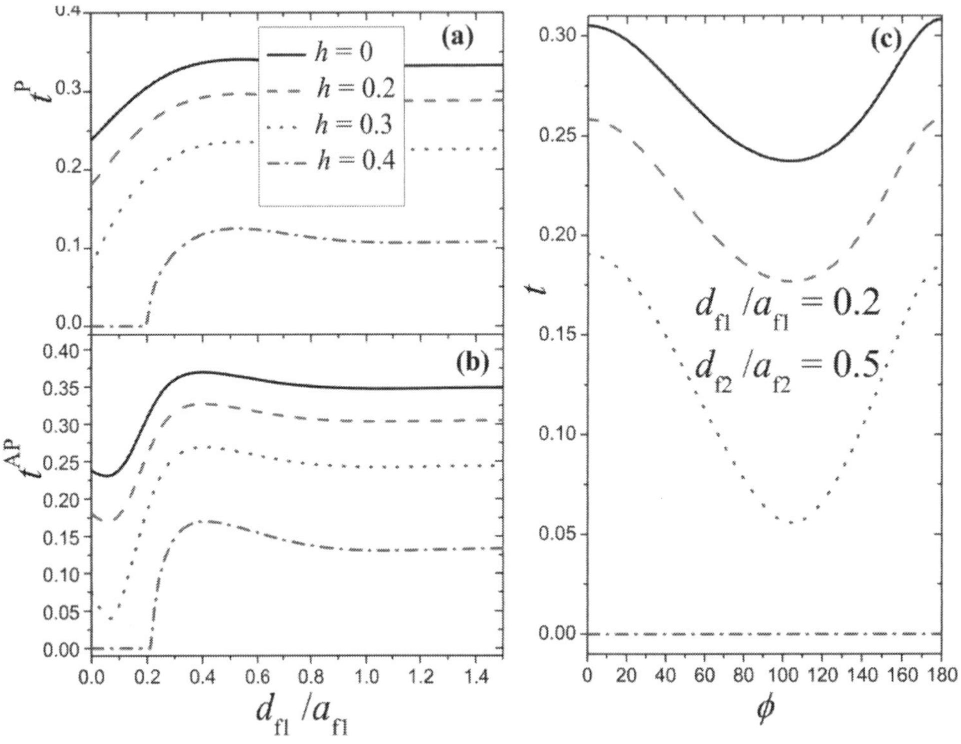

Figure 4. The phases diagrams displayed for asymmetrical $F_1/F_2/S$ system in external magnetic field h. In panel (a) shown the dependence t^P (for the parallel state) on the reduced thickness d_{f1}/a_{f1} and fixed value $d_{f2}/a_{f2} = 0.5$. In panel (b) shown the dependence $t^{AP}(d_{f2})$ for the antiparallel state. In panel (c) displayed the angular dependence $t(\phi)$ for fixed values $d_{f1}/a_{f1} = 0.2$ and $d_{f2}/a_{f2} = 0.5$. Other parameters are the same as in Figure 2(b).

The solitary re-entrant superconductivity in the clean four-layered superconductor/ferromagnet system

A.N. Minnullin, M.M. Khusainov, Yu.N. Proshin
Kazan Federal University, Kazan, 420008, Kremlevskaya 18, Russia.
E-mail: minnullin_artur@mail.ru

Abstract. The superconducting critical temperature T_c for asymmetric four-layered system ferromagnet/superconductor/ferromagnet/superconductor (F′/S/F/S′) in the clean limit using the boundary value problem for Eilenberger function is investigated. An electron-electron coupling constant in ferromagnetic metals and pair amplitude changes along the F/S interfaces are taken into account. It is shown that 0- and π - phase superconducting states of pure thin F′/S/F/S′ fourlayers are controlled by the magnitude and sign of electron correlations in the F and F′ layers, as well as by the competition between homogeneous Bardeen-Cooper-Schrieffer (BCS) pairing and inhomogeneous Larkin-Ovchinnikov-Fulde-Ferrell (LOFF) pairing. A solitary re-entrant superconductivity for the 00π state is predicted. The results of numeric calculations allow to explain the absence of the suppression of three dimensional superconductivity in short period Gd/La superlattices.

1. Introduction

One of unconventional superconducting correlations of electrons different from BCS pairing is pairing through the Larkin-Ovchinnikov-Fulde-Ferrell (LOFF) mechanism with a nonzero 3D momentum of pairs $k \neq 0$ [1, 2], which can appear in ferromagnet/superconductor (F/S) nanostructures. The appearance of this states leads to the nonmonotonic dependencies of the critical temperature T_c, electron density of states, and the Josephson current on the thickness d_f of the F layers [3, 4]. The recently revealed absence of the suppression of three dimensional (3D) superconduc tivity in short period Gd/La superlattices [5, 6] constitutes a challenge for the existing one dimensional (1D) theory of the proximity effect [3, 4]. Superconductivity in F/S nanostructures usually occurs only if the F layer thickness d_f is much smaller than the S layer thickness d_s. However, 3D superconductivity in a Gd/La superlattice not only exists at $d_f > d_s$, but appears at T_c equal to the critical temperature of a bulk lanthanum sample. Based on obtained equations, our theory explains this suprising effect.

Provided the solutions of the Eilenberger equations in the F and S layers are known, the critical temperature T_c may be found from the self-consistency equation for the pair potential $\Delta(\mathbf{r})$

$$\Delta_{s(f)}(\mathbf{r}) = 2\lambda_{s(f)}(\mathbf{r})\pi T \mathrm{Re} \sum_{\omega>0}{}' \left\langle \Phi_{s(f)}(\mathbf{p}, \mathbf{q}_{s(f)}, z, \omega) \right\rangle, \tag{1}$$

where $\lambda_{s(f)}$ is the electron-electron interaction parameter, $\omega = \pi T(2n+1)$ is the Matsubara frequency, the prime means a cutoff at the Debye frequency ω_D, T is the temperature, and $\hbar = k_B = \mu_B = 1$ is taken hereinafter.

The Eilenberger function $\Phi(\mathbf{p},\mathbf{q},z,\omega)$ can be separately defined for layers F and S using following differential equations

$$\left[2\tilde{\omega}_{s(f)} - v_{sz(fz)}\xi_{sz(fz)}\frac{\partial^2}{\partial z^2}\right]\Phi_{s(f)}(\mathbf{p},\mathbf{q}_{s(f)},z,\omega) = 2\Delta_{s(f)}(\mathbf{q}_{s(f)},z), \quad (2)$$

$v_{sz(fz))}$ and $\xi_{s(f)}(z)$ are the z components of the Fermi velocity and correlation length, respectively, which are related as

$$\xi_{sz} = \frac{v_{sz}}{2\tilde{\omega}_s}; \tilde{\omega}_s = 2\omega + i\mathbf{q}_s\mathbf{v}_{s\perp} \quad (3)$$

for superconducting layer and

$$\xi_{fz} = \frac{v_{fz}}{2\tilde{\omega}_f}; \tilde{\omega}_f = 2\omega + i(2I + \mathbf{q}_f\mathbf{v}_{f\perp}) \quad (4)$$

for ferromagnet layer, respectively. The subscripts s and f mark the parameters and functions in the S and F layers. The boundary conditions at the planar F/S contact $z = 0$, where the 2D projection of the Fermi momentum $\mathbf{p}$, is conserved have the form

$$(1-\sigma)\xi(z)\frac{\partial\Phi\left(\mathbf{p},\mathbf{q},z,\omega\right)}{\partial z}\Big|_{z=\pm 0} = \frac{\sigma}{2}\left[\Phi\left(\mathbf{p},\mathbf{q},+0,\omega\right) - \Phi\left(\mathbf{p},\mathbf{q},-0,\omega\right)\right], \quad (5)$$

where σ is transparency, ξ is correlation length.

The results of this work develop the theory of proximity effect of recent works [7, 8, 9] for F/S/F′ trilayers on the asymmetrical four-layered system. We also take into account the magnitude and sign of the electron-electron interaction in a ferromagnet and spatial variations of the Eilenberger function along F/ S contact.

2. Four-layered system

Let us consider a system consisting of four layers with different thickness: two pure ferromagnets F and F′ filling the regions $-d_f < z < 0$ and $d_s < z < d_s + d'_f$ and two layers of a pure BCS superconductors S and S′ with areas $0 < z < d_s$ and $-d_f - d'_s < z < -d_f$, respectively. The boundary conditions at the F/S contact $z = 0$ for the case of ideal transparency have the form

$$\begin{gathered}\Phi_s\left(\mathbf{p},\mathbf{q}_s,+0,\omega\right) = \Phi_f\left(\mathbf{p},\mathbf{q}_f,-0,\omega\right)\ ; \\ \xi_{sz}\frac{\partial\Phi_s\left(\mathbf{p},\mathbf{q}_s,z,\omega\right)}{\partial z}\bigg|_{z=+0} = \xi_{fz}\frac{\partial\Phi_f\left(\mathbf{p},\mathbf{q}_f,z,\omega\right)}{\partial z}\bigg|_{z=-0}.\end{gathered} \quad (6)$$

The boundary conditions at the S/F′ contact $z = d_s$ have the form

$$\begin{gathered}\Phi_s\left(\mathbf{p},\mathbf{q}_s,d_s-0,\omega\right) = (\pm)\ {\Phi_f}'\left(\mathbf{p},\mathbf{q}_f,d_s+0,\omega\right)\ ; \\ \xi_{sz}\frac{\partial\Phi_s\left(\mathbf{p},\mathbf{q}_s,z,\omega\right)}{\partial z}\bigg|_{z=d_s-0} = (\pm)\ {\xi_{fz}}'\frac{\partial{\Phi_f}'\left(\mathbf{p},\mathbf{q}_f,z,\omega\right)}{\partial z}\bigg|_{z=d_s+0}.\end{gathered} \quad (7)$$

The last boundary conditions at the S′/F contact $z = -d_f$ have the form

$$\begin{gathered}\Phi_f(\mathbf{p},\mathbf{q}_f,-d_f+0,\omega) = [\pm]\,{\Phi_s}'(\mathbf{p},\mathbf{q}_s,-d_f-0,\omega); \\ \xi_{fz}\frac{\partial\Phi_f(\mathbf{p},\mathbf{q}_f,z,\omega)}{\partial z}\Big|_{z=-d_f+0} = [\pm]\,\xi_{sz}\frac{\partial{\Phi_s}'(\mathbf{p},\mathbf{q}_s,z,\omega)}{\partial z}\Big|_{z=-d_f-0}\end{gathered} \quad (8)$$

The "+" sign in the boundary conditions corresponds to the usual 0-phase superconductivity with the coinciding phases of the order parameters Δ_f and Δ'_f, which are nonzero despite to the overwhelming majority of preceding theories [3, 4] for the F/S systems, in the F and F′ layers or Δ_s and Δ'_s in the S and S′ layers, respectively. The lower sign "-" corresponds to the nontraditional π-phase superconductivity in which the transition through the layer is accompanied with a change in the sign of the order parameter Δ_f or Δ_s [10]. Thus, we have four different states: (00),(0π),$(\pi 0)$ and $(\pi\pi)$. First symbol relates to order parameter in superconductors, the second symbol refers to the ferromagnetic layers.

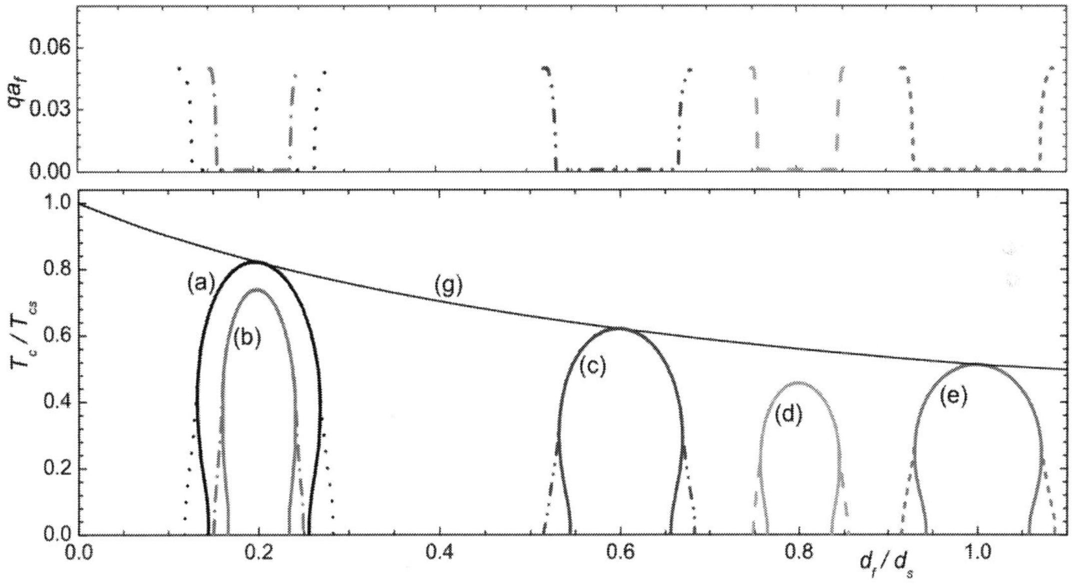

Figure 1. Phase diagrams (lower panel) $T_c(d_f)$ and (upper panel) $q(d_f)$ for the 00π states of F′/S/F/S′ fourlayers. The solid lines correspond to the lines of the transition to the BCS states with $q = 0$. Other lines are the LOFF states with $q \neq 0$. In the region of the competition between BCS and LOFF states on the wings of the peaks, the realized LOFF state has a higher T_c value than the BCS state. The thickness of layers are following: a) $d'_f = 0.2d_s$, $d'_s = d_s$; b) $d'_f = 0.2d_s$, $d'_s = 0.2d_s$; c) $d'_f = 0.6d_s$, $d'_s = d_s$; d) $d'_f = 0.8d_s$, $d'_s = 0.2d_s$; e) $d'_f = d_s$, $d'_s = d_s$; g) $d'_f = d_f$, $d'_s = d_s$;

We seek for the solutions of Eqs.(2) in the respective layers in the form that excludes fluxes through the outer boundaries of the contact, i.e.,

$$\begin{aligned}
\Phi_s &= \frac{\Delta_{s0}}{\omega_s} + A\frac{\mathrm{ch}(z-d_s/2)/\xi_{sz}}{\mathrm{ch}\,(d_s/2\xi_{sz})} + C\frac{\mathrm{sh}\,(z-d_s/2)\,/\xi_{sz}}{\mathrm{sh}\,(d_s/2\xi_{sz})}, \quad 0<z<d_s,\\
\Phi_f &= \frac{\Delta_{f0}}{\omega_f} + B\frac{\mathrm{ch}\,(z+d_f/2)\,/\xi_{fz}}{\mathrm{ch}\,(d_f/2\xi_{fz})} + D\frac{\mathrm{sh}\,(z+d_f/2)\,/\xi_{fz}}{\mathrm{sh}(d_f/2\xi_{fz})}, -d_f<z<0,\\
\Phi'_s &= \frac{\Delta_{s0}}{\omega_s} + A'\frac{\mathrm{ch}\,(z+d'_s+d_f)\,/\xi_{sz}}{\mathrm{ch}\,(d'_s/\xi_{sz})}, \quad -\left(d'_s+d_f\right)<z<-d_f,\\
\Phi'_f &= \frac{\Delta_{f0}}{\omega'_f} + B'\frac{\mathrm{ch}\,(z-d_s-d'_f)\,/\xi'_{fz}}{\mathrm{ch}\left(d'_f/\xi'_{fz}\right)}, \quad d_s<z<\left(d_s+d'_f\right).
\end{aligned} \tag{9}$$

The constants A, B, B', C can be defined from boundary conditions (6), (7) and (8). Owing to the collectivization of electron correlations and the paramagnetic effect of the exchange field, the mutual effect of F, S, F′ and S′ metals is particularly strong in the Cooper limit when their thicknesses are small, $d_{f(s)} \ll \xi_{f(s)}, a_f$, where $\xi_{f(s)} = v_{f(s)}/2\pi T$ is the coherence length and $a_f = v_f/2I$ is the spin stiffness length. The set of equations on these constants have following form

$$\begin{aligned}
&\frac{\Delta_{s0}}{\omega_s} + A - C = \frac{\Delta_{f0}}{\omega_f} + B + D,\\
&-A\frac{d_s}{2\xi_{sz}} + C\frac{2\xi_{sz}}{d_s} = B\frac{d_f}{2\xi_{fz}} + D\frac{2\xi_{fz}}{d_f},\\
&\frac{\Delta_{s0}}{\omega_s} + A + C = (\pm)\left(\frac{\Delta_{f0}}{\omega_f{}'} + B'\right),\\
&A\frac{d_s}{2\xi_{sz}} + C\frac{2\xi_{sz}}{d_s} = (\pm)\left(-B'\frac{d_f{}'}{\xi_{fz}{}'}\right),\\
&\frac{\Delta_{f0}}{\omega_f} + B - D = [\pm]\left(\frac{\Delta_{s0}}{\omega_s} + A'\right),\\
&-B\frac{d_f}{2\xi_{fz}} + D\frac{2\xi_{fz}}{d_f} = [\pm]\,A'\frac{d_s{}'}{\xi_{sz}}.
\end{aligned} \tag{10}$$

Then, we solve the set of Eqs.(10). Due to the Eilenberger functions depend on the polar angles in velocity space, the analysis of the obtained relations for the critical temperature is very complicated. But in case of component equality of the Fermi velocity (the same electronic structure of metals in the nanocontact) equations are strongly simplified by using the summation rule with the digamma function Ψ.

As a result, taking into account the two possible combinations of the exchange field I in ferromagnets, we obtain eight different states $\alpha_s\alpha_f\beta$, where the meaning of the first two symbols is explained above, and β indicates the parallel (0) or antiparallel (π) orientation of the magnetizations of the neighboring F layers.

Let us analyze the case of (00π) as the most interesting for an experimental realization, because π-magnetic states have higher critical temperature then 0-magnetic states as it was shown for four-layered system F/S/F/S [11] and for superlattices [12]. Simplified equation $(v_s = v_f)$ has form:

$$\begin{aligned}
\ln t = \frac{(c_f + c_f')(\lambda_f - \lambda_s)}{\lambda_s((c_s + c_s')\lambda_s + \lambda_f(c_f + c_f'))} + \Psi\left(\frac{1}{2}\right) - \mathrm{Re}\int\frac{d\Omega}{4\pi}\psi\left(\frac{1}{2} + \frac{\Gamma}{4\pi T_{cs}t}\right),\\
\Gamma = 2i\left(\frac{1}{2}(c_s + c_s')\mathbf{q}_s\mathbf{v}_{s\perp} + (c_f - c_f')(I + \frac{1}{2}\mathbf{q}_f\mathbf{v}_{f\perp})\right); t = \frac{T_c}{T_{cs}}.
\end{aligned} \tag{11}$$

where T_{cs} is the critical temperature of the isolated S layer, $\Psi(x)$ is the digamma function and c_x in case of similar electronic structure are the relative weights of layers, i.e $c_x = d_x/(d_f + d_f' + d_s + d_s')$. For certainty 2D momenta of LOFF pairs q_f and q_s are taken to be identical $(q_f = q_s = q)$ [7].

The exchange fields I and $-I$ of the neighboring F layers in symmetric F/S/F/S fourlayers with $d_f = d_f'$ $(c_f = c_f')$ and $d_s = d_s'$ $(c_s = c_s')$ completely compensate each other in Γ. Then, from the condition of the maximum of T_c, we obtain $q = \Gamma = 0$ and the critical temperatures T_c is determined only by the first terms in Eqs.(11), i.e

$$\ln t = -\frac{(c_f + c'_f)(\lambda_s - \lambda_f)}{\lambda_s(c_s\lambda_s + (c_f + c'_f)\lambda_f)}. \tag{12}$$

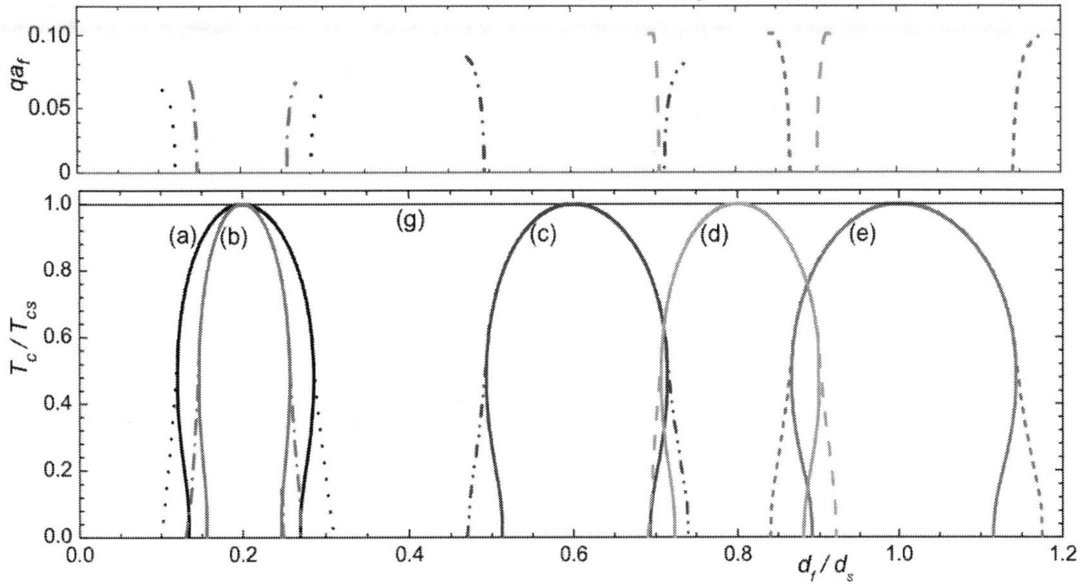

Figure 2. Phase diagrams (lower panel) $T_c(d_f)$ and (upper panel) $q(d_f)$ for the 00π states of $F'/S/F/S'$ at $\lambda_s = \lambda_f = 0.2$. The thickness of layers are following: a) $d'_f = 0.2d_s$, $d'_s = d_s$; b) $d'_f = 0.2d_s$, $d'_s = 0.2d_s$; c) $d'_f = 0.6d_s$, $d'_s = d_s$; d) $d'_f = 0.8d_s$, $d'_s = 0.2d_s$; e) $d'_f = d_s$, $d'_s = d_s$; g) $d'_f = d_f$, $d'_s = d_s$;

It depends on the difference between the parameters of electron-electron interactions λ_f and λ_s.

The diagrams of state $T_c(d_f/d_s)$ for $\lambda_s = 0.3$ and $\lambda_s = 0.2$, which are calculated from Eq.(11), consist of separate peaks with reentrant superconductivity. These peaks are characterized by the LOFF-BCS-LOFF competition for various d_f/d_s values are shown in the lower panel in Fig. 1. The g-line is the curve of the maxima of T_c that is calculated by the formula in Eqs.(12) and this line describes T_c behavior of the symmetric F/S/F/S four-layered system with $d_f = d'_f$ and $d_s = d'_s$. The upper panel in Fig. 1 shows the corresponding $q(d'_f/d_s)$ dependences in the units of a_f^{-1}.

According to Fig. 1, the LOFF superconducting state with nonzero momentum of pairs ($q \neq 0$) appears in the fourlayer at small T_c and the transition of the $F'/S/F/S'$ contact to the BCS state with $q = 0$ and pair amplitude oscillating in the interface plane becomes favorable with an increase in T_c.

A similar set of peaks of $T_c(d_f/d_s)$ with reentrant superconductivity and the LOFF-BCS-LOFF competition at $\lambda_s = \lambda_f = 0.2$ calculated from the equation in Eqs.(11) is shown in the lower panel of Fig. 2. The peaks possess the same height because the line of the maxima of $T_c(d_f/d_s)$ for the π-phase superconductivity, which corresponds to the symmetric F/S/F/S fourlayers, is determined from relation $\ln t = 0$, as it follows from Eqs.(12).

The existence of such states with π-phase magnetism explains the absence of the suppression of 3D superconductivity in short period Gd/La superlattices in experimental works [5] and [6]. The critical temperature T_c measured for this superlattices was 5 K under cooling in zero field; this value almost coincides with the critical temperature of bulk lanthanum! This means that the 00π state with $\lambda_f = \lambda_s$ occurs in the Gd/La superlattices.

Solutions obtained above for four-layered system were verified for the limiting case when

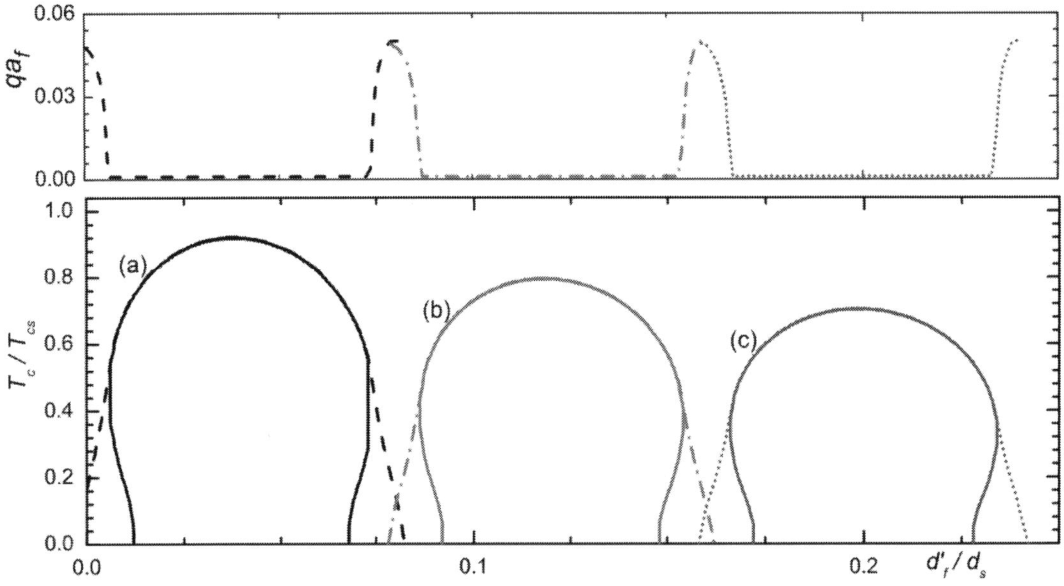

Figure 3. The limit transition into trilayer system at $d'_s = 0$. The thickness of layers are following: a) $d'_f = 0.04d_s$; b) $d'_f = 0.12d_s$; c) $d'_f = 0.2d_s$.

$d'_s = 0$, (Fig. 3). We have the complete agreement with results obtained recently for trilayer F/S/F′ [9]. Here first "0" symbol indicates the sign constancy of the superconducting order parameter at the transition through the S layer, and the symbol "π" relates to antiparallel configuration of the exchange fields in ferromagnets. As expected, set of diagrams has the form of solitary peaks with re-entrant superconductivity and competition LOFF-BCS-LOFF. We obtained interesting result: analytical solutions (11) do not coincide with previous expressions from [9], but the numerical solutions demonstrate the similar plots.

References

[1] Larkin A I and Ovchinnikov Yu N 1964 *Zh. Eksp. Teor. Fiz.* **47** 1136
[2] Fulde P and Ferrell R A 1964 *Phys. Rev. A* **135** 550
[3] Izyumov Yu A, Proshin Yu N and Khusainov M G 2002 *Usp. Fiz. Nauk* **172** 113
[4] Buzdin A I 2005 *Rev. Mod. Phys* **77** 935
[5] Goff J P, Deen P P, Ward R C et al. 2002 *J. Magn. Magn. Mater.* **240** 592
[6] Deen P P, Goff J P, Ward R C et al. 2005 *J. Phys.: Condens. Matter* **17** 3305
[7] Khusainov M G, Khusainov M M, Ivanov N M and Proshin Yu N 2009 *JETP Letters* **89** 626
[8] Khusainov M G, Khusainov M M, Ivanov N M and Proshin Yu N 2009 *JETP Letters* **90** 124
[9] Khusainov M G, Khusainov M M, Ivanov N M and Proshin Yu N 2009 *JETP Letters* **90** 359
[10] Bulaevskii L N, Kuzii V V and Sobyanin A A 1977 *JETP Lett.* **25** 290
[11] Proshin Yu N, Zimin A, Fazleev N G and Khusainov M G 2006 *Phys. Rev. B* **73** 184514
[12] Proshin Yu N, Izyumov Yu A and Khusainov M G 2001 *Phys. Rev. B.* **64** 064522

The superconducting phase transitions for the asymmetrical FS superlattices with interelectronic interaction in ferromagnet layers

M.M. Khusainov, Yu.N. Proshin

Kazan Federal University, Kazan, 420008, Kremlevskaya 18, Russia

E-mail: yu.proshin@gmail.com

Abstract. The superconducting and magnetic states coexistence in the FS superlattice, where F is ferromagnetic metal and S is superconductor, is investigated on base of microscopically derived boundary value problem for the Eilenberger function. The asymmetrical four-layered structure FSF′S′ is considered as elementary cell. The second order phase transitions are explored for case of ideal boundary transparencies and clean Cooper limit. Each layer is characterized by its own thickness and electronic structure. Materials are also differed by electron-electron interaction constants (note these constants for ferromagnets are nonzero!). It is shown, that 0- and π-phase superconducting states of clean thin superlattices FS are defined by value and sign of electronic correlations in all four layers of elementary cell. The competition between uniform BCS pairing and non-uniform Fulde-Ferrell-Larkin-Ovchinnikov pairing are also taken into account. We predict that the complex system under consideration may have up to 8 different states which are characterized by phase shifts between superconducting order parameters in S(F) and S′(F′) and mutual orientation of magnetizations in the F and F′ layers. The states with π-phase magnetism can fully explain surprising experimental behavior of short-range Gd/La superlattice, i.e the coincidence of superlattice critical temperature T_c with $T_c(La) = 5\,\text{K}$ for different thickness of the Gd layers exceeding the La layer thickness.

1. Introduction

The ferromagnet metal/superconductor (FS) nanostructures represent artificial layered systems in which various not trivial phenomena may be observed (see reviews [1, 2, 3] and references therein). For example, the superconductivity of the Fulde-Ferrell-Larkin-Ovchinnikov (FFLO) type [4, 5], triplet superconducting correlations, various types of dependencies of critical temperature from thickness of the ferromagnetic layers (monotonous decrease, output on plateau, oscillations, re-entrant superconductivity), π-phase superconductivity and π-phase magnetism. It is caused by a proximity effect due to electrical and quantum contact of ferromagnets and the superconductors.

In our recent works [6, 7] we have developed the proximity effect theory for the FS bilayers and symmetric FSF′ trilayers taking into account the electron-electron interaction in a ferromagnet (for the first time) and spatial variations of the Eilenberger function along FS border. The last one leads to appearance of two-dimensional FFLO states. Four various states which differ phases of superconducting Δ and magnetic I order parameters in adjacent F layers were found for the three-layered nanostructures FSF′ [7, 8]. For asymmetric trilayer FSF′ [8], where the thickness of layer F′ was used as the controlling parameter, the solitary re-entrant superconductivity was predicted.

As a rule a superconductivity occurs in the layered FS nanostructures at $d_f \ll d_s$, where $d_{f(s)}$ is the F(S) layer thickness. The typical case is the layered Gd/Nb and Fe/V structures [1, 2, 3]. The recently observed behavior of critical temperature in short period Gd/La superlattices [9, 10] also motivate our work. The critical temperature T_c measured for the short-period Gd/La superlattice with different ratio of Gd and La layers thicknesses ($d_f \approx 2d_s$ [10] and $d_f \approx 3d_s$ [9]) was 5 K under cooling in zero field; this value almost coincides with the T_c of bulk lanthanum. This can not be explained by existing one-dimensional theory of the proximity effect [1, 2].

We develop the proximity theory for the thin pure SF superlattice and we try to explain these facts. The unit cell of the asymmetric superlattice is four-layered structure SFS′F′ in which all four layers may have different thicknesses and other parameters.

2. Mathematical formalism

In vicinity of the second order phase transition point the critical temperature of the nonuniform superconductor may be found from the Gor'kov self-consistency equation [11] for an order parameter Δ

$$\Delta(\mathbf{r}) = \lambda(\mathbf{r})\pi T \sum_{\alpha\neq\beta}\sum_{\omega}{}' F_{\alpha\beta}(\mathbf{r},\omega), \tag{1}$$

where λ is the interelectronic interaction parameter, $\omega = \pi T(2n+1)$ is the Matsubara frequency, the prime at the second sum means a cutoff at the Debye frequency ω_D, T is the temperature, and $\hbar = k_B = \mu_B = 1$ is taken hereinafter. The Gor'kov function F is defined [11]

$$F_{\alpha\beta}(\mathbf{r},\omega) = \frac{1}{\pi N(r)} \iiint_V \langle G_{\alpha\alpha}(\mathbf{r},\mathbf{r}',\omega) G_{\beta\beta}(\mathbf{r},\mathbf{r}',-\omega)\rangle_{imp} \Delta(\mathbf{r}')dV'. \tag{2}$$

Here $G_{\alpha\beta}(\mathbf{r},\mathbf{r}',\omega)$ is the Fourier transform of the Green function of normal metal, instead of frequency is substituted the Matsubara one, and averaging is conducted on all configurations of nonmagnetic impurities. The integral boundary-value problem for the Gor'kov function in the FS nanocontact may be obtained by the diagrams technics [1, 6, 11]

$$F_{\alpha\beta}(\mathbf{r},\omega) = \frac{1}{\pi N(\mathbf{r})} \iiint_V K_{\alpha\beta}(\mathbf{r},\mathbf{r}',\omega)\left[\Delta(\mathbf{r}') + \frac{1}{2\tau(\mathbf{r}')} F_{\alpha\beta}(\mathbf{r}',\omega)\right] dV'. \tag{3}$$

Here $K_{\alpha\beta}(\mathbf{r},\mathbf{r}',\omega) = G_{\alpha\alpha}(\mathbf{r},\mathbf{r}',\omega)G_{\beta\beta}(\mathbf{r},\mathbf{r}',-\omega)$, integration is conducted on all contact volume, the scattering velocity on impurities is equal to zero ($\tau^{-1} \equiv 0$).

The Eilenberger function Φ is connected with the Gor'kov function F by a relation [12]

$$F(\mathbf{q},z,\omega) = \frac{1}{4\pi}\iint \Phi(\mathbf{p},\mathbf{q},z,\omega)\, d^2\mathbf{p} \tag{4}$$

Here $\mathbf{p}$ is the unit vector moving at integration on a solid angle of the Fermi sphere. The Eilenberger function satisfies to the differential boundary-value problem [6, 7] including the equation of diffusion type

$$\left[2\omega + i\Gamma - v_z(z)\xi_z(z)\frac{\partial^2}{\partial z^2}\right]\Phi(\mathbf{p},\mathbf{q},z,\omega) = 2\Delta(z) \tag{5}$$

and boundary conditions at the sharp FS borders

$$\xi_z(z)\frac{\partial\Phi(\mathbf{p},\mathbf{q},z,\omega)}{\partial z}|_{z=\pm 0} = \frac{\sigma}{2(1-\sigma)}\left[\Phi(\mathbf{p},\mathbf{q},+0,\omega) - \Phi(\mathbf{p},\mathbf{q},-0,\omega)\right]. \tag{6}$$

Here σ is transparency of the boundary, $\mathbf{v}$ is Fermi velocity, the 2D wave vector $\mathbf{q}$ lies in the FS contact plane, the pair-breaking parameter Γ will be determined below, ξ is correlation length

$$\xi_z = \frac{v_z}{2\omega}; \quad \omega = 2\omega + i(2I + \mathbf{q}\mathbf{v}_\perp). \tag{7}$$

3. Asymmetric SF superlattice: the basic assumptions and the boundary-value problem solution

Let's consider the SF superlattice which consist of repeating four-layered block SFS′F′. In common asymmetrical case all layers of this elementary cell have own parameters: thickness d_ζ, Fermi velocities $\mathbf{v}_\zeta$, interelectronic interaction constants λ_ζ, exchange fields I_ζ, coherence lengths ξ_ζ, etc ($\zeta = s, f, s', f'$ numbers layers in unit cell). To solve analytically the Eilenberger equations we follow the papers [6, 7] and make next assumptions for all metals and boundaries. First of them is ideal transparency (σ=1), the second is the Cooper limit ($\xi_{\zeta z}/d_i \gg 1$), at last there are no impurities (a clean limit). The novel assumption is the account of interelectronic interaction in the ferromagnet which in preceding theories [1, 2, 3] was supposed zero owing to a powerful exchange field. The constant λ_f can be either negative and positive. The last case of concealed electron-electron interaction corresponds to the latent superconductivity suppressed in bulk ferromagnet by the strong exchange field.

The superlattice case differs from four-layered case, which will be considered in separate paper, by periodical translational conditions on the Eilenberger function (L is period of the FS superlattice)

$$\Phi(\mathbf{p}, \mathbf{q}, z, \omega) = \Phi(\mathbf{p}, \mathbf{q}, z + L, \omega); \quad L = d_s + d_f + d_{s'} + d_{f'}. \tag{8}$$

Boundary conditions into elementary cell SFS′F′ for the second FS′ boundary (let $z = 0$) are given by the equations

$$\Phi_{s'}(\mathbf{p}, \mathbf{q}_{s'}, +0, \omega) = \Phi_f(\mathbf{p}, \mathbf{q}_f, -0, \omega); \ \xi_{sz}\frac{\partial \Phi_s(\mathbf{p}, \mathbf{q}_s, z, \omega)}{\partial z}\bigg|_{z=+0} = \xi_{fz}\frac{\partial \Phi_f(\mathbf{p}, \mathbf{q}_f, z, \omega)}{\partial z}\bigg|_{z=-0}. \tag{9}$$

Boundary conditions at third S′F′ boundary ($z = d_{s'}$) look like

$$\begin{gathered} \Phi_{s'}(\mathbf{p}, \mathbf{q}_{s'}, d_{s'} - 0, \omega) = e^{i\alpha_f}\,\Phi_{f'}(\mathbf{p}, \mathbf{q}_{f'}, d_{s'} + 0, \omega); \\ \xi_{s'z}\frac{\partial \Phi_{s'}(\mathbf{p}, \mathbf{q}_{s'}, z, \omega)}{\partial z}\bigg|_{z=d_{s'}-0} = e^{i\alpha_f}\,\xi_{f'z}\frac{\partial \Phi_{f'}(\mathbf{p}, \mathbf{q}_{f'}, z, \omega)}{\partial z}\bigg|_{z=d_{s'}+0}. \end{gathered} \tag{10}$$

Boundary conditions at the SF border ($z = -d_f$) looks like

$$\begin{gathered} \Phi_f(\mathbf{p}, \mathbf{q}_f, -d_f + 0, \omega) = e^{i\alpha_s}\Phi_s(\mathbf{p}, \mathbf{q}_s, -d_f - 0, \omega); \\ \xi_{fz}\frac{\partial \Phi_f(\mathbf{p}, \mathbf{q}_f, z, \omega)}{\partial z}\Big|_{z=-d_f+0} = e^{i\alpha_s}\xi_{sz}\frac{\partial \Phi_s(\mathbf{p}, \mathbf{q_s}, z, \omega)}{\partial z}\Big|_{z=-d_f-0}. \end{gathered} \tag{11}$$

The phase shifts α_s and α_f can be equal only 0 and π, and, correspondingly, they determine the cases with 0- and π-types of superconductivity [1, 2, 6]. Firstly π-types of superconductivity was proposed by Bulaevskii et al. [13], we following paper [6] introduce π-phase of superconductivity on the F layers, this case corresponds $\alpha_f = \pi$. The first three conditions (9)-(11) are equivalent to the analogous relations for asymmetrical four-layered system. The last boundary conditions connect the Eilenberger function and its derivative of fourth layer (F′) and the same values of first layer (S) owing to periodical translational conditions (8) and look as (here $\alpha_3 = 0, \pi$ is the phase shift)

$$\begin{gathered} \Phi_s(\mathbf{p}, \mathbf{q}_s, -d_f - d_s + L + 0, \omega) = e^{i\alpha_3}\Phi_{f'}(\mathbf{p}, \mathbf{q}_{f'}, d_{s'} + d_{f'} - 0, \omega); \\ \xi_{sz}\frac{\partial \Phi_s(\mathbf{p}, \mathbf{q}_s, z, \omega)}{\partial z}\Big|_{z=-d_f-d_s+L+0} = e^{i\alpha_3}\xi_{f'z}\frac{\partial \Phi_{f'}(\mathbf{p}, \mathbf{q}_{f'}, z, \omega)}{\partial z}\Big|_{z=d_{s'}+d_{f'}-0}. \end{gathered} \tag{12}$$

For simplicity below we will use that the S and S′ metals are the same superconductors as well as the F and F′ metals are the same ferromagnets. As trial solutions of the Eilenberger equations we use the four common solutions centered at the middle of corresponding layer z_ζ

$$\Phi_\zeta = \frac{\Delta_{s0}}{\omega_s} + A_\zeta \frac{\cosh[(z - z_\zeta)/\xi_{sz}]}{\cosh(z_\zeta/\xi_{sz})} + C_\zeta \frac{\sinh[(z - z_\zeta)/\xi_{sz}]}{\sinh(z_\zeta/\xi_{sz})}, \tag{13}$$

where Δ_{f0} and $\Delta_{f'0}$ are initial virtual superconducting order parameters in metal F and F′, correspondingly, if they were nonferromagnetic, i.e. $I = I' = 0$.

Owing to the collectivization of electron correlations and the paramagnetic effect of the exchange field, the mutual effect of F, S, F′ and S′ metals is essentially strong in the Cooper limit when their thicknesses are small, $d_{f(s)} \ll \xi_{f(s)}, a_f$, where $\xi_{f(s)} = v_{f(s)}/2\pi T$ is the coherence length and $a_f = v_f/2I$ is the spin stiffness length. Substituting the trial Eilenberger functions in boundary conditions (9)-(12) and using periodical condition (8), we gain the nonuniform set of the linear equations on eight constants A_ζ and C_ζ

$$\begin{aligned}
\frac{\Delta_{s'0}}{\omega_{s'}} + A_{s'} - C_{s'} &= \frac{\Delta_{f0}}{\omega_f} + A_f + C_f, \\
-A_{s'}\frac{d_{s'}}{2\xi_{s'z}} + C_{s'}\frac{2\xi_{s'z}}{d_{s'}} &= A_f\frac{d_f}{2\xi_{fz}} + C_f\frac{2\xi_{fz}}{d_f}; \\
\frac{\Delta_{s'0}}{\omega_{s'}} + A_{s'} + C_{s'} &= e^{i\alpha_f}\left(\frac{\Delta_{f'0}}{\omega_{f'}} + A_{f'} - C_{f'}\right), \\
A_{s'}\frac{d_{s'}}{2\xi_{s'z}} + C_{s'}\frac{2\xi_{s'z}}{d_{s'}} &= e^{i\alpha_f}\left(-A_{f'}\frac{d_{f'}}{2\xi_{f'z}} + C_{f'}\frac{2\xi_{f'z}}{d_{f'}}\right); \\
\frac{\Delta_{f0}}{\omega_f} + A_f - C_f &= e^{i\alpha_s}\left(\frac{\Delta_{s0}}{\omega_s} + A_s + C_s\right), \\
-A_f\frac{d_f}{2\xi_{fz}} + C_f\frac{2\xi_{fz}}{d_f} &= e^{i\alpha_s}\left(A_s\frac{d_s}{2\xi_{sz}} + C_s\frac{2\xi_{sz}}{d_s}\right); \\
\frac{\Delta_{s0}}{\omega_s} + A_s - C_s &= e^{i\alpha_3}\left(\frac{\Delta_{f'0}}{\omega_{f'}} + A_{f'} + C_{f'}\right), \\
-A_s\frac{d_s}{2\xi_{sz}} + C_s\frac{2\xi_{sz}}{d_s} &= e^{i\alpha_3}\left(A_{f'}\frac{d_{f'}}{2\xi_{f'z}} + C_{f'}\frac{2\xi_{f'z}}{d_{f'}}\right).
\end{aligned} \tag{14}$$

Note that, in four-layered nanostructure case we had six unknown constants and six equations. After solving the set of Eqs.(14) we can get the Eilenberger functions. Due to given assumptions (Cooper limit, ideal transparencies, etc) and collectivization of electron correlations and exchange fields we obtain the same the Eilenberger function for all layers, but they do not similar for different set of phase shifts. In common case we have $2^3 = 8$ different solution $(\alpha_f\alpha_s\alpha_3)$. Owing to translation symmetry (see Eq. (8)) only four states of them are realized. The total phase shift on elementary cell $(\alpha_f+\alpha_s+\alpha_3)$ should be equal to either zero or 2π. Hence α_3 can be called a compensation phase shift. So, we have only four different solution having physical meaning: (000), $(0\pi\pi)$, $(\pi0\pi)$ $(\pi\pi0)$. Direct calculations confirm this consideration.

Taking into account the two possible combinations of the exchange field I in ferromagnets

$$I' = \exp(i\chi)I, \tag{15}$$

where $\chi = 0(\pi)$ is the magnetic phase shift [14] corresponding to (anti)parallel mutual orientation of magnetizations in adjacent F and F′ layers, we obtain **eight** different states $[\alpha_f\alpha_s\alpha_3\chi]$.

The expression for the Eilenberger functions in the common case $[\alpha_f \alpha_s \alpha_3 \chi]$ follows

$$\Phi_f = \Phi_{s'} = e^{i\alpha_f}\Phi_{f'} = e^{i\alpha_s}\Phi_s = \frac{2\left(c_f + c_{f'}\exp(i\alpha_f)\right)\Delta_{f0} + 2\left(c_s\exp(i\alpha_s) + c_{s'}\right)\Delta_{s0}}{2\omega + i\Gamma}, \quad (16)$$

where Γ is pair-breaking factor and c_ζ are the relative weights of layers, i.e

$$\Gamma = (c_s + c_{s'})\,\mathbf{q}_s\mathbf{v}_{s\perp} + \left(c_f + e^{i\chi}c_{f'}\right)(2I + \mathbf{q}_f\mathbf{v}_{f\perp}), \quad c_\zeta = \frac{d_\zeta/v_{\zeta z}}{\sum\limits_\eta d_\eta/v_{\eta z}}. \quad (17)$$

Note, we obtained surprising result that the states for asymmetric infinite superlattice $[\alpha_f\alpha_s\alpha_3\chi]$ coincide completely with states for finite asymmetric four-layered structure FSF′S′ $\{\alpha_f\alpha_s\chi\}$. Because of this, we will below use the truncated four-layered classification $\{\alpha_f\alpha_s\chi\}$

The superlattice critical temperature T_c in the common case is determined from integral equation, because the Eilenberger functions depend on the polar angles in velocity space,

$$\begin{aligned}\ln\frac{T_c}{T_{cs}} = &\left\{\lambda_s \iint \frac{d\Omega}{4\pi}\left[\lambda_s\left(c_{s'} + e^{i\alpha_s}c_s\right) + \lambda_f\left(c_f + e^{i\alpha_f}c_{f'}\right)\right]\right\}^{-1} * \\ &* \left\{\iint \frac{d\Omega}{4\pi}\left[\lambda_s\left(c_{s'} + e^{i\alpha_s}c_s - 1\right) + \lambda_f\left(c_f + e^{i\alpha_f}c_{f'}\right)\right] + \right. \quad (18)\\ &\left. + \lambda_s \iint \frac{d\Omega}{4\pi}\left[\lambda_s\left(c_{s'} + e^{i\alpha_s}c_s\right) + \lambda_f\left(c_f + e^{i\alpha_f}c_{f'}\right)\right]\mathrm{Re}\left[\Psi\left(\frac{1}{2}\right) - \Psi\left(\frac{1}{2} + \frac{\Gamma}{4\pi T_c}\right)\right]\right\},\end{aligned}$$

where $\Psi(x)$ is the digamma function and T_{cs} is the critical temperature of the isolated S layer.

The analysis of the obtained equations for the critical temperature is very complicated. So, we present the essentially simplified expression for the case when all metals have the same electronic structure (i.e. there is component equality of the Fermi velocity of metals), then $c_\zeta = d_\zeta/(d_f + d_{f'} + d_s + d_{s'})$ and

$$\ln\frac{T_c}{T_{cs}} = \frac{\lambda_s\left(c_{s'} + e^{i\alpha_s}c_s - 1\right) + \lambda_f\left(c_f + e^{i\alpha_f}c_{f'}\right)}{\lambda_s\left[\lambda_s\left(c_{s'} + e^{i\alpha_s}c_s\right) + \lambda_f\left(c_f + e^{i\alpha_f}c_{f'}\right)\right]} + \Psi\left(\frac{1}{2}\right) - \mathrm{Re}\int\frac{d\Omega}{4\pi}\Psi\left(\frac{1}{2} + \frac{\Gamma}{4\pi T_{cs}t}\right). \quad (19)$$

Note, that in symmetrical case which was examined in experiments for the Gd/La superlattice [9, 10] we have $d_f = d_{f'} \Rightarrow c_f = c_{f'}$ and $d_s = d_{s'} \Rightarrow c_s = c_{s'}$ then expression (19) is further simplified. Especially it takes place for the π-magnetic states ($\chi = \pi$) when the exchange fields I and $-I$ of the adjacent F layers completely compensate each other in Γ. Then, from the condition of the maximum of T_c, we obtain $q = \Gamma = 0$ and the critical temperatures T_c is determined only by the first term in Eq.(19).

Observed unexpected facts for the zero field cooled samples of the symmetric Gd/La superlattice with antiparallel mutual magnetizations orientation [9, 10] (an existence of T_c for $d_f > d_s$, an independence of T_c in regard to the d_f/d_s and extremely high value equal to critical temperature of bulk La) can be completely explained by the our formula for the case $\{00\pi\}$

$$\ln t = -\frac{c_f\left(\lambda_s - \lambda_f\right)}{\lambda_s\left(c_s\lambda_s + c_f\lambda_f\right)}. \quad (20)$$

So, we see that T_c coincides with $T_{cs} = T_{\mathrm{La}}$ if $\lambda_f = \lambda_s$. The value $\lambda_s = \lambda_{\mathrm{La}} \simeq 0.28$ can be estimated from $T_{\mathrm{La}} \simeq 5\,\mathrm{K}$ using the standard BCS expression [7].

Corresponding phase diagrams are shown in Fig. 1. Here we use that 2D momenta of FFLO

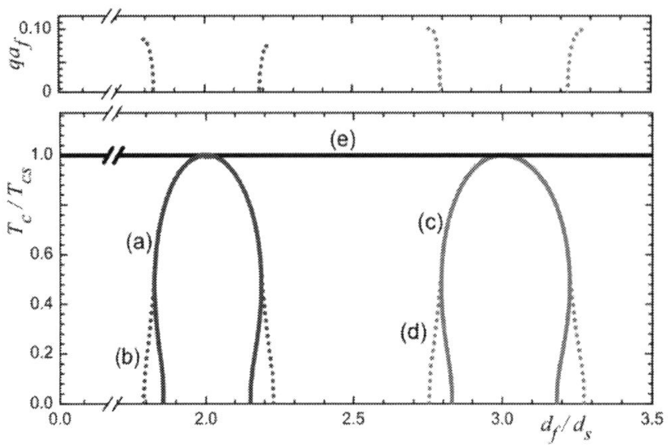

Figure 1. (Color on-line) Phase diagrams (lower panel) $T_c(d_f)$ and (upper panel) $q(d_f)$ for the $\{00\pi\}$ states of the FS superlattice with $d_{s'} = d_s$ at $\lambda_s = \lambda_f = 0.28$. The thickness of F layers are following: (a,b) $d_{f'} = 2d_s$; (c,d) $d_{f'} = 3d_s$; (e) symmetrical case $d'_f = d_f$

pairs q_f and q_s are identical ($q_f = q_s = q$) [6]. The straight bold black horizontal line (e) is calculated from Eq. (20).

For asymmetric case ($d_{f'}$ is fixed) the samples with varied F layer thickness d_f are investigated by Eq. (19). The phase diagram $T_c(d_f)$ looks like as separate peak with re-entrant superconductivity. Both peaks shown in the lower panel in Fig. 1 are characterized by the FFLO-BCS-FFLO competition for various d_f/d_s values. The dotted lines (b,d) correspond to the FFLO states with pair wave vector $q \neq 0$ (see the upper panel for the corresponding $q(d_f)$ dependencies). The solid color curves (a,c) correspond to the BCS state with $q = 0$. We see that the FFLO state with pair amplitude oscillation in the FS interface plane is favorable on peak wings.

In conclusion, the proximity effect theory for clean asymmetric FS superlattice is build. The unconventional experimental T_c behavior of the short-range Gd/La superlattice [9, 10] is explained and the value of interelectronic interaction for the Gd is estimated. The separate re-entrant superconductivity for asymmetric FS superlattice is predicted (the Gd/La superlattice is best candidate for experimental checking this effect).

References

[1] Izyumov Yu A, Proshin Yu N and Khusainov M G 2002 *Physics-Uspekhi* **45** 109
[2] Budin A I 2005 *Rev. of Mod. Phys.* **77** 3
[3] Efetov K B, Garifullin I A, Volkov A F, Westerholt K 2008 *Magnetic Heterostructures\Springer Tracts in Modern Physics*, **227** 251
[4] Larkin A I and Ovchinnikov Yu N 1964 *Zh. Eksp. Teor. Fiz.* **47** 1136
[5] Fulde P and Ferrell R A 1964 *Phys. Rev. A* **135** 550
[6] Khusainov M G, Khusainov M M, Ivanov N M and Proshin Yu N 2009 *JETP Letters* **89** 626
[7] Khusainov M G, Khusainov M M, Ivanov N M and Proshin Yu N 2009 *JETP Letters* **90** 124
[8] Khusainov M G, Khusainov M M, Ivanov N M and Proshin Yu N 2009 *JETP Letters* **90** 359
[9] Goff J P, Deen P P, Ward R C et al. 2002 *J. Magn. Magn. Mater.* **240** 592
[10] Deen P P, Goff J P, Ward R C et al. 2005 *J. Phys.: Condens. Matter* **17** 3305
[11] Abrikosov A A, Gorkov L P and Dzjaloshinsky I E 1965 *Quantum field theoretical methods in statistical physics* (New York: Pergamon)
[12] Demler E A, Arnold G B and Beasley M R 1997 *Phys. Rev. B* **55** 15174
[13] Bulaevskii L N, Kuzii V V and Sobyanin A A 1977 *JETP Lett.* **25** 290
[14] Proshin Yu N, Izyumov Yu A and Khusainov M G 2001 *Phys. Rev. B.* **64** 064522

On the nature of ferromagnetism in oxide semiconductors doped with 3d-elements

A A Achkeev[1*], I R Vakhitov[1], R I Khaibullin[1,2] and L R Tagirov[1,2]

[1] Institute of physics, Kazan (Volga region) Federal University, Kazan, Russia
[2] Laboratory of Radiation Physics, Kazan Physical Technical Institute of Russian Academy of Science, Kazan, Russian

*E-mail: fermion@list.ru

Abstract. The origin and mechanisms of ferromagnetism in the new class of magnetic materials, oxide-diluted magnetic semiconductors (ODMS), are examined in a framework of the Stoner-Anderson model. Within the Green function formalism, a condition (the Stoner criterion) for nucleation of ferromagnetism is obtained for itinerant electrons in the narrow defect (vacancy) band, and an additional contribution due to interaction with 3d magnetic ions is derived. The "trigger" character of the transition to the ferromagnetic state in ODMS is discussed in its dependence on the type and concentration of 3d magnetic impurity dopant. The results of calculations are compared with the experimental data for the spontaneous magnetic moment in semiconducting titanium dioxide (TiO_2) doped with 3d magnetic ions and containing various concentrations of oxygen vacancies.

1. Introduction

At present, there is an elevated interest to new class of magnetic materials – oxide diluted magnetic semiconductors which exhibit the room temperature ferromagnetism [1], [2]. Diluted magnetic semiconductors could be a key spintronic material intended to control both spin and charge degrees of freedom of carriers in semiconductor devices.

Nowadays there exist several models of ferromagnetism in ODMS [3]–[5]. Most of researchers associate ferromagnetism with the localized magnetic moments on 3d-impurities which order by indirect exchange interactions. However, there are studies in which ferromagnetism is ascribed to electrons of defect band of the oxide crystal [5]. Which mechanism of magnetic ordering is realized in a particular system remains an open question. Therefore, the aim of this study is to clarify mechanisms of the long-range magnetic order in defective oxide semiconductors for the case when impurities are distributed in a sample in the form of a solid solution by substituting cations of the matrix.

2. Basics of ferromagnetism model in diluted magnetic oxide semiconductors.

We took titanium dioxide (TiO_2) rutile structure as a model ODMS which will be considered in this paper. Band structure and magnetic properties of rutile doped by 3d-elements are shown in figure 1a. It is well known that conductivity of pure titanium dioxide is provided by ionized point defects (as a rule by oxide vacancies) which are electronic donors [6], [7]. Energy levels of oxide defects lie in the band-gap of TiO_2 at 0.3 eV below the bottom of conduction band (figure 1a). This levels form narrow quasi-continuous band with 0.3-0.4 eV width. We will argue below what the narrow defect band can be responsible for the ferromagnetism nucleation.

Our experimental studies revealed abrupt rise of the saturation magnetic moment of 3d-implanted TiO_2 films with increasing the concentration of 3d-elements. These findings are shown in figure 1b. The figure presents dose dependence of the magnetic moment per impurity atom in the single-crystal TiO_2 films. An important circumstance is that the above feature is observed in samples where the impurity is distributed in the form of dilute solution with substitution concentration of a few percent [8]. This allows to conclude that direct exchange can be neglected. Then, the origin of the ferromagnetism should be searched for in the host, but not in the magnetic moments of impurity alone.

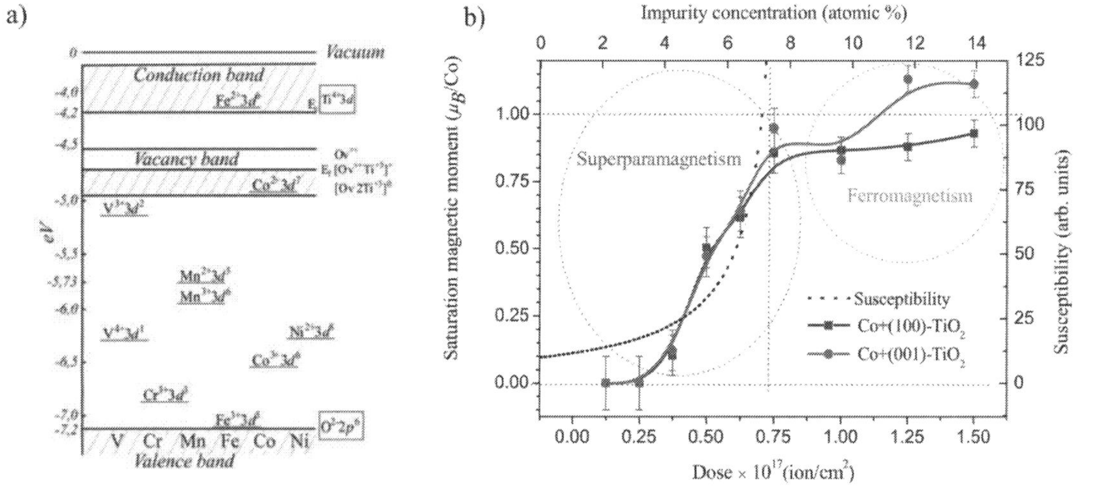

Figure 1. Electrical and magnetic properties of rutile doped by 3d-elements: a) band structure and d-levels energy of 3d-elements [6], [7]; b) dose dependence of the spontaneous magnetic moment of Co-implanted single-crystal TiO_2 films. Theoretical concentration dependence of the magnetic susceptibility (dash line) is calculated using expression (6). Transition to the ferromagnetic state is observed at dose of 0.75×10^{17} ion/cm^2.

Let's consider in detail transition from the paramagnetic state to the ferromagnetic in ODMS under the influence of dopant 3d-elements. Our hypothesis is that the magnetic transition is caused by spin-splitting of the vacancy band. A mechanism of the magnetism implies itinerancy of electrons in the vacancy band, and existence of the exchange-type interaction I (analog of the Stoner exchange) between them. We assume that the vacancy band is narrow with high density of states $G(E_F)$ in vicinity of the Fermi level E_F (the Stoner criterion is close to fulfillment $IG(E_F)\simeq1$), and completed defects percolation in the sample (overlapping of the electron states otherwise localized on the defects).

The model we use can be formalized by the Hamiltonian:

$$H = H_s + H_d + H_{s-d}. \tag{1}$$

Here H_s is the Stoner Hamiltonian of the free-electrons systems in the second-quantization notations:

$$H_s = \sum_{k\sigma} \varepsilon_k c^+_{k\sigma} c_{k\sigma} + \frac{I}{2N_c} \sum_{m,k',q',\sigma} c^+_{m-q'\sigma} c^+_{k'+q',-\sigma} c_{k',-\sigma} c_{m\sigma} , \tag{2}$$

ε_k is the energy of the free-electron state with momentum k, $c^+(c)$ – the creation (annihilation) operators, I is the exchange integral, N_c – the number of electron states in the vacancy band.

The second term H_d describes electrons localized on the impurity atoms [9]:

$$H_d = \sum_{i,\sigma} \varepsilon_d d^+_{i\sigma} d_{i\sigma} + \frac{U}{2} \sum_{i,\sigma} d^+_{i\sigma} d_{i\sigma} d^+_{i\bar{\sigma}} d_{i\bar{\sigma}} , \tag{3}$$

here E_d is the energy of the impurity d-level, $d^+(d)$ is the creation (annihilation) operators at impurity cites, U is parameter of the Coulomb repulsion of electrons at a d-orbital. The summation on the index i runs over all impurities the number of which is N_d.

The third term H_{s-d} is the *s-d* hybridization energy of the localized and the itinerant electrons [10], [11]:

$$H_{s-d} = \sum_{ik\sigma} \left(V_{ki} c^+_{k\sigma} d_{i\sigma} + V^*_{ki} d^+_{i\sigma} c_{k\sigma} \right), \tag{4}$$

where V_{ki} is the potential of hybridization.

3. Ferromagnetism criterion

To find the modified criterion of ferromagnetism, we can calculate the transverse dynamic susceptibility of electrons in the vacancy band using the formalism of double-time thermodynamic Green functions. The Fourier component of the transverse dynamic susceptibility of the electron system is expressed in terms of Fourier component of the Green function $\langle\langle B(q) \mid B^+(q) \rangle\rangle$ [12]:

$$\chi^{\perp}(q,\omega) = 2\pi i \langle\langle B(q) \mid B^+(q) \rangle\rangle_{\omega} , \tag{5}$$

where $B(q) = \sum_k c^+_{k\uparrow} c_{k+q\downarrow}$ и $B^+(q) = \sum_k c^+_{k+q\downarrow} c_{k\uparrow}$.

We solve the problem up to the second order term in V_k, which is a small parameter compared with the energy of the Coulomb repulsion U and the exchange integral I. In the limit $q \to 0$, $\omega \to 0$ we have the following expression for the susceptibility:

$$\chi^{\perp}(0,0) = \frac{G(E_F)(1 + \dfrac{N_d |V_F|^2}{N_c (E_F - E_d + I/2 - Un/2)^2})}{1 - IG(E_F)(1 + \dfrac{N_d |V_F|^2}{N_c (E_F - E_d + I/2 - Un/2)^2})} . \tag{6}$$

Then, the criterion of ferromagnetism takes the following form:

$$IG(E_F)(1 + \frac{N_d |V_F|^2}{N_c (E_F - E_d + I/2 - Un/2)^2}) = 1 . \tag{7}$$

Here E_F is the energy of the itinerant electron state on the Fermi level, V_F is the matrix element of the hybridization on the Fermi level.

4. Discussion

Expression (6) describes paramagnetic susceptibility of electrons in the vacancy band of ODMS. The susceptibility indefinitely increases at a certain concentration N_d / N_c of magnetic impurity. This moment correspond to the spontaneous polarization of electrons in the vacancy band ("trigger" behavior). The calculated concentration dependence of the susceptibility is shown on figure 1b. Fitted to the experimental data (figure 1b) susceptibility goes to infinity at dose ~ 0.7×10^{17} ion/cm^2 that corresponds to the mean concentration of 6-7 atomic percent.

First of all, ferromagnetism of ODMS is determined in the first place by parameters of the narrow vacancy band but not by magnetic parameters of the dopant. It is just we observe in our experiments: the magnetic ordering temperature (700-800 K for cobalt) and the spontaneous magnetization are almost independent on the impurity concentration.

The nucleation of the ferromagnetic state depends on the impurity concentration, square of the hybridization matrix element V_F at the Fermi level, and location of the localized energy level of the impurity E_d relative to the Fermi level. In fact, it means that the transition threshold depends on the type of impurity. For example, figure 2 presents values of the spontaneous magnetic moments per impurity for different 3d-elements: Ti, V, Cr, Mn, Fe, Co, Ni, Cu. There is a wide range of magnetic moments for the particular 3d-impurity, however, we can single out general trend for the data set (see the solid curve in figure 2). Notably, V and Co have highest values of the mean magnetic moment. This fact, it is in agreement with our mechanism of ferromagnetism.

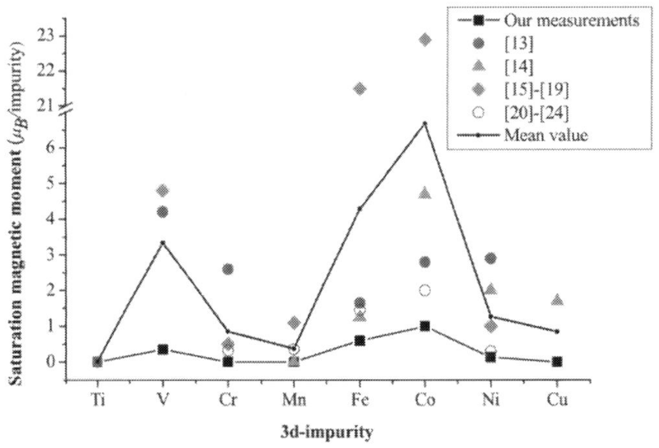

Figure 2. Value of the magnetic moment of TiO_2 single-crystalline plates implanted by 3d-elements (Ti, V, Cr, Mn, Fe, Co, Ni, Cu) at room temperature. For comparison, the data of papers [13]-[24] are shown.

According to the suggested mechanism, ferromagnetism of ODMS is spatially inhomogeneous and confined to areas of percolated impurities. Consequently, increase in the concentration of defects (actually, the oxygen vacancies) leads to escalation of the ferromagnetic areas and hence, to increase of the total magnetic moment. Particularly, this explains a wide spread of the saturation magnetic moments in the samples prepared by different techniques. Our measurements show that the oxide vacancies injection strongly affects the saturation magnetic moment. Figure 3 shows hysteresis loops before and after the vacancy injection from copper electrode by an electric field [25]. The saturation magnetic moment increases several times after the oxygen vacancies injection. At the same time, the temperature of the magnetic ordering (~720 K) and the coercive field remain almost steady. Therefore, the mechanism of the magnetic ordering is the same for both cases.

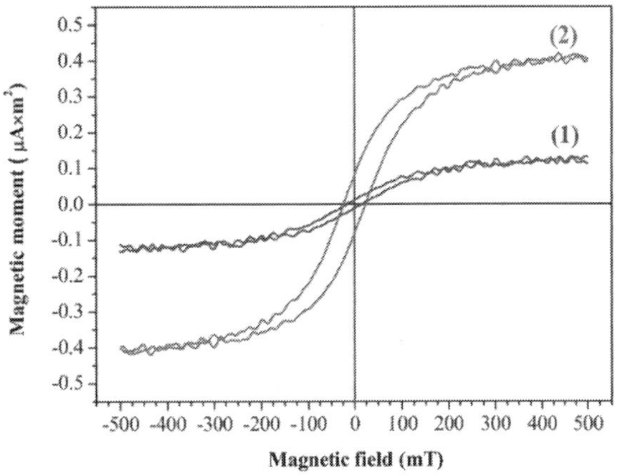

Figure 3. Hysteresis loops of the Co-implanted TiO_2 films before (1) and after (2) the oxygen vacancies injection [25].

5. Conclusion

The long-range magnetic ordering mechanism in ODMS with high concentration of defects is suggested. Spin-splitting of the vacancy band and the "trigger" (on the concentration of the defects) behavior of transition to the ferromagnetic state are the main features of the suggested mechanism. The qualitatively new result is that ferromagnetic properties of ODMS are determined mainly by exchange interaction of electrons in narrow vacancy band. This result is in a qualitative agreement with our experimental data.

This study was supported by the Russian Foundation for Basic Research (grant no. 10-02-01130) and by Ministry of Education and Science of the Russian Federation (contract no. 02.740.11.0797).

References

[1] Janisch R, Gopal P and Spaldin N 2005 *J. Phys.: Cond. Matter.* **17** R657
[2] Chambers S 2006 *Surf. Sci. Reports.* **61** 345
[3] Coey J, Venkatesan M and Fitzgerald C 2005 *Nat. Mater.* **4** 173
[4] Dietl T 2010 *Nat. Mater.* **9** 965
[5] Coey J., Stamenov P, Gunning R, Venkatesan M and Paul K 2010 *New J. Phys.* **12** 11
[6] Breckenridge R and Hosler W 1953 *Phys. Rev.* **91** 793
[7] Malati M and Wong W 1984 *Surf. Technol.* **22** 305
[8] Achkeev A, Khaibullin R, Tagirov L, Mackova A, Hnatowicz V and Cherkashin N 2011 *Phys. Sol. State.* **53** 508
[9] Hubbard J 1963 *Proc. Roy. Soc. Lond. A.* **276** 238
[10] Anderson P 1961 *Phys. Rev.* **124** 41
[11] Smith D 1968 *J. Phys. C (Proc. Phys. Soc.).* **21** 1263
[12] Kuzmin E, Petrakovskiy G, Zavadskiy Je 1976 *Physics of magnetically ordered materials* (Novosibirsk: Nauka) p 287
[13] Hong N, Sakai J, Prellier W, Hassini A, Ruyter A, Gervais F 2004 *Phys. Rev. B* **70** 195204
[14] Duhalde S, Torres R, Vignolo M, Golmar F, Chillote C, Cabrera A and Sánchez F 2007 *J. Phys.: Conf. Series.* **59** 479
[15] Orlov A, Kulemanov I, Parhomenko Yu, Perov N and Semisalova A 2011 *Mater. Jelektron.*

Journal of Physics: Conference Series **394** (2012) 012018 doi:10.1088/1742-6596/394/1/012018

Tehn. **3** 4
[16] Kaspar T, Heald S, Wang C, Bryan J, Droubay T, Shutthanandan V, Thevuthasan S, McCready D, Kellock A, Gamelin D and Chambers S 2005 *Phys. Rev. Lett.* **95** 217203
[17] Xu J, Lin Y, Lu Z, Liu X, Lu Z, Wang J, Zou W, Lv L, Zhang F and Du Y 2006 *Sol. State Commun.* **140** 514
[18] Orlov A, Balagurov L, Konstantinova A, Perov N and Yarkin D 2008 *J. Magn. Magn. Mater.* **320** 895
[19] Hou D, Meng H, Jia L, Ye X, Zhou H and Li X 2007 *Phys. Lett. A.* **364** 318
[20] Wang Y, Hui Liua, Li Z, Zhang X, Zheng R and Ringer S 2006 *Appl. Phys. Lett.* **89** 042511
[21] Xu J, Wang J, Lin Y, Liu X, Lu Z, Lu Z, Lv L, Zhang F and Du Y 2007 *J. Phys. D: Appl. Phys.* **40** 4757
[22] Kim K, Park Y, Ahn G, Kim Ch and Park J 2006 *J. Appl. Phys.* **99** e152
[23] Han G, Luo P, Guo Z, Fiaz Un Nahar, Tay M, Wu Y and Wang S 2006 *Thin Solid Films* **505** 137
[24] Cho J, Hwang T, Joh Y, Kim E, Kim D, Lee K, Park H, Ri H, Kim J and Cho C 2006 *Appl. Phys. Lett.* **88** 092505
[25] Khaibullin R, Bazarov V, Osin Yu, Ibragimov Sh and Tagirov L 2009 *The influence of oxygen vacancies on ferromagnetism oxide magnetic semiconductors* (Nizhny Novgorod, Russia, 16-20 March 2009) (Izd. Centr. IFM RAN Materials XIII Symposium " Nanofizika i Nanojelektronika " vol 1) pp 90

Ab-initio investigation of spin states of sodium cobaltate $Na_{2/3}CoO_2$

Y V Lysogorskiy[1,2], O V Nedopekin[2], S A Krivenko[2], B Minisini[1], and D A Tayurskii[2]

[1] Institut Supérieur des Matériaux et Mécaniques Avancés du Mans, 72000, 44 Av. Bartholdi, Le Mans, France

[2] Institute of Physics, Kazan Federal University, 420008, Kremlyovskaya St. 18, Kazan, Russia

E-mail: yura.lysogorskii@gmail.com

Abstract. Resent experiments in the lamellar system Na_xCoO_2 detected a transition of Co planes into a puzzling metallic state at $x \geq 2/3$, which co-exists with a robust arrangement of the 3*d* cobalt electrons: The triangular Co lattices are disproportionated in the spinless Co^{3+} sites (Co1), and $Co^{3.44+}$ sites (Co2) with enhanced magnetism forming conducting sublattices. This textures concur with a tightening of the ferromagnetic (FM) interaction in planes, and emerge when the sodium ions become arranged in layers in between the CoO_2 slabs. In the present research we have investigated *ab-initio* the appearance of such state in $Na_{2/3}CoO_2$. Towards this end in view we studied an interplay between the electronic coupling to the superstructure of the Na^+ ions and local correlations of the itinerant *d* electrons treated within the GGA+U approximation. Employing the exact crystallographic supercell, the electronic organization has been analyzed upon increasing the energy U of the Coulomb repulsion within the 3*d* shells at $T = 0$. The metallic ground state, being a spin density wave with the inplane FM and antiferromagnetic interplane correlations, has been obtained and established to posses two regimes. When $U > 2$ eV, a crossover develops from a uniform state of the *d*-lattice to the regular phase with the spin/charge disproportionation between the sites. In particular at the representative value $U = 5$ eV, the $Co1^{3+}$ sites with suppressed magnetism appears, while the spin-active Co^{4+} holes are accumulated by the Co2 sites. A related formation of an isolated, narrow conduction band at the Fermi level implies a considerable enhancement of the electron correlations in the crystal field imposed by the Na^+ patterns.

1. Introduction

Intriguing behavior of the lamellar sodium cobaltates Na_xCoO_2 is in focus of numerous resent researches. Large thermopower and relatively low resistivity make these material promising for thermoelectronics [1, 2]. The pronounced magnetic field effect on the thermopower implies considerable spin entropy and an importance of Co(3*d*) electronic correlations in the ground state. A discovery of superconductivity ($T_c \approx 5K$) in $Na_xCoO_2 \cdot 1.3\ H_2O$ ($x \approx 0.33$) farther promotes the interest to the electron physics of the lamellar sodium cobaltates, providing an example of a layered non-cuprate oxide superconductor [3].

Structurally, Na_xCoO_2 consists of quasi-two-dimensional CoO_2 slabs stacked along the crystallographic *c*-axis and separated by Na layers [Fig. 1(a)]. Forming the triangular lattice, the planar Co ions are surrounded by edge-sharing O_6 octahedra. To the first approximation, the octahedral crystal field splits the local 3*d* levels into the lower t_{2g} orbital triplets and upper e_g

THE GRAPHIC (FIGURE 1) ON THIS PAGE WAS NOT AVAILABLE

ON THE E-MEDIA VERSION OF THIS PAPER.

Figure 1. (Color online) Structure of $Na_{2/3}CoO_2$ presented in accordance with NMR measurements of Alloul *et al.*, see [10]. (a) Elementary cell with the respective sodium and cobalt ion positions. (b) Sketch of the cobalt plane: The violet (red) open circles specify the inert ions Co1a (Co1b) being Co^{3+}. The light (dark) blue circles stand for the magnetically-active Co2a (Co2b) ions $Co^{3.44+}$. The conducting kagome sublattice is indicated by the thin solid lines.

doublets with the gap about 2 eV between them [4]. This suggests the ground state configuration t_{2g}^6 $(S = 0)$ and t_{2g}^5 $(S = 1/2)$ for the Co^{3+} and Co^{4+} ion, respectively.

Sodium ions Na_x^+ donate x electrons into the $3d$ shells of the $Co^{4+}O_2^{2-}$ planes. Contr-intuitively, in the metallic cobaltates the correlations *enhance* upon increasing x [i.e. upon decreasing the density $1 - x$ of the itinerant holes $Co^{4+}(t_{2g}^5)$] in particular in the range $0.65 < x < 0.85$ about the band-insulator limit $x = 1$. The experimental indications of this are the following:

(i) Sommerfeld constant γ is markedly increased from the low-x value, about 11 mJ/mol K^2, to $30 - 50$ mJ/mol K^2 [5], demonstrating the heavy-fermion behavior [6].

(ii) Localized spin states emerge, which are manifested by the Curie-Weiss behavior of magnetic susceptibility [7].

(iii) Simultaneously, intersite ferromagnetic (FM) spin correlations become strengthened at $x > 0.65$ [7], and the inplane FM order [more precisely A-AFM (antiferromagnetic) spin structure] is formed when $0.75 < x < 0.85$ [8, 9].

(iv) In addition, imprints of robust electronic arrangements, irrelevant for Fermi liquids, appear in the metal [10]: In the triangular lattice, the cobalt $3d$-electrons become segregated between the Co1 and Co2 sites, being in different magnetic and valence states. In particular, at $x = 2/3$ the extended spin-active holes, repelled from the inert sites $Co1^{3+}$ are almost confined in the kagome sublattice $Co2^{3.44+}$ [10], see Fig. 1(b).

Experimentally, for $0.65 < x < 0.85$ the Na^+ ions become arranged in the positions Na1 and Na2 between the Co planes, when the temperature decreases below few hundred Kelvins [11], see Fig. 1(a). Then, a tunneling of the holes Co^{4+} to the Co1 sites could be suppressed due to their electric repulsion from the adjacent cations $Na1^+$.

A relevant description of this puzzling metal represents a problem, which we address in the present study. Previous theoretical approaches [4, 12–15] to the collective states of CoO_2 slabs disregarded the sodium ions superstructure, relying on a strong on-site/inter-site Coulomb interaction between the $3d$ electrons. However, the repulsion energy of the itinerant holes from the Na^+ ions is considerable [16] and should be also included into the description of the electron physics. Within an approximation of a chaotic distribution of the Na^+ cations, sodium were established to markedly enhance the correlations in the metallic cobaltates [17, 18].

In the present work, we study the role of the realistic sodium *arrangement* in the formation of the unconventional metallic state of the CoO_2 planes. To this end, we consider the representative system $Na_{2/3}CoO_2$ and investigate an interplay between the two relevant interactions, (i) the on-site Coulomb repulsion with the energy U between the $3d$ electrons and (ii) the electron coupling to the regular superstructure of the Na^+ ions. Starting with the experimental crystal supercell [10], we have described the electronic states at $T = 0$ in a dependence on U within the GGA+U approximation. Our approach is detailed in Sec. 2, the electronic structure is deduced in Sec. 3, and results are discussed in Sec. 4.

2. *Ab-initio* approach

2.1. GGA+U method

We performed the first-principle calculations with the spin-polarized density functional theory (DFT). The generalized gradient approximation (GGA) was implemented with the functional form of the Perdew–Burke–Ernzerhof exchange correlation revised for solids (PBEsol), see Ref. [19]. The electron-ion interaction was represented with the projector augmented-wave method (PAW) [20] implemented in the Vienna Ab Initio Simulation Package (VASP 5.2.2) [21]. The plane-wave cutoff energy was 500 eV.

As the internal coordinates and so the crystal cell parameters were fully relaxed. The Brillouin zone (BZ) was sampled with a $5 \times 5 \times 5$ mesh, containing the gamma point; the corresponding density of the mesh was about one point per $0.004\AA^{-3}$. The energy convergence criteria 10^{-5} eV has been chosen. The strong electron correlations were treated with the simplified (rotationally invariant) LSDA+U approach, introduced by Dudarev *et al.* [22], and $U_{\text{eff}} = U - J$ was used as the Hubbard parameter U.

Two PAW potentials have been tested for sodium: one with seven valence electrons ($2p^6 3s^1$) and another with one valence electron ($3s^1$). In the first case, all the p-electrons were found to be localized at the sodium cations, signifying a sufficiency of the Na^+ potential with one valence electron for the calculations.

2.2. Supercell structure

In Ref. [10], using the NMR/NQR technique, the crystallographic positions of the sodium and cobalt ions in $Na_{2/3}CoO_2$ were identified and the unit cell with the symmetry group $R-3c$ was established for this compound, see Fig. 1(a). In this cell the cobalt ions occupy four distinct positions [Fig. 1(b)]: The Co1a and Co1b sites have the minimal distance from the sodium ions $Na1^+$, suggesting the larger crystal potential, ejecting the holes to the remaining Co2a and Co2b sites forming the kagome sublattice.

We performed a geometry optimization of such cell in the range from $U = 0$ eV to 6 eV, employing the GGA and GGA+U technique. The arrangement of the sodium ions remained robust even without symmetrization of charge densities and forces. Thus, the stability of the sodium pattern, reported in Ref. [10], was confirmed. Again, we have calculated the energies of the system as for FM and so for AFM interlayer spin configurations. The AFM order was found to have less energy (≈ 7.5 meV per one irreducible layer) and, thus, to be more favorable.

3. Results

First, we calculated electronic states for the hypothetic case $U = 0$, when the correlations are "switched off" in the cobalt $3d$ shells in $Na_{2/3}CoO_2$. Then, the GGA provides a uniform Fermi liquid state of the Co planes with the small FM spin-polarization $\approx 0.3\ \mu_B$ of the band due to

THE GRAPHIC (FIGURE 2) ON THIS PAGE WAS NOT AVAILABLE

ON THE E-MEDIA VERSION OF THIS PAPER.

Figure 2. Magnetic structure of the cobalt plane calculated for $Na_{2/3}CoO_2$ within the GGA+U approximation. Panel (a): Magnetic moments of the Co sites in a dependence on U, the energy of the Coulomb repulsion within the d-levels. The results for the Co1a (Co1b) ions are indicated by the short (long) dash line, while the thin solid (bold solid) line corresponds to the Co2a (Co2b) ions. Panel (b): Schematic view of the triangular Co plane. Filled circles represent the sites Co2b with the substantial magnetic moments ($\approx 0.6\ \mu_B$ at $U = 5$ eV).

weak correlations remaining in the system, see Fig. 2(a). The electronic spectrum, described in terms of the density of states (DOS), is presented in Fig. 3(a). The wide conduction band (with the width ≈ 1.6 eV) mostly originates from the t_{2g} levels, providing the relatively small DOS $\rho_F \approx 1.1\ \text{eV}^{-1}$ at the Fermi energy E_F. In turn the lower, valence band comes from the occupied O($2p$) states, while the upper band corresponds to the e_g excitations, see Fig. 3(a). The coupling to the Na^+ does not crucially affects the CoO_2 planes when $U = 0$: The DOS resembles to that, obtained with the LSDA+U model, disregarding sodium effect [23].

When the correlations appear within the $3d$ shells ($U > 0$) the obtained spin-density wave in the Co plane gradually transforms to a non-uniform regime: The marked disproportionation develops between the Co1 and Co2 sites when $U > 3$ eV, see Fig. 2(a). The effect is particularly strong when $U \geq 5$ eV: The enhanced moments $\mu_{\text{Co2b}} \approx 0.6\mu_{\text{B}}$ at the Co2b sites form the FM hexagon patterns, while the magnetism of the Co1 and Co2a sites becomes almost suppressed, see Fig. 2(b).

With the realistic value $U = 5$ eV (taken in between the estimations 4.91 eV for the Co^{3+} ion and 5.37 eV for the Co^{4+} ion in the lithium cobaltate [24]), the effect of the sodium superstructure becomes crucial for the spectrum of the CoO_2 planes, see Fig. 3(b). In particular, at the Fermi level the isolated narrow d-band is formed, indicating a considerable enhancement of the correlations in the CoO_2 planes. This sharp feature has the bandwidth ≈ 125 meV and $\rho_F \approx 5.42\ \text{eV}^{-1}$. Just the electron coupling to the sodium-cation system promotes the correlations, which otherwise remains weak: $\rho_F \approx 0.5$ eV for $U = 5$ eV [23].

As evident from Fig. 3(c), at the Co2b sites the DOS (taken in the average below E_F) becomes less than the respective DOS of the other cobalt sites, indicating that the spin separation [Fig. 2(b)] goes hand-in-hand with the charge redistribution in the Co lattice. The non-magnetic state of the Co1 and Co2a ions (having more populated $3d$-shells) implies their closeness to the Co^{3+} ($t_{2g}^6, S = 0$) configuration. Avoiding these sites, the magnetic holes $Co^{4+}(t_{2g}^5)$ tend to the Co2 sublattice, thereby forming the localized spin states in the metal [Fig. 2(b)]. We obtain the FM interaction between spins in plane and the AFM coupling between planes in accordance with the experimentally observed A-AFM correlations in Na_xCoO_2 ($0.65 < x < 0.85$) [8, 9].

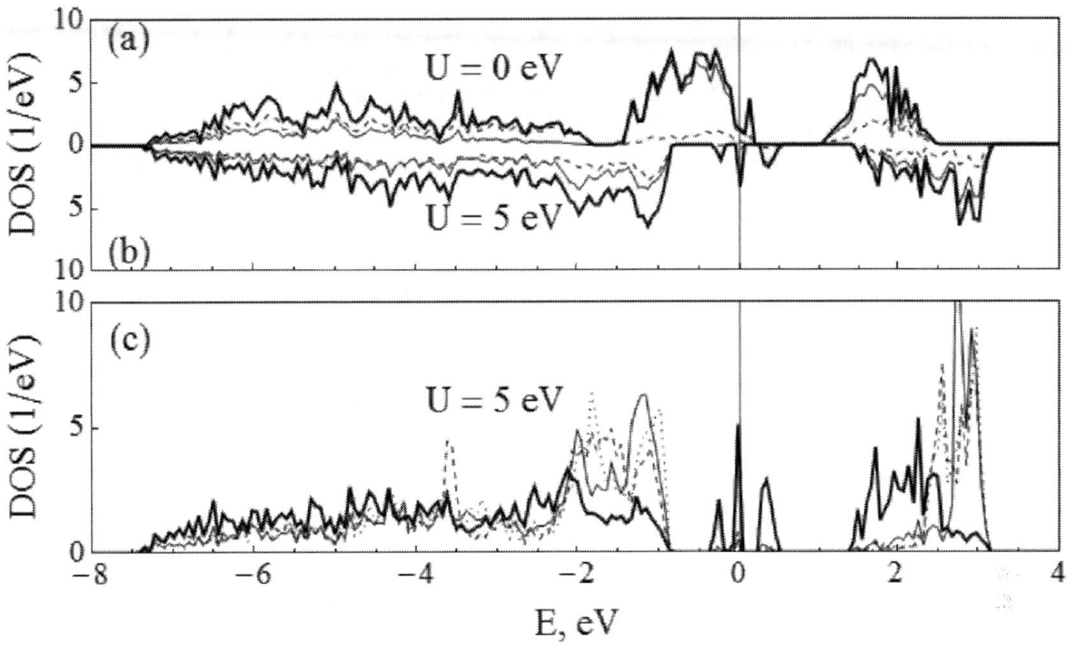

Figure 3. Energy dependence of the electronic density of states (DOS) within the GGA+U approximation. Plots (a) and (b): DOS for $Na_{2/3}CoO_2$ (bold line) calculated with $U = 0$ eV (a) and $U = 5$ eV (b). Corresponding cobalt (oxygen) contributions to the DOS are given by the solid (dash) lines. Plot (c): Electronic DOS at the cobalt sites obtained with $U = 5$ eV. The short (long) dash line represents DOS at the Co1a (Co1b) sites, and the thin solid (bold solid) line corresponds to the Co2a (Co2b) ions.

The narrow band obtained in the DOS at E_F indicates, that the electronic disproportionation coexists with the robust metallicity ($\rho_F > 0$) of the Co planes up to $U \approx 6$ eV. Experimentally, the metallic state of the cobaltate was found, e.g., in Ref. [7].

The calculated Fermi surface (FS) has seven pockets. The central one, located about Γ symmetric point of the BZ of the triangular lattice, always has a hole character (convex). The six pockets, found nearby the K points, gradually change their nature from an electron-like at $U = 1$ eV to the hole-like at $U = 6$ eV. With $U = 1$ eV the volume V_F of the FS encompasses 32% of the BZ volume (V_{BZ}), in close correspondence with the concentration of the holes ($1-x = 1/3$) in the d-lattice of $Na_{x=2/3}CoO_2$. For $U = 6$ eV, we obtain $V_F = 0.31V_{BZ}$. Thus, the Luttinger's theorem is satisfied with a good accuracy for such electronic liquid. The central pocket comprises 7% (12.5%) of V_{BZ} for $U = 1$ eV ($U = 6$ eV).

We estimated the cobalt magnetic moment $\mu_{\mathrm{Co2}} = 0.4\mu_B$ per Co2 site for $U = 5$ eV. Again, measurements of magnetic susceptibility of $Na_{2/3}CoO_2$ provided the larger value $\mu_{\mathrm{Co2}}^{(exp)} \approx 1.2\mu_B$ in the temperature range from 50 K to 250 K [8]. A plausible way to resolve this discrepancy is to take into consideration excitations of Co ions towards higher spin states.

4. Discussion

The strong FM-correlations in the triangular Co planes are known to appear at $x \geq 2/3$ within the Hubbard-model approach for the cobaltates, which disregards the Na superlattice [12, 18, 25]. Then, a delicate treatment of the nearest and next-nearest neighbor correlations enabled to obtain imprints of the kagome-like charge/spin density wave (the M-point pike of the charge susceptibility) in the metallic state [25]. This instability is related with a pronounced van Hoove singularity, existing in the DOS nearby the Fermi level in an effective kagome lattice [26]. However, such approach provided an electronic disproportionation, only when the 3*d* electron intersite repulsion ($V \approx 0.1 - 0.4$ eV) between the nearest-neighbor sites of the lattice was included into the model [18].

The present approach does not require an intersite electronic repulsion to produce the effect: The obtained metallic state, combining the strong spin/charge disproportionation and intersite FM spin correlations, results from a peculiar optimization of the double-exchange interaction [27, 28] in the crystal field imposed by the Na^+ arrangement. In more details, we shall discuss this physics elsewhere.

The deduced electronic structure of the metal (magnetic Co2b-hexagons) partially reproduces the realistic magnetic kagome Co2 lattice, cf. Fig. 2(b) with Fig. 1(b). However, the magnetism of the Co2a sites is suppressed [see Fig. 2(b)] indicating, that these ions are close to the Co^{3+} state likewise the sites Co1a and Co1b. Virtual transitions of the $Co2^{3+}$ ions to their high spin (HS) configuration $t_{2g}^5 e_g$ ($S = 1$) could enhance their moments and proliferate the magnetism to the Co2a sites, reproducing the kagome lattice. We intend to clarify this issue in a future investigation. Resembling HS fluctuations were implemented, e.g., in the spin-polaron theories of the CoO_2 planes by Khaliullin and Chaloupka [13].

An existence of six small FS pockets, obtained nearby the K-points of the BZ in our study, were not revealed by ARPES experiments so far [29]. In principle, they could be hidden for ARPES, because this technique mostly detects surface electronic states, which could differ significantly from those in a bulk of a sample [30]. In turn, transport measurements, employing the Shubnikov–de Haas effect, corroborated the presence of the FS pockets in Na_xCoO_2 at $x = 0.71$ and 0.84 [31]. Another evidences in favor of an existence of these features are provided by bulk measurements with the x-ray Compton scattering technique, performed in the unhydrated and hydrated sodium cobaltates [32]. On the theoretical ground, a more accurate treatment of the 3*d* correlations beyond the mean-field GGA+U approximation could verify a robustness of the FS pockets within the present approach.

5. Conclusion

In the present work we have calculated *ab-initio* electronic and magnetic properties of sodium cobaltate $Na_{2/3}CoO_2$ within GGA+U approximation. In contrast to previous studies [12, 23, 33], we have used the supercell of this compound, deduced from the experiments in Ref. [10]. We have established the sodium arrangement to remain stable, as it does not change after the full structure relaxation.

Electronic state of the FM metal $Na_{2/3}CoO_2$ has been described microscopically in a dependence on the energy U of the Coulomb repulsion within the 3*d* shells. The metallic ground state, being a spin density wave with A-AFM correlations (FM-inplane and AFM-interplane), has been obtained. The state was found to posses the two regimes: A uniform phase of the d-lattice for $U < 2$ eV, and the regular phase with the spin/charge disproportionation between the sites for $U > 2$ eV. In particular at the representative value $U = 5$ eV, the $Co1^{3+}$ sites with suppressed magnetism appears, while the spin-active Co^{4+} holes are accumulated by the Co2 (more exactly Co2b) sites in accordance with the experimental behavior [34]. A related formation of an isolated, narrow conduction band at the Fermi level implies a considerable enhancement of the electron correlations in the crystal field imposed by the Na^+ patterns.

Such a peculiar metallic electronic structure appears due to the specific interplay between the electrostatic potential induced by the periodic Na^+ arrangement and the strong correlations in the 3*d* orbitals.

Acknowledgments

Authors would like to thank I. Mukhamedshin for many stimulating discussions, critical reading of the manuscript, and for providing us with the crystallographic data. Also, Y.L. would like to thank Region Pays de la Loire, France for the partial financial support (MOA119). The partial support by the RFBR under the project No. 10-02-01005-a is acknowledged as well.

References

[1] Fujita K, Mochida T and Nakamura K 2001 *Jpn. J. Appl. Phys.* **40** 4644
[2] Terasaki I, Sasago Y and Uchinokura K 1997 *Phys. Rev. B* **56** 12685
[3] Takada K, Sakurai H, Takayama-Muromachi E, Izumi F, Dilanian R and Sasaki T 2003 *Nature* **422** 53–55
[4] Singh D J 2000 *Phys. Rev. B* **61** 13397
[5] Schulze T, Brühwiler M, Häfliger P, Kazakov S, Niedermayer C, Mattenberger K, Karpinski J and Batlogg B 2008 *Phys. Rev. B* **78** 205101
[6] Miyoshi K, Morikuni E, Fujiwara K, Takeuchi J and Hamasaki T 2004 *Phys. Rev. B* **69** 132412
[7] Foo M, Wang Y, Watauchi S, Zandbergen H, He T, Cava R and Ong N 2004 *Phys. Rev. Lett.* **92** 247001
[8] Alloul H, Mukhamedshin I R, Collin G and Blanchard N 2008 *EPL (Europhysics Letters)* **82** 17002
[9] Bayrakci S P, Mirebeau I, Bourges P, Sidis Y, Enderle M, Mesot J, Chen D P, Lin C T and Keimer B 2005 *Phys. Rev. Lett.* **94** 157205
[10] Alloul H, Mukhamedshin I R, Platova T A and Dooglav A V 2009 *EPL (Europhysics Letters)* **85** 47006
[11] Zandbergen H, Foo M, Xu Q, Kumar V and Cava R 2004 *Phys. Rev. B* **70** 024101
[12] Korshunov M, Eremin I, Shorikov A, Anisimov V I, Renner M and Brenig W 2007 *Phys. Rev. B* **75** 094511
[13] Khaliullin G and Chaloupka J 2008 *Phys. Rev. B* **77** 104532
[14] Shorikov A, Korshunov M M and Anisimov V I 2011 *JETP Lett.* **93** 80
[15] Motrunich O and Lee P 2004 *Phys. Rev. B* **69** 214516
[16] Roger M, Morris D J P, Tennant D A, Gutmann M J, Goff J P, Hoffmann J U, Feyerherm R, Dudzik E, Prabhakaran D, Boothroyd A T, Shannon N, Lake B and Deen P P 2007 *Nature* **445** 631
[17] Marianetti C and Kotliar G 2007 *Phys. Rev. Lett.* **98** 176405
[18] Piefke C, Boehnke L, Georges A and Lechermann F 2010 *Phys. Rev. B* **82** 165118
[19] Perdew J, Ruzsinszky A, Csonka G and Vydrov O 2008 *Phys. Rev. Lett.* **100** 136406
[20] Blochl P 1994 *Phys. Rev. B* **50** 17953
[21] Kresse G and Furthmuller J 1996 *Phys. Rev. B* **54** 169
[22] Dudarev S, Botton G, Savrasov S, Humphreys C and Sutton A 1998 *Phys. Rev. B* **57** 1505
[23] Zhang P, Luo W, Cohen M and Louie S 2004 *Phys. Rev. Lett.* **93** 236402
[24] Zhou F, Cococcioni M, Marianetti C A, Morgan D and Ceder G 2004 *Phys. Rev. B* **70** 235121
[25] Boehnke L and Lechermann F 2012 *Phys. Rev. B* **85** 115128
[26] Peil O, Georges A and Lechermann F 2011 *Phys. Rev. Lett.* **107** 236404
[27] Zener C 1951 *Phys. Rev.* **82** 403
[28] Anderson P W and Hasegawa H 1955 *Phys. Rev.* **100** 675
[29] Hasan M Z, Chuang Y D, Qian D, Li Y W, Kong Y, Kuprin A, Fedorov A V, Kimmerling R, Rotenberg E, Rossnagel K, Hussain Z, Koh H, Rogado N S, Foo M L and Cava R J 2004 *Phys. Rev. Lett.* **92** 246402
[30] Pillay D, Johannes M and Mazin I 2008 *Phys. Rev. Lett.* **101** 246808
[31] Balicas L, Jo Y, Shu G, Chou F and Lee P 2008 *Phys. Rev. Lett.* **100** 126405
[32] Laverock J, Dugdale S B, Duffy J A, Wooldridge J, Balakrishnan G, Lees M R, Zheng G q, Chen D, Lin C T, Andrejczuk A, Itou M and Sakurai Y 2007 *Phys. Rev. B* **76** 052509
[33] Zhou S, Gao M, Ding H, Lee P A and Wang Z 2005 *Phys. Rev. Lett.* **94** 206401
[34] Mukhamedshin I R and Alloul H 2011 *Phys. Rev. B* **84** 155112

Journal of Physics: Conference Series **394** (2012) 012020 doi:10.1088/1742-6596/394/1/012020

Ab-initio investigation of $GdLiF_4$ structure under pressure

A Petrova[1], B Minisini[2], O Nedopekin[1] and D Tayurskii[1]
[1)]Institute of Physics, Kazan Federal University, 420008, 18 Kremlevlskaya, Kazan, Russia
[2)]Institut Supérieur des Matériaux et Mécaniques Avancés du Mans, 72000, 44 Av.Bartholdi, Le Mans, France

E-mail: npetrova131@gmail.com

Abstract. This work is devoted to the *ab-initio* studies of rare-earth double fluoride $GdLiF_4$ under external hydrostatic pressure. Structural and mechanical properties were considered by means of Vienna Ab-initio Simulation Program (VASP). A good agreement of lattice parameters and bulk modulus with experimental data has been obtained. The seven independent elastic constants of $I4_1/a$ $GdLiF_4$ structure were calculated from stress-strain method. The provideab-initio studies has revealed the instability of $GdLiF_4$ crystal structure above 10 GPa in accordance with available experimental findings.

1. Introduction

It has recently been concluded that pressure-induced post-scheelite (or post-fergusonite) structures in oxides, fluorides, and any otherABX_4-type compounds are not really well known [1]. That is why the investigation of high-pressure behaviour of $GdLiF_4$by measurements of its angle-dispersive x-ray powder diffraction patterns as a function of pressure was provided. Particularly the pressure-induced decomposition of $GdLiF_4$sheelite has been found.

The goal of the present paper is to study the evolution of structural and mechanical parameters with a pressure by means of *ab-initio* calculation and to compare the obtained dependencies with experimental data.

2. Calculations

$GdLiF_4$ is a sheelite compound.Scheelite crystals have a tetragonal $I4_1/a$ symmetry. It can be presented as dipyramidal pseudo-octahedra. In this structure, the primitive unit cell has two ABX_4 units.The crystal structure $GdLiF_4$ (see figure 1) can be described as highly ionic with Gd^{3+}cations and tetrahedral LiF_4 anions. Each Li^{1+}cation is surrounded by four equivalent fluorines in tetrahedral symmetry. Each Gd^{3+}cation shares corners with eight adjacent LiF_4 tetrahedra and the Gd^{3+}cations are surrounded by eight fluorines in approximately octahedral symmetry forming bisdisphenoids.

The crystallographic directions in $GdLiF_4$ consist of two equivalent directions (the a-axes), and one unique direction (the c-axis).

During the calculations we considered $GdLiF_4$ structure with initial lattice parameters (A= 5.235(1) Å, C= 11.01(2) Å, V= 302.0 Å^3) taken from the experiment [2].

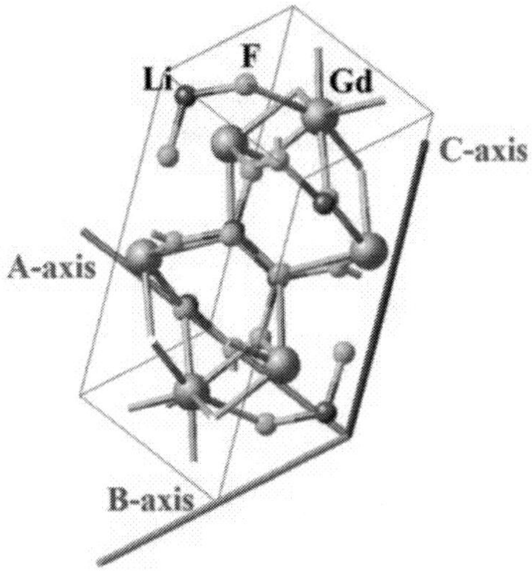

Figure 1. A tetragonal structure of $GdLiF_4$.

For numerical simulations we used Density Functional Theory [3, 4] as implemented in Vienna Ab-initio Simulation Package [5], part of MedeA© modelling interface. The exchange–correlation functional was approximated by the gradient corrected form proposed by Perdew-Burke-Ernzerhof [6]. The electronic degrees of freedom were described using the projector augmented wave method.

Three stages of structure optimization were performed in the range from 0 to 20 GPa with step 2 GPa. The atom positions, shape and volume of cell can be changed at first and second stages. At the third stage only atom positions can be changed. Such scheme of optimization was used for more accurate calculation of lattice parameters and energy of the structure. Full optimization of the structure was performed until the maximum force dropped below 0.002 eV/A^{-1} whereas the self-consistent field convergence criterion was set at 10^{-6}eV. A k-mesh of 0.5 $Å^{-1}$ leading to 12 symmetry-unique k-points was used to sample the Brillouin zone. For each k-point the wave functions were expanded in terms of plane wave until a cutoff energy 500 eV. To smooth the Fermi step function the Gaussian smearing with σ-parameter 0.1 eV was used.

Different methods were considered to investigate the influence of 4f-electrons on the structural and mechanical properties of $GdLiF_4$ structure. In this work non-magnetic, spin-polarized, spin-orbit coupling methods were used as well as the approximation that 4f-electrons of gadolinium are «kept frozen» in the core or are «included in the valence states».

3. Results

3.1 Structural properties

The obtained evolutions of the lattice parameters of $GdLiF_4$ as a function of pressure are presented on Figure 2 for different methods of taking into account the influence of f-electrons.

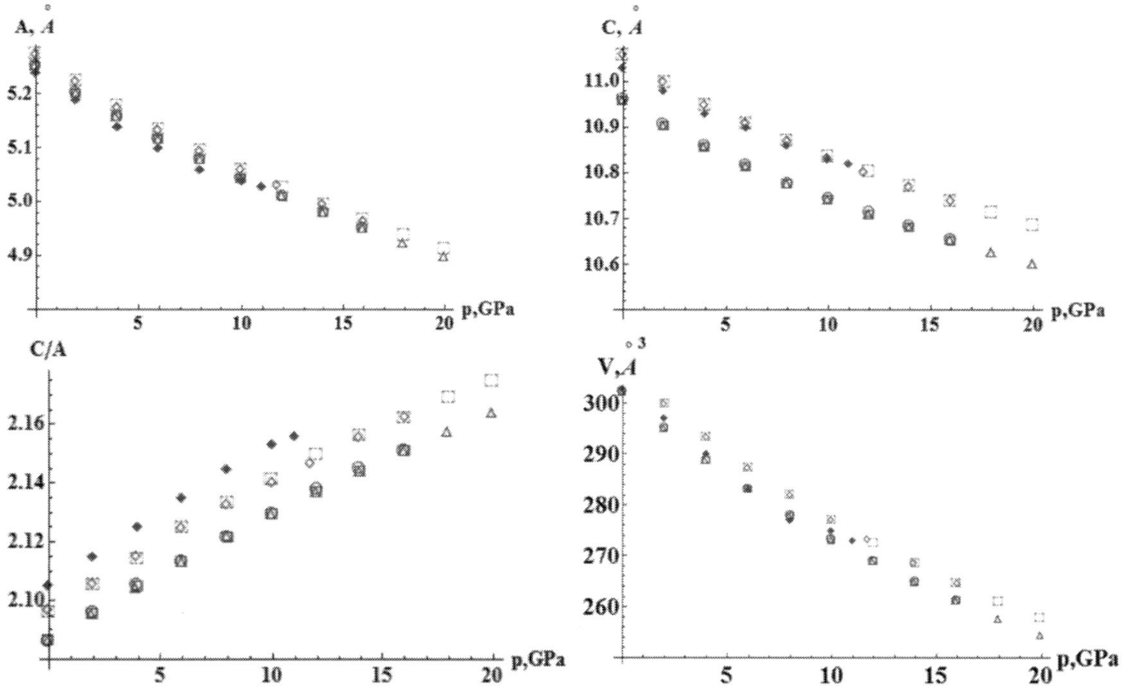

Figure 2.Evolution of structural parameter versus pressure for different methods of taking into account the influence of f-electrons (f-electrons are «kept frozen in the core»: non-magnetic state-empty square, spin-polarized state-empty triangle, spin-orbit -empty circle; f-electrons «included in the valence states»: spin polarized-chain line square, spin orbit-empty rhomb; experimental data [2]-solid rhomb).

The graphics demonstrate a good agreement between experimental results and calculated lattice parameters. Based on these results we can conclude thattaking into account spin degree of freedoms makes not too much influence on the structural parameters. The case of f-electrons «kept frozen in the core» gives the best agreement with experimental results for *A* parameter. A visible difference between the cases of f-electrons considered as a core and as a valence appears on pictures, which show the evolution of *C* lattice parameter and *C*/*A* axial ratio. As a pressure increases, the *C*/*A* ratio is increasing, indicating that the distortions in a crystal are enhanced at high pressures. The dependence obtained by case of f-electrons «included in the valence states» is closer to the experimental data. The analysis of volume evolution shows the best agreement with experimental data for the approximation when f-electrons are «kept frozen in the core». Since further we will be interested in the mechanical properties, we can restrict our consideration by the approximation of non-magnetic f-electrons «kept frozen in the core».

3.2 Mechanical properties

We have evaluated the elastic constants under pressure and have compared the values of bulk modulus and volume with experimental results. Mechanical-thermal analysis of $GdLiF_4$ was made by Mechanical-Thermal (MT) module of VASP 5.2 [7].

For a crystal belonging to the TII Laue group, the elastic stiffness constant matrix referred to the crystallographic axis (*X*, *Y*, *Z*) reference frame is

$$C_{ij} = \begin{pmatrix} c_{11} & c_{12} & c_{13} & 0 & 0 & c_{16} \\ c_{12} & c_{22} & c_{13} & 0 & 0 & -c_{16} \\ c_{13} & c_{13} & c_{33} & 0 & 0 & 0 \\ 0 & 0 & 0 & c_{44} & 0 & 0 \\ 0 & 0 & 0 & 0 & c_{44} & 0 \\ c_{16} & -c_{16} & 0 & 0 & 0 & c_{66} \end{pmatrix}. \quad (1)$$

The seven independent elastic constants of $GdLiF_4$ structure versus pressure were calculated from stress-strain method using strain 0.002. The calculation parameters were the same as in section 2. On the figure 3 the elastic constants C_{11}, C_{12}, C_{13} and C_{33} show strong pressure dependence, in contrast to C_{16} and C_{44}. The value of C_{66} elastic constant decreases as a function of pressure.

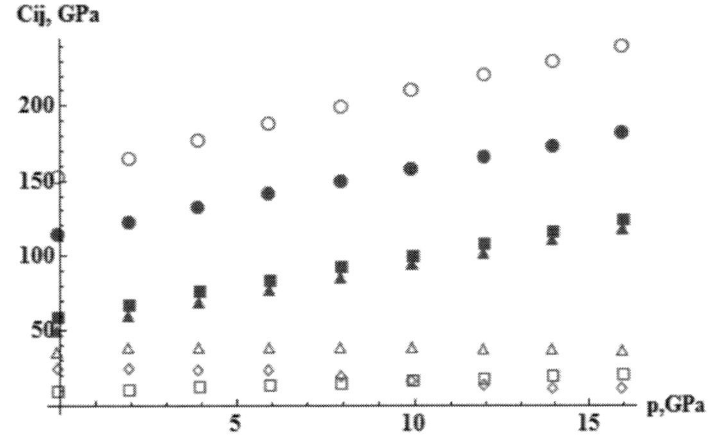

Figure 3. Evolution of elastic constants versus pressure (C_{11}-solid circle, C_{12}-solid triangle, C_{13}-solid square, C_{16}-empty square, C_{33}-empty circle, C_{44}-empty triangle, C_{66}-empty rhomb).

From the constants of elasticity the following modulus were calculated: shear module is equal to 34 GPaand Young's module is equal to 89 GPa.

By the fitting P-V curves with a Birch-Murnanghan equation of state,

$$P(V) = \frac{3B_0}{2}\left[\left(\frac{V_0}{V}\right)^{\frac{7}{3}} - \left(\frac{V_0}{V}\right)^{\frac{5}{3}} \right]\left\{ 1 + \frac{3}{4}(B'_0 - 4)\left[\left(\frac{V_0}{V}\right)^{\frac{2}{3}} - 1 \right] \right\} \quad (2)$$

bulk modulus B_0,the first pressure derivative of the bulk modulus B'_0 and equilibrium volume V_0 were evaluated.The results are presented in the table 1.

Table 1. The comparison of the mechanical parameters.

Methods	**V_0, volume, $Å^3$**	**B_0, Bulk modulus, GPa**	**B'_0, the first pressure derivative of the bulk modulus**
B-M	302.4	79	3.9
MT	302.4	80	-
B-M experiment[2]	302.9±0.3	76±4	5.01±1.03

A good agreement is observed between the bulk modules obtained by MT calculations, by Birch-Murnaghan (BM) approximation and experimental values (B-M approximation of experimental data).

It is well known that a Born condition of the stability of crystals demands the positive sign ofthe eigenvalues of the elastic constants matrix [8, 9]. MT calculations of $GdLiF_4$ structure give the negative eigenvalue above 10 GPa. It gives the evidence of the instability of $GdLiF_4$compound above this value of pressure as observed in the experiment [2] where decomposition of $GdLiF_4$ sample has been found at 13.1 GPa.

4. Conclusion

We obtained a good agreement of structural parameters (*A*, *C*, *C*/*A*, *V*) with experimental values. The pressure dependence of volume has been fitted with the Birch–Murnaghan equation of state to evaluate the equilibrium volume and the bulk modulus. Mechanical parameters (bulk modulus, B_0 and equilibrium volume, V_0) show a good agreement with the relevant values obtained in the experiment and directly calculated with using MT module. The seven independent elastic constants of $I4_1/a$ $GdLiF_4$ structure were evaluated versus pressure. Mechanical-thermal analysis of $GdLiF_4$ structure revealed the instability of this material above 10 GPa.

Acknowledgements

This work was supported in part by the Ministry of Education and Science of Russian Federation. A.P. thanks Region Pays de la Loire for partial individual financial support (MOA00120).

References

[1] Errandonea D and Manjorn F J 2008 *Progress in Materials Science* **53** 711
[2] Grzechnik A, Crichton WA, Bouvier P, Dmitriev V, Weber H P and Gesland J Y 2004 *J Phys. Condens. Matter* **16** 7779
[3] Hohenberg P, Kohn W 1964 *Phys. Rev.* B **136** 864
[4] Kohn W, Sham L J 1965 *Phys. Rev.* A**140** 1133
[5] Kresse G and Furthm¨uller J 1996 *Phys. Rev.* B**54**11169
[6] Perdew J P, Burke K and Wang Y 1996 *Phys. Rev.* B **54** 16533
[7] Le Page Y, Saxe P, 2002 *Phys. Rev. B***65** 104104
[8] Vsin'ko G and Smirnov N A 2002 *J. Phys.: Condens. Matter***14** 6989
[9] Born M and Huang K 1954 Dynamical Theory of Crystal (London: Oxford University Press) p.129

SPMCS 2012 IOP Publishing
Journal of Physics: Conference Series **394** (2012) 012021 doi:10.1088/1742-6596/394/1/012021

Ab-initio investigations of $LuLiF_4$ compound under pressure

A Petrova[1], B Minisini[2], O Nedopekin[1] and D Tayurskii[1]

[1)]Institute of Physics, Kazan Federal University, 420008, 18 Kremlevskaya, Kazan, Russia
[2)]Institut Supérieur des Matériaux et Mécaniques Avancés du Mans, 72000, 44 Av.Bartholdi, Le Mans, France

E-mail: npetrova131@gmail.com

Abstract. We have performed ab-*initio* calculations for the structural and mechanical properties of sheelite compound$LuLiF_4$under pressure. A good agreement with experimental results has been obtained.

1. Introduction

A great progress has been done in the last years in the study of pressure-effects on the structural and mechanical properties of sheelite compounds $MLiF_4$, where M stands for lanthanides [1]. These compounds have the $CaWO_4$-type or scheelite-type structure which is a superstructure of fluorite CaF_2 (Fm3m, Z = 4). The fluorine atoms are situated in a distorted simple cubic arrangement and the Li^{1+} and M^{3+}cations are fourfold and eightfold coordinated by fluorines, respectively.

The interest to the $LuLiF_4$ compound was invoked by the fact that this material possesses the own electrical dipole moment in some temperature range. $LuLiF_4$ structure has been investigated using synchrotron angle-dispersive x-ray powder diffraction in a diamond anvil cell at high pressures and room temperature [2]. The authors investigated structural and mechanical data that allows us to compare our results with experimental one.

Earlier Minisini et al. [3] have studied thermo-mechanical properties of $LuLiF_4$sheeliteat ambient pressure by means of density functional theory (DFT) was performed. Particularly, elastic stiffness coefficients of $LuLiF_4$ sheelite have been estimated by stress–strain method.

2. Calculation parameters

We performed *ab-initio* calculations by means of DFT [4, 5] with using VASP 5.2 [6] (Vienna Ab-Initio Simulation Package), the part of the MedeA® interface. The exchange–correlation functional was approximated by the gradient corrected form proposed by Perdew-Burke-Ernzerhof[7]. The electronic degrees of freedom were described using the projector augmented wave method and basis of plane waves [8, 9].

Based on the results obtained [10] only non magnetic method was considered and f-electrons of Lu were considered as the «kept frozen in the core». It was made due to the fact that taking into account spin degree of freedom makes no any essential influence on structural and mechanical parameters. During calculations we considered $LuLiF_4$ structure with symmetry $I4_1/a$ and C12/c1, Z = 4 in the range of pressure from 0 to 20 GPa with step 2GPa. Full optimization of the structure was performed until the maximum force dropped below 0.005 eV, whereas the self-consistent field convergence criterion was set at 10^{-6}eV. Next parameters have been chosen to obtain the converging results. A k-mesh is equal to 9*9*9 for the structure with symmetry $I4_1/a$ and 7*7*9 for the structure with symmetry C12/c1. For each k-point the wave functions were expanded in terms of plane waves until a cutoff energy of 796 eV for $I4_1/a$ structure and until 576 eV for C12/c1 structure. For forces and energy optimization Methfessel-Paxton scheme was used with smearing 0.12 eV for $I4_1/a$ structure and 0.17 eV for C12/c1 structure. The method of Methfessel-Paxton allows to describe very accurate the total energy of the system.

The structure optimization was made by three stages. The atom positions, shape and volume of cell can be changed at the first and second stages. At the third stage only atom positions can be changed.

It is well known that Lu has $4f^{14}$ $5d^1$ $6s^2$ electronic configuration. GGA (generalized gradient approximation) +U scheme was applied namely to d-electrons of Lu, because of f-orbital is completed. This method was used in order to treat strong correlation for d-electrons. Different Habbard parameters in the range from 1 to 3 eV were considered, but the most suitable parameter is equal to 1 eV.

For investigation mechanical properties Mechanical-Thermal (MT) module of VASP 5.2 was used. The calculation parameters were the same as above; the value of strain is 0.002.

3. Results

3.1 Structural properties

We performed the analysis of structural parameters of$LuLiF_4$ compound with two different symmetries $I4_1/a$ and C12/c1. The structure parameters ($a^{t/m}$, $b^{t/m}$, $c^{t/m}$, where t-tetragonal and m-monoclinic symmetries) and experimental points [2] versus pressure were plotted on the figure 1.

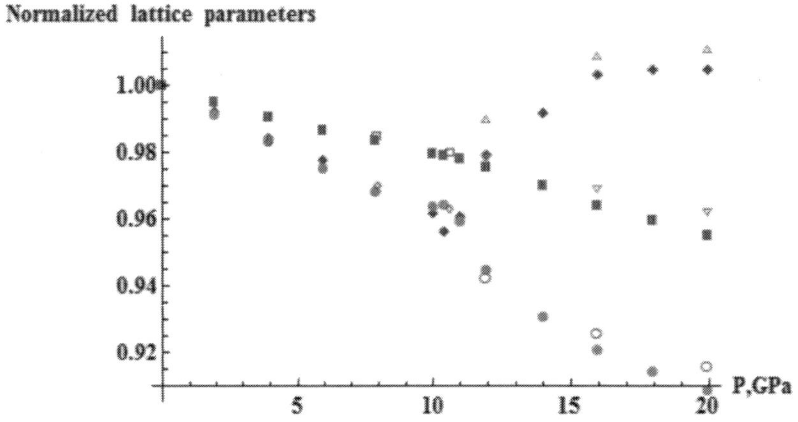

Figure 1. Lattice parameters dependence on pressure normalized to the respective value at ambient pressure (experimental values: empty square- c^T / c_0 , empty rhomb- a^T / a_0 , empty triangle- a^m / a_0 , empty inverse triangle- b^m / b_0 , empty circle- c^m / c_0 ; *ab-initio* calculations: solid rhomb- a^m / a_0 , solid square- b^m / b_0 , solid circle- c^m / c_0).

We can see a good agreement with experimental results. *Ab-initio* calculations show that the structural parameters of $LuLiF_4$ with two symmetries are mostly the same below 10.5 GPa. It can be explained by the fact that C12/c1 is a subgroup of $I4_1/a$ symmetry. Two symmetries structural parameters diverge after 10.5 GPa.

The figure 2 presents the angle β (between a and c axes) dependence on pressure for $LuLiF_4$ structure with C12/c1 symmetry. This parameter also changes sharply at 10.5 GPa.

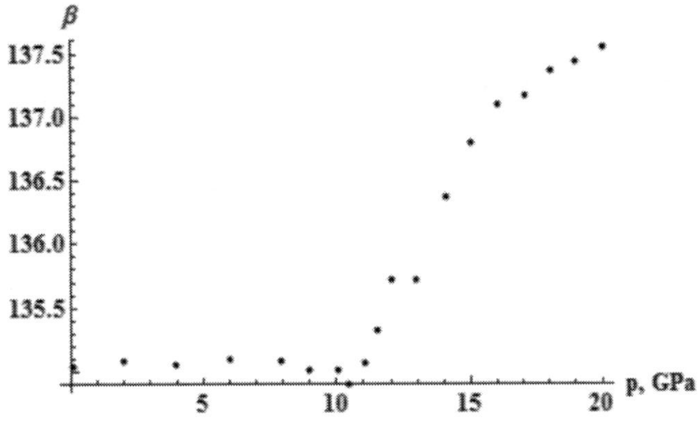

Figure 2. The angle of C12/c1 structure versus pressure.

3.2 Mechanical properties

The behavior of unit-cell volumes of $LuLiF_4$ were investigated for two symmetries until 20 GPa. The corresponding figure is presented below:

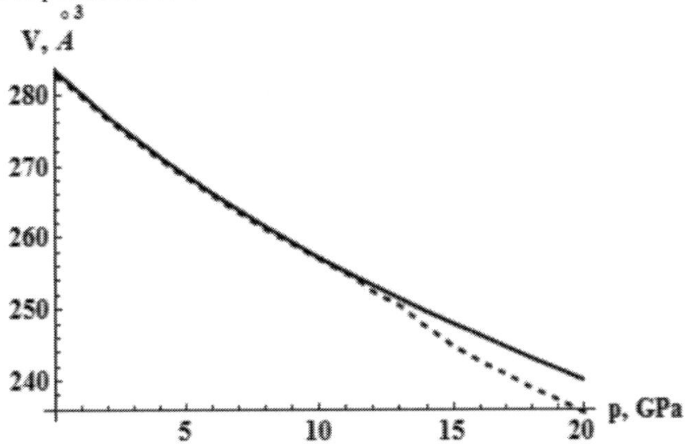

Figure 3. The solid line presents the unit-cell volume of $I4_1/a$ structure, the dashed line presents the unit-cell volume of C12/c1 structure versus pressure.

Based on this curve we can conclude that volumes are practically the same and change smoothly with pressure increasing up to 10.5 GPa.

The P-V data was approximated by Birch-Murnanghan equation of state in the range from 0 GPa to 10 GPa:

$$P(V) = \frac{3B_0}{2}\left[\left(\frac{V_0}{V}\right)^{\frac{7}{3}} - \left(\frac{V_0}{V}\right)^{\frac{5}{3}}\right]\left\{1 + \frac{3}{4}(B'_0 - 4)\left[\left(\frac{V_0}{V}\right)^{\frac{2}{3}} - 1\right]\right\} \tag{1}$$

As a result bulk modulus B_0 and equilibrium volume V_0were evaluated. Also the values of equilibrium volume and bulk module were obtained by M-T module of VASP 5.2.

The results presented in the table 1 demonstrate a good agreement between the bulk module obtained by MT calculations, by Birch-Murnaghan (B-M) approximation and experimental results(B-M approximation of experimental data) [2].

Table 1. Equilibrium volume V_0, bulk modulus B_0 and the first pressure derivative of the bulk modulus B'_0

Methods	V_0,volume, Å^3	B_0, Bulk modulus, GPa	B'_0, The first pressure derivative of the bulk modulus
B-M	283.4	83	4.50
MT	283.3	85	-
B-M experiment [2]	280.7±0.2	85±3	5.59±0.55

Table 2 presents the values of elastic stiffness coefficients obtained by MT calculations and the results obtained in the paper [3].

Table 2. Elastic stiffness coefficients (GPa) estimated by strain-stress method at 0 GPa

Elastic stiffness Coefficients,GPa	Paper [1]	MT calculations
C_{11}	131	128
C_{12}	54	54
C_{13}	62	59
C_{16}	13	9
C_{33}	168	165
C_{44}	40	40
C_{66}	29	24

Shear module (38 GPa) and Young's module (98GPa) were obtained from the elasticity constants.

4. Conclusion

Thus, in this work we presented a density functional theory analysis of $LuLiF_4$ structure under pressure. A good agreement of structural parameters (a, b, c) for $I4_1/a$ and C12/c1 symmetries with experimental results has been obtained. The values of bulk module and equilibrium volume obtained by MT calculations and by B-M approximation agree well with experimental results. The comparison of the elastic constants obtained by M-T module with relevant results of the paper [3] gives similar values.

Acknowledgements

This work was supported in part by the Ministry of Education and Science of Russian Federation.

References

[1] Errandonea D and Manjorn F J 2008 *Progress in Materials Science* **53** 711
[2] Grzechnik A, Friese K, Dmitriev V, Weber H-V, Gesland J-Y and Crichton W A 2005 *J. Phys.: Condens. Matter* **17** 763
[3] Minisini B, Bonnaud P, Wang Q A and Tsobnang F 2008 *Computational Materials Science* **42** 156
[4] Hohenberg P and Kohn W 1964 *Phys. Rev.* B **136** 864
[5] Kohn W and Sham L J 1965*Phys. Rev.* A **140** 1133

[6] Kresse G and Furthmuller J 1996 *Phys. Rev.* B **54** 11169
[7] Perdew J P, Burke K and Wang Y 1996 *Phys. Rev.* B **54** 16533
[8] Blöchl P E 1994 *Phys Rev*.B**50** 17953
[9] Methfessel M and Paxton A 1989 *Phys. Rev.* B **40** 3616
[10] Petrova A, Minisini B, Nedopekin O and Tayurskii D, 2012 *Ab-initio investigations of* $GdLiF_4$ *compound under pressure,* see this volume

Ab initio studying of topological insulator Bi_2Se_3 under the stress

Y V Lysogorskiy, A G Kijamov, O V Nedopekin and D A Tayurskii

Institute of Physics, Kazan Federal University, Kazan 420008, Russia

E-mail: yura.lysogorskii@gmail.com

Abstract. A topological insulator is an unusual state of quantum matter which, while being an insulator in the bulk, has topologically protected electronic states at the surface. These states could be used in different applications, such as spintronics and quantum computing. However, it is difficult to distinguish the surface and bulk contributions into transport properties, such as conductivity. In order to distinguish surface and bulk contributions an external pressure could be applied. In the present work we have performed *ab initio* calculations of topological insulator Bi_2Se_3 under the stress for bulk and surface models. Calculations have been made by means of density functional theory within generalized gradient approximation, the spin-orbit interaction was taken into account as well.

It was found that topologically protected surface states remains robust under the stress. Moreover, pressure tends to increase the Fermi velocity of surface electrons, as well as increase electronic density of states at the bottom of the conduction band of the bulk of Bi_2Se_3. Thus, the results of *ab initio* calculations could complement the experimental investigations of high pressure transport properties of topological insulators. The experimentally detected increase of carrier density could be related to the effects of the bulk.

1. Introduction

Topological insulators (TI) are the electronic materials that have a bulk band gap like an ordinary insulators, but have conducting states on their edges. Unlike fragile surface states in ordinary metals, the surface states of TIs are topologically protected, i.e. non-magnetic impurities can neither destroy nor impact the topological nature of the surface states. These states are possible due to the combination of strong spin-orbit (SO) interaction and time-reversal symmetry [1, 2]. The best candidate for the study of topological surface states (TSS) is Bi_2Se_3 thanks to its large bulk insulating gap ($E_G \sim 300$ meV).

TIs could be used for technological applications in spintronics and quantum computing [2]. In order to exploit these materials in electronics, the detailed study of their electronic and transport properties is required. In a recent years, a numerous investigations of different properties of topological insulators, such as conductivity, magnetoresistance and optical properties [3–7] as well as *ab initio* simulations of TIs compounds [8–10] have been performed.

High-pressure studies are useful to understand the properties of the existing topological insulators and to design new materials because the increase of pressure reduces the interatomic distances and finely tunes materials properties. In order to complement the experimental investigations, which aimed to distinguish surface and bulk contributions into transport properties, *ab initio* calculations could be of great help.

The present paper is organized as follows: in Section 2 a short overview of experimental results about the transport properties of TIs under the pressure is given; Section 3 is devoted to the details of *ab initio* calculations as well as calculated electronic properties; Section 4 contains discussion and conclusion.

2. Transport properties at high pressure

Bi_2Se_3 is a narrow band gap semiconductor with a tetradymite crystal structure (space group $R\bar{3}m - D^5_{3d}$). Its rhombohedral layered structure is formed by quintuple layers (QL), consisted from five hexagonal close-packed atomic sublayers (Se(1)-Bi-Se(2)-Bi-Se(1)), see Fig. 1. Inside the QL the strong covalent bonds are formed, while QLs are linked together by the weak van der Waals interaction.

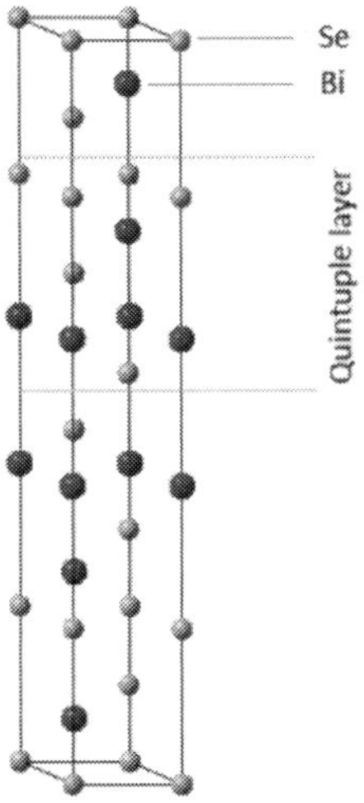

Figure 1. The crystal structure of Bi_2Se_3.

Results of the electrical transport measurements under the high pressure was reported in the Ref. [3]. The experiments were performed in pressure range from zero to 20.8 GPa at the temperature as low as 20 K. The authors pointed out the increase of resistance in the pressure range from 0 to 8 GPa. The resistivity drops substantially as the pressure increases above the 8 GPa, when structural transition occurs. The carrier density increases by more than two orders of magnitude in the measured pressure range. All measurements were carried out on high mobility, nominally stoichiometric single crystals of Bi_2Se_3. The substantial increase of carrier density with pressure might be expected to further obscure the surface conduction.

Measurements of the transport and optical properties of Bi_2Se_3 under hydrostatic pressure as well as *ab initio* calculations of its electronic properties were reported in Ref. [6]. The authors observed the decrease of Hall electron mobility as well as increase of Hall electron concentration for several Bi_2Se_3 samples with different thicknesses. The corresponding *ab initio* calculations indicated an increasing of direct band gap at Γ point which is coherent with the carried out measurements of optical band gap.

In order to match an experimentally observed increase of carrier density with respect to the pressure with the results of *ab initio* simulations, the density of state (DOS) for bulk and surface Bi_2Se_3 should be calculated.

3. Calculations

We have performed *ab initio* calculations of the bulk and surface electronic properties of topological insulator Bi_2Se_3 in the pressure range from 0 to 8 GPa. Calculations have been performed by means of density functional theory (DFT) with exchange-correlation functional PBEsol [11] and with taking into account the spin-orbit interaction, which is realized in the VASP code [12] embedded into MedeA© interface. A plane-wave cutoff energy was 256 eV, and $8 \times 8 \times 8$ k-mesh in the Brillouin zone (BZ) for the bulk structures and $8 \times 8 \times 1$ for the surface structures was used. In order to simulate in a right way the hexagonal cell K-meshes was centered at the Γ point. The hydrostatic pressure to the bulk was applied. All atoms in bulk were fully relaxed until forces become less than 0.005 eV/Å.

Band structures and DOS for bulk Bi_2Se_3 at three different pressures are depicted on the Fig. 2(a). As it can be seen, the increase of pressure leads to the increase of direct band gap at Γ point (which correlates with data presented in Ref. [6]) while the total band gap remains almost constant and equal to 300 meV. Electronic density of states at the bottom of conduction

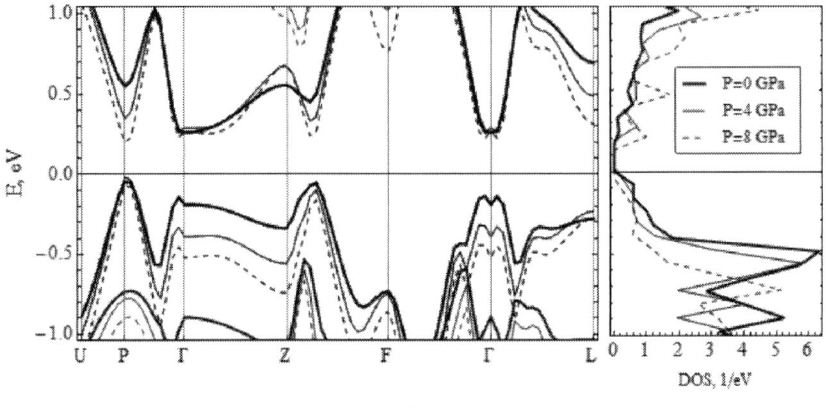

(a)

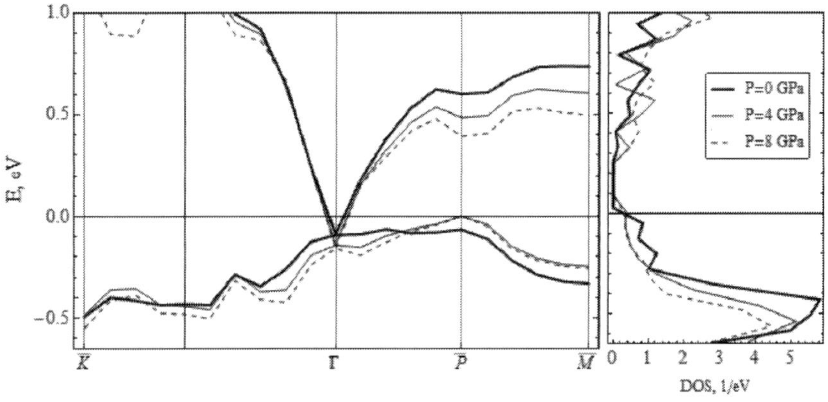

(b)

Figure 2. Calculated band structure and density of state (DOS) near the Fermi level for (a) bulk and (b) 7 QL slab of Bi_2Se_3 at three different pressures. P and $\overline{P}$ k-points are in a middle on $\Gamma - U$ and $\Gamma - \overline{M}$ lines respectively. Notation for the BZ points is the same as in Ref. [9].

band (BCB) slightly increases with respect to pressure due to the decrease of conduction band near the L and P points in BZ, as can be seen from Fig. 2(a).

To simulate effect of the high pressure on the surface states, we used optimized bulk structure at given pressure and made from it the 7 QL slab surrounded by a 20 Å vacuum layer in the direction perpendicular to plane in order to avoid the influence of periodic boundary condition. That is, all interatomic distances as well as lattice parameters in *ab*-plane remains the same as for bulk. In order to make sure in the topological character of surface electronic states, we have determined Z_2-topological invariants according to the method described in Ref. [8]. Topological invariant for a surface models was found to be equal to unity for all considered pressures, that confirms the system does indeed have the TSS which remains intact by pressure. However, it was established, that the slabs thinner than 5 QL do not form a TSS, probably due to the strong interaction between top and bottom edges of slab. This fact is coherent to previously reported experimental results [13], where gapped states for the Bi_2Se_3 films thinner than 5-6 QL were detected. *Ab initio* calculations for similar compound Bi_2Te_3 indicate that the minimal thickness necessary to form TSS is equal to 5 QL [8].

The calculated band structures along $\overline{K} - \Gamma - \overline{M}$ path in hexagonal BZ for a slab of Bi_2Se_3 consisted of 7 QL under different pressures are presented on the Fig. 2(b). Here we present only

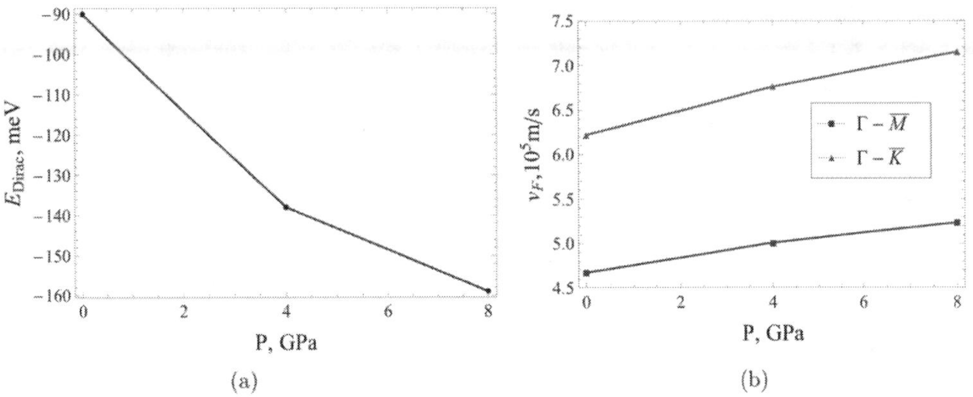

Figure 3. (b) Dirac point energy and (a) velocity at Fermi level along $\Gamma - \overline{K}$ and $\Gamma - \overline{M}$ directions in BZ for Bi_2Se_3 in dependence on pressure, calculated fot 7 QL slab.

two bands close to the Fermi level and responsible for TSS formation. In opposite to the bulk, one can find a nonzero DOS at Fermi level, that produced by TSS. However, DOS at the BCB has the similar tendency to increase as for a bulk.

Two bands, depicted on the Fig. 2(b), form the so called Dirac cone. The crossing of these bands at Γ point is a Dirac point. Pressure dependence of Dirac point depth with respect to Fermi energy is depicted on the Fig. 3(a). Calculated Dirac point depth at zero pressure (90 meV) well agrees with the experimentally observed one (160 meV) [13]. The dependence of Fermi velocity on pressure is given on the Fig. 3(b). The order of Fermi velocity magnitude is in a good agreement with the experimental value for Bi_2Se_3 $v_{\rm F} \simeq 5.0 \times 10^5 m/s$ [1] and $v_{\rm F} \simeq 6.0 \times 10^5 m/s$ along $\Gamma - \overline{M}$ direction and $v_{\rm F} \simeq 8.0 \times 10^5 m/s$ along $\Gamma - \overline{K}$ direction for Bi_2Te_2Se compound which is similar to the Bi_2Se_3 [14].

4. Discussion and conclusion

The carrier concentration could be calculated as

$$n = \int_{E_C}^{\infty} f(E) g_C(E) dE, \quad (1)$$

where E_C - energy of the BCB, $f(E)$ - Fermi-Dirac distribution and $g_C(E)$ is density of states for conduction band. That is, DOS at the BCB should make influence on the carrier concentration considerably. The *ab initio* calculations indicate the increase of this quantity both for a bulk and a slab models. This fact allows us to make an assumption that experimentally observed increase of carrier concentration could be results of a change in *bulk* electronic properties. On the other hand, authors in Ref. [6] supposed that bulk carrier density decreases, which is in contradiction with our results.

The external pressure does not make any significant influence on the TSS. Indeed, surface states are topologically protected against any perturbations which conserve time-reversal symmetry, such as defects, distortions and non-magnetic impurities. Moreover, TSS are low-sensitive to elastic scattering on impurities as well as scattering on low-energy phonons, as was reported in Ref. [5]. Thus, external pressure stands alongside with other types of disturbances that do not affect the surface states.

On the surface band structure (Fig. 2(b)) one can see the peculiarity at the $\overline{P}$ point. This k-point corresponds exactly to the P k-point on the bulk band structure. One can suggest that this peculiarity could be the precursor for the insulator-to-metal phase transition of Bi_2Se_3 [3]

above 8 GPa. Nevertheless, the wave function of this state is distributed in the slab and has no any relationship with TSS.

Calculations indicate that external pressure tends to increase the surface electron velocities at the Fermi level. In its turn, surface electron mobility should significantly depend on the Fermi velocity, because the scattering mechanisms for TSS are rather weak. So, the compression of the Bi_2Se_3 could be proposed as a mechanism to increase the mobility of surface electron.

In conclusion, we have performed an *ab initio* calculations of band structure and DOS for both bulk and surface of Bi_2Se_3 with taking into account the spin-orbit interaction. Topologically protected surface states remain intact by pressure. It was established that increase of pressure in the range from 0 to 8 GPa leads to increase of DOS in the bottom of the conduction band, whereas the indirect band gap remains almost constant and equal to $E_g = 300$ meV. Hereby, we suppose, that increase of the Hall electron concentration with respect to external pressure observed in [3, 6] could be caused by this increase of DOS.

Acknowledgement
The work was supported in part by the Ministry of Education and Science, Russian Federation.

References

[1] Zhang H J, Liu C X, Qi X L, Dai X, Fang Z and Zhang S C 2009 *Nature Physics* **5** 438–42
[2] Moore J E 2010 *Nature* **464** 194–8
[3] Hamlin J J, Jeffries J R, Butch N P, Syers P, Zocco D a, Weir S T, Vohra Y K, Paglione J and Maple M B 2012 *J. Phys.: Cond. Matt.* **24** 035602
[4] Di Pietro P, Vitucci F, Nicoletti D, Baldassarre L, Calvani P, Cava R, Hor Y, Schade U and Lupi S 2012 *Phys. Rev. B* **86** 045439
[5] Pan Z H, Fedorov A, Gardner D, Lee Y, Chu S and Valla T 2012 *Phys. Rev. Lett.* **108** 187001
[6] Segura A, Panchal V, Sánchez-Royo J, Marín-Borrás V, Muñoz Sanjosé V, Rodríguez-Hernández P, Muñoz a, Pérez-González E, Manjón F and González J 2012 *Phys. Rev. B* **85** 195139
[7] Alegria L D, Schroer M D, Chatterjee A, Poirier G R, Pretko M, Patel S K and Petta J R 2012 *Nano Lett.* **12** 4711–4
[8] Park K, Heremans J, Scarola V and Minic D 2010 *Phys. Rev. Lett.* **105** 186801
[9] Larson P, Greanya V, Tonjes W, Liu R, Mahanti S and Olson C 2002 *Phys. Rev. B* **65** 085108
[10] Vilaplana R, Santamaría-Pérez D, Gomis O, Manjón F J, González J, Segura A, Muñoz A, Rodríguez-Hernández P, Pérez-González E, Marín-Borrás V, Muñoz Sanjose V, Drasar C and Kucek V 2011 *Phys. Rev. B* **84** 184110
[11] Perdew J, Ruzsinszky A, Csonka G, Vydrov O, Scuseria G, Constantin L, Zhou X and Burke K 2008 *Phys. Rev. Lett.* **100** 136406
[12] Kresse G and Furthmüller J 1996 *Phys. Rev. B* **54** 11169–86
[13] Taskin A A, Sasaki S, Segawa K and Ando Y 2012 *Phys. Rev. Lett.* **109** 066803
[14] Neupane M, Xu S Y, Wray L, Petersen A, Shankar R, Alidoust N, Liu C, Fedorov A, Ji H, Allred J, Hor Y, Chang T R, Jeng H T, Lin H, Bansil A, Cava R and Hasan M 2012 *Phys. Rev. B* **85** 235406

SPMCS 2012 IOP Publishing
Journal of Physics: Conference Series **394** (2012) 012023 doi:10.1088/1742-6596/394/1/012023

Growth Kinetics of the Homogeneously Nucleated Water Droplets: Simulation Results

Anatolii V. Mokshin and Bulat N. Galimzyanov
Department of Physics, Kazan Federal University, 420008 Kazan, Russia
E-mail: anatolii.mokshin@mail.ru

Abstract. The growth of homogeneously nucleated droplets in water vapor at the fixed temperatures $T = 273$, 283, 293, 303, 313, 323, 333, 343, 353, 363 and 373 K (the pressure $p = 1$ atm.) is investigated on the basis of the coarse-grained molecular dynamics simulation data with the mW-model. The treatment of simulation results is performed by means of the statistical method within the mean-first-passage-time approach, where the reaction coordinate is associated with the largest droplet size. It is found that the water droplet growth is characterized by the next features: (i) the rescaled growth law is unified at all the considered temperatures and (ii) the droplet growth evolves with acceleration and follows the power law.

1. Introduction

Nucleation is one of the possible ways, by means of which a phase transition can be initiated. According to the classical nucleation theory, it is assumed that nucleus (cluster) of a new phase becomes stable to grow in a 'mother phase', when its size reaches some critical value – the critical size n_c. Consequently, both related processes, nucleation and nuclei growth, belong to the initial stages of phase transition. The character of phase transition depends directly on the nucleation rate J_s and the growth rate G. Appearance of liquid droplets (bubbles) as origins of condensation (vaporization) and emergence of crystallite grains at crystallization are the typical examples of nucleation. A feature of homogeneous nucleation[1] for most of molecular systems at the natural conditions is that the characteristic spatial scale [1], which is associated with the linear critical size of the nucleated phase, takes the values $10^{-9} \div 10^{-7}$m. Thereby, the processes of nucleation and of nuclei growth are excellent candidates to be studied by means of the methods of molecular dynamics simulation. However, the probabilistic character of nucleation (the values of nucleation rate, critical size, growth rate and others) prompts applying the statistical treatment of the simulation results.

Recently, it was shown that the mean-first-passage-time (MFPT) approach can be successfully used to define all the characteristics of nucleation kinetics from the data of molecular dynamics simulations [2, 3]. In the idealized case of the symmetric nucleation barrier, the MFTP-distribution obtained from the set of independent simulations or from experiments is interpolated in the nucleation regime by the error-function, and such parameters as the steady-state nucleation rate J_s (or the nucleation time scale $\tau_s = (J_s V)^{-1}$, V is the system volume), the critical size n_c and the curvature of the nucleation barrier associated with the Zeldovich factor

[1] Homogeneous scenario means the same probability for the nucleation event emergence over the whole system.

Z can be extracted by numerical fitting. An extended MFTP approach allows one to define all these parameters for a wider class of nucleation scenario by numerical analysis of the first derivative of the MFPT-curve with respect to the reaction coordinate (say, nucleus size). In the present work we show that the same MFPT-approach can be also applied to define the characteristics of the nuclei growth regime. We demonstrate this analysis as applied for the case of the water droplet growth.

2. Methods and results

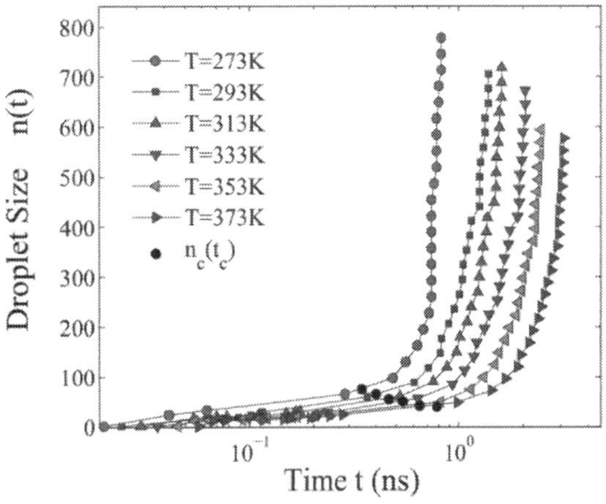

Figure 1. Growth curves of the nucleated water droplet as obtained from NpT-simulations within the MFPT-approach. The pressure is the same for all the considered cases, $p = 1$ atm. Black circles indicate the critical sizes n_c.

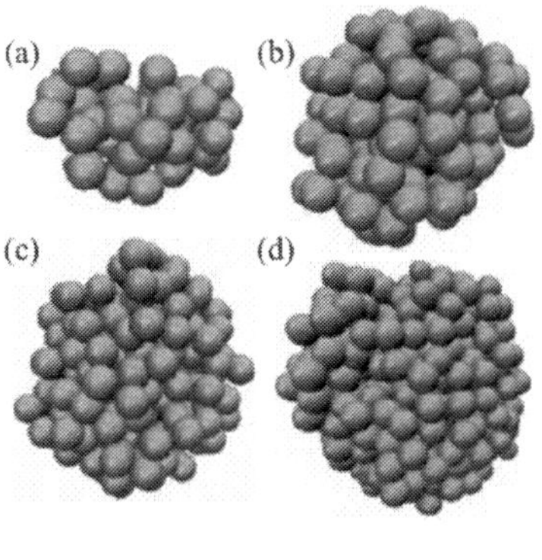

Figure 2. Water droplet at the temperature T = 293 K and at the different time after nucleation: (a) n = 55 molecules, $t(n)$ = 0.75 ns, (b) n = 123 molecules, $t(n)$ = 0.83 ns, (c) n = 197 molecules, $t(n)$ = 0.88 ns, (d) n = 308 molecules, $t(n)$ = 0.91 ns. Circles are the conventional symbols of water molecules.

The basic idea of the MFPT-approach is the next. First, it is necessary to focus on the reaction coordinate and to define the time scales of the first appearance of its values, $\tau(n)$. As a reaction coordinate for the case of the droplet nucleation, it is convenient to utilize the droplet size n, which defines the number of molecules (particles) involved in the nucleus of new (liquid) phase. Second, by defining these time scales from M independent 'experiments' and by averaging results, one obtains the MFPT-distribution:

$$\tau^{\mathrm{MFPT}}(n) = \sum_{i=1}^{M} \tau_i(n).$$

In fact, the MFPT-distribution contains the information about nucleation and nucleus growth kinetics. For example, the first inflection points in the MFPT-curve defines the critical size n_c with the lag-time $t_c = \tau^{\mathrm{MFPT}}(n_c)$ (for details, see [2]). Then, the part of the inverted MFPT-distribution $n(\tau^{\mathrm{MFPT}})|_{n \geq n_c}$ corresponds to the most probable growth law of the nucleated droplet. Let us suppose the growth law in the following generalized form [4]:

$$n(t, t_c) = n_c + c_g \rho_c \mathcal{G}_c^{d\nu} (t - t_c)^{d\nu}, \tag{1}$$

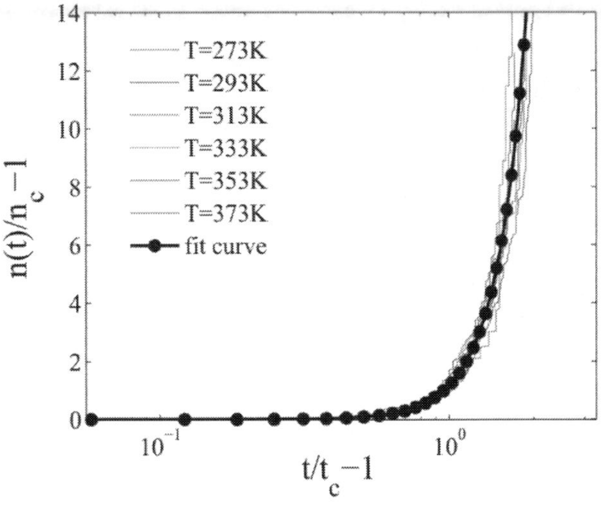

Figure 3. Rescaled growth curves at the different temperatures $T \in [273;\ 373]$ K (the pressure $p = 1$ atm); t_c is the lag-time of the droplet with the critical size n_c. Full circles correspond to the fit (see text for details).

where c_g is the shape-factor, ρ_c is the density of the nucleated phase, ν is the growth exponent – it is equal to unity in the case of constant growth rate, $\mathcal{G}_c$ is the growth parameter and d is the system dimension. Without losing the generality in such a definition, the growth rate can be time-dependent term and takes the next form : $G(t) = \nu\mathcal{G}_c^{\nu}t^{\nu-1}$ (see also Ref. [1]). By fitting equation (1) to the distribution $n(\tau^{\mathrm{MFPT}})|_{n\geq n_c}$, one can extract the values of the growth characteristics.

Molecular dynamics NpT-simulations were performed with the coarse-grained mW-model [5] of water vapor at the temperatures $T = 273$, 283, 293, 303, 313, 323, 333, 343, 353, 363 and 373 K at the pressure $p = 1$ atm.[2] For an each (p, T)-point the set from 100 independent runs at the identical initial conditions was carried out, and the time evolutions of the largest nucleated (liquid) droplet were stored for every run. Identification of liquid clusters was performed on the basis of the cluster analysis with the Stillinger rule [6, 7].

Figure 1 shows the growth curves of water droplet at the different temperatures. These results are obtained from molecular dynamics simulations within the MFTP-approach. The MFPT-treatment of these results reveals the interesting features of the nucleation-growth kinetics. The critical size $n_c(T)$ has a weak decay with the temperature (for the considered temperature range and the fixed pressure), that contradicts the classical nucleation theory predictions [1]. On other hand, the lag-time t_c increases with temperature from $t_c(T = 273\ \mathrm{K}) = 0.69$ ns to $t_c(T = 373\ \mathrm{K}) = 1.56$ ns. The nucleated droplets have shapes, which are closed to spherical ones (see figure 2). Further, the growth parameter $\mathcal{G}_c$ decreases with temperature from $\mathcal{G}_c(T = 273\ \mathrm{K}) = 576\ \mathrm{m}^{\nu}/\mathrm{s}$ to $\mathcal{G}_c(T = 373\ \mathrm{K}) = 242\ \mathrm{m}^{\nu}/\mathrm{s}$, whilst the growth exponent is independent on the temperature and takes the value $\nu = 1.3$. This indicates on the next time dependence of the growth rate: $G(t) \sim t^{0.3}$. Other result is that the growth curves rescaled to

[2] The initial configurations were obtained by the isobaric cooling from the equilibrated vapor at the temperature $T = 900$ K and at the pressure $p = 1$ atm.

the form

$$\frac{n(t)}{n_c} - 1 = \frac{c_g \rho_c (\mathcal{G}_c t_c)^{3\nu}}{n_c} \left(\frac{t}{t_c} - 1 \right)^{d\nu}$$

are independent on the temperature and collapse onto a single growth curve (see figure 3).

3. Conclusions

The following results are presented in this work.

(i) The treatment of the experimental data and of the simulation results on the nucleation-growth kinetics can be performed on the basis of the MFPT-approach, which allows one to extract the main characteristics of the cluster growth process.

(ii) The growth of water droplets was studied on the basis of coarse-grained molecular dynamics simulations for a wide range of the temperature.

(iii) It is found that the rescaled growth law of water droplets within the mW-model is unified at all the considered temperatures.

(iv) It is established that the droplet growth evolves with acceleration and follows the power law.

References

[1] Kashchiev D 2000 *Nucleation: Basic Theory with Applications* (Oxford: Butterworth Heinemann)
[2] Wedekind J, Strey R, Reguera D 2007 *J. Chem. Phys.* **126** 134103
[3] Mokshin A V and Barrat J.-L. 2009 *J. Chem. Phys.* **130** 034502
[4] Mokshin A V and Barrat J.-L. 2010 *Phys. Rev. E* **82** 021505
[5] Molinero V, Moore E B 2009 *J. Phys. Chem. B* **113** 4008.
[6] Mokshin A V and Galimzyanov B N 2012 (in preparation)
[7] Stillinger F H 1963 *J. Chem. Phys.* **38** 1486.

Systems and models with anticipation in physics and its applications

A Makarenko

Institute for Applied System Analysis at National Technical University of Ukraine "KPI' Peremogy Avenue 37, 03056, Kiev-56, Ukraine

E-mail: makalex@i.com.ua

Abstract. Investigations of recent physics processes and real applications of models require the new more and more improved models which should involved new properties. One of such properties is anticipation (that is taking into accounting some advanced effects).It is considered the special kind of advanced systems – namely a strong anticipatory systems introduced by D. Dubois. Some definitions, examples and peculiarities of solutions are described. The main feature is presumable multivaluedness of the solutions. Presumable physical examples of such systems are proposed: self-organization problems; dynamical chaos; synchronization; advanced potentials; structures in micro-, meso- and macro- levels; cellular automata; computing; neural network theory. Also some applications for modeling social, economical, technical and natural systems are described.

1. Introduction

The history of self-organization science has many brilliant achievements: dissipative structures theory; oscillations in distributed systems; auto-waves; deterministic chaos; collapses; synchronization etc. The ordinary differential equations, discrete time equations and partial differential equations with memory effects have been used as the background for such investigations [1].

But recently it has been found that also very interesting are the investigation of the systems with anticipation. The term 'anticipation' had been firstly attached to the systems with the intrinsic models for predicting the evolution of the systems and of their environment [2-5] and many others. But now since the works by Daniel M. Dubois (Liege, Belgium) the notion of 'strong anticipation' had been introduced and investigated [3-5]. In case of strong anticipation the system haven't the models for predictions but are self- making with accounting presumable future states of the system. D. Dubois also introduced the notion of incursive and hyper-incursive systems (with presumable multivaluedness of solutions). Also D. Dubois firstly described the elementary single element with hyper-incursion. One of the most important examples of the models with anticipation follows from modeling of large social systems. Some examples of investigations of the systems with anticipation had been described in [6-8]. Remark that one of the most new and interesting properties in such systems is presumable multivaluednes of the solutions (that is existing of some values of the solution at each moment of time). Because of this it is very prospective to consider the properties of multivalued solutions from the point of view of synchronization investigations.

So in the proposed paper the next issues will be represented: The description of strong anticipation following D.Dubois, general example of network of elements with strong anticipation; examples of multivalued solutions in different systems with anticipation and some presumable types of solutions behavior; presumable possibilities of manifestation of such kind of solutions and further research problems. Thus the given paper should be considered as the beginning the investigations of new class of systems namely the systems of coupled strong anticipated elements.

2. Strong anticipation

Since the beginning of 90-th in the works by D.Dubois – see [3-5] the idea of strong anticipation had been introduced: "Definition of an incursive discrete strong anticipatory system …: an incursive discrete system is a system which computes its current state at time t, as a function of its states at past times $,...,t-3,t-2,t-1$, present time, t, and even its states at future times $t+1,t+2,t+3,...$

$$x(t+1)=A(...,x(t-2),x(t-1),x(t),x(t+1),x(t+2),...,p) \tag{1}$$

where the variable x at future times $t+1,t+2,t+3,...$ is computed in using the equation itself.

Definition of an incursive discrete weak anticipatory system: an incursive discrete system is a system which computes its current state at time t, as a function of its states at past times $,...,t-3,t-2,t-1$, present time, t, and even its predicted states at future times $t+1,t+2,t+3,...$

$$x(t+1)=A(...,x(t-2),x(t-1),x(t),x^*(t+1),x^*(t+2),...,p) \tag{2}$$

where the variable x^* at future times $t+1,t+2,t+3,...$ are computed in using the predictive model of the system" [5], (Dubois, 2001, p. 447).

Thus as the further research problem in the field of anticipatory systems should be considered the behavior of the system with strong anticipation. The simplest but rather general counterparts for the equations (1) and (2) are the equations for continuous time case and for discrete time case accounting of strong anticipation with single anticipation time.

Remark that the common example for synchronization investigations is the chain of couple maps with strong anticipation on one forward discrete step:

$$u_i(n+1)=f\ (u_i(n),u_i(n+1),\alpha\ )+\varepsilon\ g\ (\{\vec{u}\},\{\vec{u}(n+1)\},\theta\ ) \tag{3}$$

where $u_i(n)=u_i(\gamma n)$, γ -time step.

The equations (1) - (3) are rather new mathematical objects for physics and their full considerations are the tasks for further investigations. So here we had described only first results of such objects investigations. We describe the results on complex behavior research for discrete maps, artificial neural networks, cellular automata. The complex branching non-homogeneous in space solutions also are proposed.

3. Examples of models with anticipation

In this subsection we briefly describe for illustration the possibilities of strong anticipation property manifestation in some problems.

3.1. Two-Steps Discrete-Time Anticipatory Models

The proposed model has two features (see details at [9]). The first is two-steps nature (that is passing on two steps ahead at discrete). The second is that the function f(x) in the model has a piecewise-linear character, and looks like the transition function of neurons in neuronets. Remark that the piecewise character of the nonlinearity usually allows to simplifying mathematical investigations.

We write down the proposed model as follows:

$$\begin{cases} p_{t+1} = \alpha \cdot m + (1-\alpha)p_t - \alpha \cdot f(p_{t+2} - p_{t+1}) \\ p_{t+2} = \alpha \cdot m + (1-\alpha)p_{t+1} - \alpha \cdot f(p_{t+2} - p_{t+1}) \end{cases} \tag{4}$$

The function f(x) depends on the parameter α and has the following expression:

$$\begin{cases} f(x) = 0, & x \le 0 \\ f(x) = \alpha \cdot x, & x \in (0, 1/\alpha] \\ f(x) = 1, & x > 1/\alpha \end{cases}$$

The software had been developed so that it is possible to visualize some branches of the states of map, but only those, that are not thrown out to infinity. This facilitates the understanding of the processes, which take place in the region of ambiguousness. Namely, at the Figure 1 below we can see branching of solution at discrete time steps. In other calculations we had seen the number of different period of cycles and increasing of the number of cycles in the course of time, that is we had seen a tendency to phenomena which we may titled as the analogues of usual „chaos".

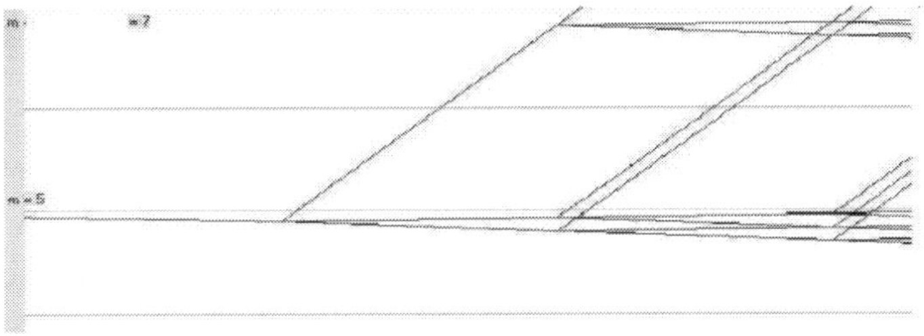

Figure 1. The graph of branching of solution. The parameters values are: $\alpha = 2$, m=0. Limitations for visualization of the values of variables are the next: maximal m+7 and minimal m-5, without representing the branches, which are thrown out to infinity.

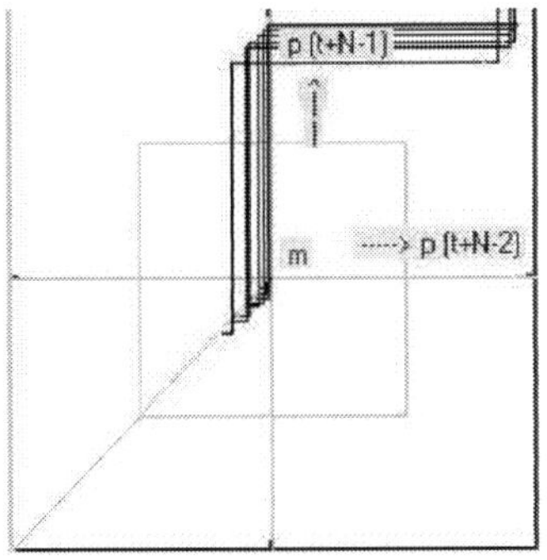

Figure 2. Map in state space. The sequence of values of (p_t, p_{t+1}) is represented for the solution from figure 1.

3.2. Cellular automata with anticipation

Recently we have investigated some kind of such models. With anticipation accounting: game 'Life' with anticipation; movement of pedestrian crowds; general problems [6].

First of all let's make a short definition of cellular automata. The cellular automata are discrete dynamic system which represents set of identical cells connected among them. Cells create a lattice of the cellular automata. Lattices can be different types, differing both on the dimension, and under the form of cells [6, 10].

The local rule for cell k on the Z_d is the transformation T_k which transforms the state $s_k(t)$ in S of cell k at moment t to the state $s_k(T+1)$ in S of the same cell at moment (t+1).

$$S_i\ (t+1) = G_i(\{s_i(t)\}, R), \tag{5}$$

where N_k – some neighborhood of cell k on the lattice Z_d; $\{s_k(t)\}$ is the set of cell's states within N_k, the transformation T_k result depends only on the states of elements within the neighborhood N_k (locality), R – some parameters for rules which define transformations. The collection of local transformations T_k define the global transformation G on the configuration space C which represents the states of all cells at each time moments.

3.2.1. Game 'Life' as example of CA

One of the basic examples of CA is well-known Game 'Life' by J.Conway [6, 10]). At a time t let some subset of the cells in the array are living. The living cells at time t+1 are determined by those at time t according to the following evolutionary rules (the details are posed at next section of paper). Remark that anticipation follows to the origin of equations of type:

$$S_i\ (t+1) = G_i(\{s_i(t)\},...,\{s_i(t+g(i))\}, R), \tag{6}$$

The main peculiarities of such equation is that such equations may have multivalued solutions (of course in some range of parameters). Next graph presents a case of well-developed solution multivaluedness within the model "LifeA".

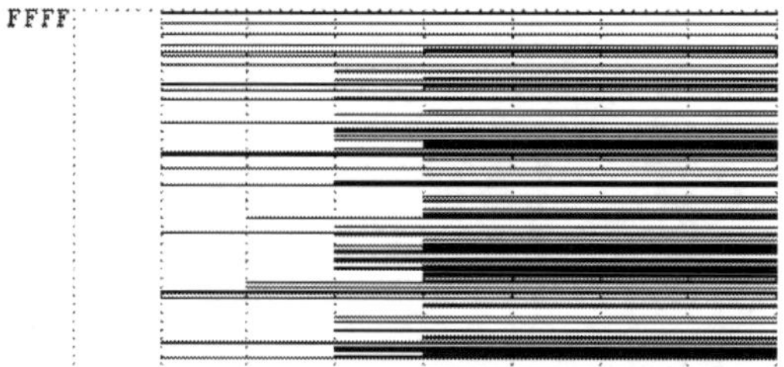

Figure 3. A large number of configurations coexisting in system (additive next state function, α = 0.5, initial state is 0000)

Such objects are very interesting and prospective for investigations of synchronizations. Remark that the main peculiarities in such investigated systems are presumable multivaluedness and space non-homogeneity of solutions behavior.

3.3. Neural networks with anticipation

Other very important object of proposed type (with anticipation and discrete time dynamic) is neural networks with anticipation. We already have investigated some such models (see [11]). Here we propose for the understanding of presumable behavior only very short description of calculations result. One of the simplest variant of such models with anticipation accounting has the form (counterpart for Hopfield's neural networks):

$$x_j(n+1) = f\left((1-\alpha)\sum_{i=1}^{N} w_{ji}x_i(n) + \alpha\sum_{i=1}^{N} w_{ji}x_i(n+1)\right) \tag{7}$$

where α is anticipation parameter. The case of $\alpha = 0$ corresponds to the absence of anticipation (with classical neuronet). Below at the picture we propose the example of presumable behavior of the model (7) at different time moments.

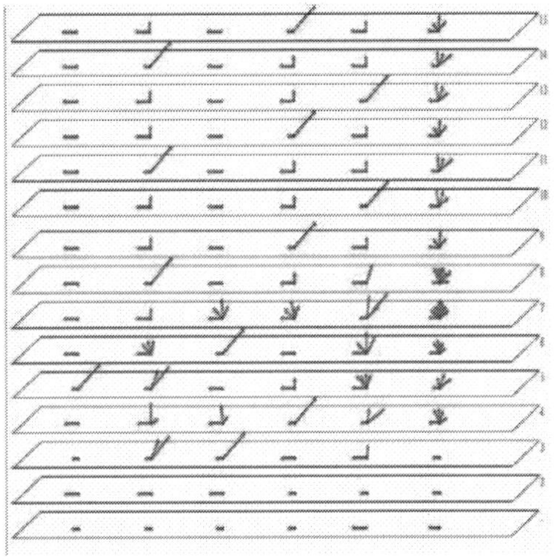

Figure 4. Behavior of the multivalued solution.

The numbers at right side of figure corresponds to the discrete tine moments. The network had 6 elements. The non-homogeneous in space solution is seen.

Two properties of artificial neural network solutions are represented at the figure 4: multivaluedness and non-homogeneities of elements behavior. Both are rather new and prospective for development of synchronization investigations as in theory as in applications.

3.4. Presumable properties of solutions and further research problems

Here we pose some comments which may be useful also for considering proposed problem. The detailed consideration of such multivalued dynamical systems may constitute the subject of further investigations. Here we only pose some general comments. Because the map acts on the multivalued functions of [0, 1] than the dynamical system is infinite-dimensional. Very important is the asymptotic solutions when $t \to \infty$. For single-valued case (without anticipation) by A. Sharkovsky and other [12] it has been found the existence of so called relaxation oscillation regimes when the solution tends to the function with finite number of discontinuities at fix length time interval and so-called turbulent solutions when the number of discontinuities tends on fix length interval tends to infinity as $t \to \infty$. Some limit objects for such solution may have fractal structure. Moreover it has been proved that the limiting regime is multivalued function for which some probability distribution corresponds. And finally very interesting for such problems is the structure of attractors in single-valued case. It had been found that attractors are infinite-dimensional and contained generalized functions (A. Sharkovsky, Yu. Maistrenko, E.Romanenko). The attractors are with simple dynamics: '… the movement on attractors is rather simple – periodical or almost periodical, also the elements of attractor are functions acting from I=[0, 1] into I, may be very complex (for example with cantorian set of multivaluedness points" [12].

Based on our results it may be proposed some generalizations for cases with accounting of anticipating. At first as we described above the solutions in such case may be multivalued. So the fractal dimension of solutions may be bigger then that in single-valued case. Moreover the behavior of the solutions may be much more complex. Different regimes on the different branches of solutions may exist (some examples of such different branches see in [9]). So each branch may have different stochastic properties, different types of relaxation behavior etc. But may be the most interesting is the common behavior of mixture of the branches in multivalued case. Especially interesting is the problem of limiting behavior as $t \to \infty$: limit solutions, attractors of such solutions, probability properties of limiting objects. But also the new problems arrows: observability of solutions, selection the single-valued trajectories of the system, its limiting objects complexity and such complexity measures, searching adequate mathematical spaces for considering the problems and solutions. Just the problems of adequate definition of periodic behavior (and moreover of dynamical chaos), Lemeray's staircase, Poincare's bifurcation diagram, ergodicity, mixing are interesting.

Conclusion

Thus in given paper we propose to consider the problem of complex behavior and self-organization for new class of objects namely for the chains and networks with strong anticipation. The presumable multivaluedness of solutions follows to new interesting properties in the frame of already existing concepts. But also new features can appear (for example non-heterogeneous multivalued solutions) which are very prospect for further investigation. We described only the first results of investigations and only some new presumable forms of research problems. But evident mathematical novelty of proposed problems and presumable great importance in applications (for examples for social systems; computation and signal processing theory; consciousness investigations etc.) follows to needs of further development of such investigations.

References:

[1] Danilenko V, Danevich T, Makarenko A, Skurativskyi S and Vladimirov V 2011 *Self-organization in nonlocal non-equilibrium media.* (Kyiv: S.I. Subbotin Institute of Geophysics).
[2] *Anticipator behavior in adaptive learning systems: from brains to individual and social behavior* 2007 ed M Butz, Sigaud O, Pezzulo G, Baldassare G (Berlin-Heidelberg: Springer-Verlag)
[3] Dubois D 1997 *BioSystems* **43** 97
[4] Dubois D 1998 *Int. J. of Computing Anticipatory Systems* **2** 3
[5] Dubois D 2001 *AIP Conference Proceedings* **573** 437
[6] Krushinskiy D, Makarenko A 2010 *AIP Conference Proceedings* **1303** 246
[7] Makarenko A 2002 *Int. J. of Computing Anticipatory Systems* **13** 77
[8] Makarenko A 2004 *Int. J. of Computing Anticipatory Systems* **15**, 217
[9] Makarenko A, Stashenko A 2006 *AIP Conference Proceedings* .**839** 265
[10] Wolfram S 2002 *New kind of science.* (USA: Wolfram Media Inc.)
[11] Makarenko A 2011 Neural networks with anticipation and some problems of complexity theory. *Proc. of Int. Conf. on Complex Systems: Synergy of Control, Communications and Computing – COSY 2011, Ohrid, Macedonia, 2011* pp. 257 – 262
[12] Scharkovski A, Maystrenko Yu and Romanenko E 1993 *Discrete equations and their applications.* (Dordrecht: Kluwer)

SPMCS 2012 IOP Publishing
Journal of Physics: Conference Series **394** (2012) 012025 doi:10.1088/1742-6596/394/1/012025

Role of Riemann's and Goldbach's hypotheses in the behaviour of complex systems: *Introduction to the concept of "Sciances"*

A Le Méhauté[1,2] and D Tayurskii[1]

[1] Institute of Physics, Kazan Federal University, Kremlevskaya 18 Kazan 420008 Russia
[2] Institut Franco-Québécois-Paris, 41 Rue Jacob 75 006 Paris France

E-mail: alm@materialsdesign.com

Abstract. The authors have already established a bi univocal correspondence between Riemann zeta functions and dynamic processes under the control of integro-differential operator of non-integer complex order. We recall that the Riemann zeta function can then be related to hyperbolic geodesics whose angles at the boundary are determined by the real part of the power laws that define the Riemann series. It is suggested that Riemann's conjecture can be reduced to a geometrical phase transition with a reduction of the parameter of order resulting from the combination of a pair symmetries associated with a quasi-self similarity of geodetics. The well-known relationship with the set of prime numbers must be considered as the result of the local existence of stationary 'state' in the dynamics.
The work is focused on the 'non stationary' behaviour of Riemann zeta function. It is shown that the main characteristic of the dynamics of complex systems may be associated to a hybridizing between a pair of states and/or processes able to give a geometrical status to the concept of the time and equally to the concept of energy. It is based on the mirror properties of complementary zeta function. It is shown also that the set of prime numbers, which controls the transitions between 'states', is the simplest form of the complexity. This analysis suggests the existence of a mathematical relationship between Riemann's and Goldbach's hypothesis. Such relationship would be the base of an extension of the principles of the science for the analysis of the complexity. According to previous proposal we name this extension that enlarges the principles of the science: sciance with 'a'.

1. Introduction

For any 'Boeotian' a Complex System is the system with respect to which the standard reason is impotent [1,2]. Complexity filters the needs of the scientific spirit for clarity within a cloud of confusion, uncertainty and, in its weak form, the randomness. The Greeks invented two approaches [3], the empiricism on the one hand (the human force for testing the beings) and on the other hand the metaphysical world of ideas (for approaching the being of the beings) because the sensitive relationship with the nature is essentially complicated but they did not imagine the complexity. In front of complexity the Greek objective posture has many weaknesses [4,5]. To overcome these weaknesses the Middle Age philosophers (Th. Aquinas, Ockham, Duns Scotus) found the sketching of the future science and the concept of modelization. The closure of our questioning by the existence of God was replaced by the transcendental ontology of the object of science. For a long time religions had searched to give meaning to a life without which it makes no sense, but as pointed by Nietzsche 'the Gods are dead in laughing' [6]. Despite a high sophistication, the hermeneutics applied to the great myths, including the myth of a pure science, the philosophical analysis did not succeed for reducing the uncertainty imposed to the human slaving, that is why from the Renaissance the thinkers attempts step by step to build the foundations for modern science based on the number and the measurement [7]. Analytic philosophy and conceptual reductionism on the one hand and experimental empiricism on the other help us to formalize in the practice the Science: a new relationship between the thinking subject and the object of its thinking. Without asserting with E. Morin that science is a source of cretinism [1] or with Bergson that science creates no future [8], we can observe the deleterious tendency of science, and especially techno sciences to enter into a dangerous ultra-specialization and the relative failure of the unification initial project for sciences [9]. The worst, in our artificial technological world, is that the conceptual limits we reach cannot be considered as methodological but epistemological [3]. This situation is infinitely deleterious with respect to the efficiency of our relationship with the world [8]. In trying to think the world as source of power, the science, and more specifically the techno-science, attempts to reduce a human to a simple 'agent of action' that cannot resist and then destroys the raft of the planet by fitting the 'quality plans' of 'competitive micro societies' put within concurrencies environment. After centuries of assurances about our ability to manage in harmony the thought and the technical power, the postmodernism [10] points out that the techno-scientific approach leads the loss of pragmatism and efficiency of the great tale of the science. Forgetting the specificities of the complexity the engineer-scientist turns the human being into the organ of self-perpetuating complicated machinery, whose teleology is reduced to the chance and/or simple local interactions. The meaning must then be suppressed from such a posture. Would it be possible to think a new scientific spirit, a new science [8], able to unlock the bar that leads to a Maelstrom? Would it be possible to balance the techno-scientific postmodern evolution and the thinking of the creative human being? Probably this issue will has to stay open even if, as it is done in this note, we try to give a scientific answer to this question by using the image of zeta function properties [11-15]. Indeed we think that this mathematical image could be a key of understanding the world of complex systems.

2. Introduction to complex systems

For ten years the Steering Committee of the French National Network on Complex Systems (RNSC) attempts to define what should be the science of the complexity. It tries to organize a scientific community around the question of a scientific access to complexity [2]. A proposal among others is that the object of this science or more precisely the approach of the complexity considered as such an object could be thought through three categories of questions:

(i) the question of the new objects characterized by multi-scale structures (Penrose-tiling, complex fractal geometry, hyperbolic, disordered networks and Gromov groups), topologies of high 'gender', living structures (cells, neurones, cities, companies, etc and their networks.)

(ii) the consideration of extreme and singular situations in experimental terms, situation concerning the treatment of huge data base, the coupling Subject-Object, the topology and the metric involves by singularities and burst of singularities, etc.

(iii) the irreducible confrontation between multiple point of views about a same object what we call a 'conceptual anisotropy'.

These proposals are obviously limited, and must be discussed within the community of Complex Systems, but it carries naturally with it many uncomfortable problems raised by the millennium confrontation between Concept and Knowledge in physics and biology. For example in physics we can look over the experimental and financial consequences of the limitations of the standard model [16], the extreme complexification of the attempts for unification of gravitation and quantum mechanics [17], the outstanding issues regarding the relevance of the concept of super-symmetry [18], the always questionable validity of the equivalence principle [17,19], etc. In this perspective we recall for example that the singularities of the equations of the general relativity theory are far from being completely known. As well, backed onto analytical philosophy, the modern biology (i) carries, far from the concept of complexity, the belief in the ability of genomic analysis of finding answers about the effectiveness of the living (why not also its meaning) and (ii) pushes implicitly a strategic of excellence leading to a dangerous eugenics. The being must then be repaired, fit like a machine and finally has to be created without defect using a perfect design. Obviously these programs are not only based on mercantile research interests on the beings. They aim also to answer to geopolitical requirements using the totalitarian phantasms of numerical controlled of the being and of the society by the mean of techno-sciences. They are also founded upon the idea that a complex system must be reduced to a complicated one, that is to say, to strictly mechanical machinery. Such a point of view ignores that the reliability of a machine is based upon the reliability of its organs and that like a living system the reliability of a Complex Systems overcomes the reliance of its components to stabilize its teleonomy via its homeostasis. The reliability of complex system must be based upon the inverse process of the one of a machine.

In contrast, it is paradoxical that with its insolence freedom, (except in financial artefact(ual) dependence [20,21]), the field of mathematics, has kept a scientific goal and an international organization according to its proximity with the philosophy. Even if the dynamic of the mathematics involves daily progresses there is no more mathematics than modern philosophy. Both are eternal because through a retro process of thinking, the object created by the mind of the thinker is the product of his thinking. This singular position of mathematics may help us to think, upon the specific basis of the singularity of the Complex systems having regard to the standard empiric. Let us also observe that a central object of mathematics is our relationship with the singularity of our position of thinker and the infinity of our environment. That could light the posture that we should have to adopt with respect to the Complex Systems, position that could justified the need for a new definition of a field of science that we suggest to call sci<u>a</u>nce [9]. This concept would be able to balance the totalitarian posture focussing on the 'bit' and the calculus. Though the generalization of freedom of scientific spirit seems always, as pointed by E. Galois having regard to the solution by radicals, the source of intrinsic 'ambiguity', the desire of freedom which is the basis for all sciences, could be to mimic the prominent efficiency of maths using the same ambiguous concerning the relationship between the subject and the object [2].

Coming back to Galois' remark, we recall that the principle of group theory applied to equation is not to examine each root, but their symmetries. This goal leads our interest to the classes of objects rather than individual objects, and to the structures of equations instead of their solutions. These principles animate the dynamics of Complex Systems. In practice as shown by the attempts of neurobiological understandings of the consciousness, the emergence of Complex System must be the result of a phase transition, a fuzzy situation of emergence that involves divergence of series over infinite number singularities. The phase transition is then led by a gain or lost of symmetries, which concerns most of the time scaling structures! But the paradigms of the standard techno-science roughly assert that the complex is mainly reducible to the mechanical entanglements that could be unravelled. Such hypothesis ignores the existence of complex knots. It also ignores the current failure of the modelization of the phase transition, the complexity of emerging phenomena such as the consciousness as well as the raising of modular structures when passing from integer automorphism to fractional ones (see [22], extension of Mandelbrot set using the automorphism $z \to z^{\alpha} + C$). It ignores the foundation of the self-organization, of the catastrophes and of the homeostasis that can be spectrally reduced only if we forget the boundary information and the long-range absence of periodicity. The state of our knowledge and the gap between these knowledge and the problems we face with Complex Systems, suggest a great modesty imposed by lake of ontology of the complex object which must be distinguished from a simple mechanical entanglement between basic components [1,3]. Nevertheless, in spite of this criticism, we believe that the major principles of science and rigor stay starting points to handle the complexity and that is why it was suggested to enlighten the complexity starting from practical scientific problems leads by zeta Riemann conjecture.

Complexity confronts out the thinking of our relationship with the nature to an epistemological difficulties leads by the role of the infinitude which is not simply combinatorial, but meta-combinatorial. How to describe this extension problem without using the words given by the standard language [4]? It is in the mathematical field that we must seek this language, not only because mathematicians are always the infinitude for purpose, but more fundamentally (i) because mathematics ignores the distinction between the subject and the object (ii) because the purpose of the mathematical research is precisely the 'brain-food' of the mathematician point of view and (iii) because the concept of forcing in set theory and of limit in categorical theory allow us to think about the extension of the thinking, we mean: the creativity. This paper will focus its proposals on an analogy based on the complexity involved by zeta Riemann function in particular because we have knowledge about its degeneracy in quantum physics [15, 23-28]. According to our proposal concerning a new approach to Riemann conjecture [29-31] we have to start from what it is probably the keystone of the problematic of complexity: the coupling between different scales.

3. Coupling within scales: Dynamics in fractal geometry

There is very interesting link between the fractional differential equations and the fractional dynamics from one side and the fractal geometry from other [31-35]. The significance of the fundamental relationship between these two branches of mathematics must relate the understanding and the mastering of what we call the space-time when the media is heterogeneous. Since the publication of the Mandelbrot's book on fractal geometries [35] the related geometrical issues about such a space-time have been extensively explored in the works done by his students, but many questions stay open and it is very interesting to note that in this matter two schools face each other:

(i) the traditional academics argue that fractal geometries do not disrupt the physical standard rules of physics. Fractal geometries are then considered as simple boundary conditions for standard and simply traditional equations [37,38], for instance, the characteristic length to scale the fractal may be the diffusion length without modification of the diffusion law. Time-space dependence of diffusion coefficient must be written using an anomalous diffusion coefficient depending on time [39];

(ii) the experts in engineering of complex systems *a contrario* assert that the convolution between scales come from the infinite behaviour, even if the scaling is not infinite and that correlations in scale contributes to the emergence of singularities of behaviour affecting deeply the physical principles themselves for instance the concept of velocity itself disappears [40-42].

Despite many experimental results that link for instance the relationship between the fractal dimension and entropy production of devices challenging nonextensive physical systems over a large range of scale the debate is unresolved. It is already unresolved in spite of many experimental data concerning the diffusion limitation in fractal media. Without going into technical details we can say that the physical standard is based on two paradigms:

(i) a time parameter must be in the set of real numbers; this parameter may disappear when it is merged with the space by using the concept of velocity [43,44] and

(ii) the homogeneity of space-time which gives rise, via Noether's theorems, to invariants such as the energy.

The both invariants of conceptual interest for this paper are the energy and his operational factotum, the velocity. It is precisely this latter that scaling coupling breaks by generalizing the harmonic form of space-time relationship. The main equation is the equation of identity: $\eta(\omega)^{d_f}(i\omega) = const$. This equation fits well the Euclidean traditional case $(d_f = 1)$ and the stochastic case $(d_f = 2)$ but it contains much more than both cases. Starting from that generalization, the concept of energy, given by a square of velocity as an invariant must be revisited in heterogeneous media. Related scientific queries lead to discover that the two paradigmatic models of the modern empiricism namely determinism and the stochastic models, are singularities into of a more general conceptualization of the exchanges of energy whose heart is none other than the class of differential equations of non-integer order [32-35]. However, these equations show, after careful analysis that standard time in the real 1D space is no less than a degenerative form of a general concept of time that must be extended all over the complex numbers [15,45]. The case of both paradigmatic models, deterministic and diffusive, appears in fact as a very specific case (although one should put a reserve for the case of Quantum Mechanics obviously related to the stochastic model [4,46,47]).

The parameterization of geodesic dynamics by complex number due to heterogeneity of the geometry is indicative of the close relationship between fractal and hyperbolic geometry. But precisely this link hence points out the issue of ambiguity in the frame of a rigorous and rational posture, that is to say the competition between many stable solutions. In many complex problems this competition explains the total absence of stable solutions even if the problem may be reduced to a deterministic geodesic on Riemann manifold. We recall that hyperbolic geometry, especially through its folding, involves multiple solutions associated a same state of definition. The usual reductionist intuition related to Euclidean and extensive systems, does not render easy the perception of the general character of such a situation. The need for thinking of linear causality confronted to the hyperbolic character of the reality may be the origin of catastrophic misunderstandings when we face complex systems (4): What is the purpose of our understanding in this case? What is a status of what we call the state of the system? What is the status of the time? To give a simple image of the ambiguity from the dynamics we suggest using the transfer function of a canonical fractional differential equation [35,48].

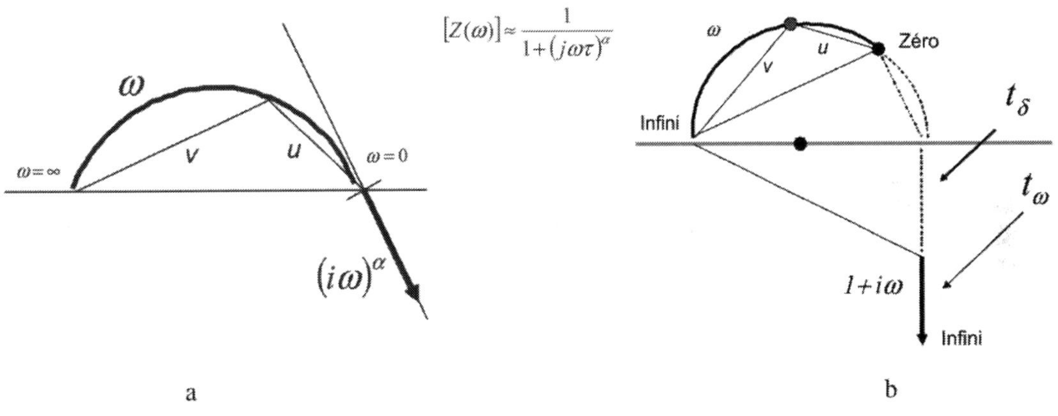

Figure 1. See the text for explanations

The figures 1a left and 1b right reveals two modalities for the conscious representation of an object (related to two types of referential) of the underlying complex phenomenon in fractal media

- The first 1a backed upon internal variables, is directly related to what Poincare called the 'analysis situs' and must be related to a fundamental automorphism domain in space-time which is a punctuated torus characterized by an angle at infinity. The automorphism is based upon a set of (i) a singular universal point at infinity in time ($\omega = 0,\infty$) and a finite distance η in space [35,53]. The relationship between the two limits, $\omega = 0,\infty$ namely the subject and the object -the sole couple of ontologic reality-, is here represented as dynamic process. This relationship points out the internal correlations of the geometry. The angle at infinity attached to the internal long-range correlations is given via the alpha parameter (see transfer equation). If the process is deterministic the space-time dimension is two, long-range correlation is given by $\alpha = 1$ and the angle at infinity is 90° (subject and object are plunged within two orthogonal space and time: Poincaré automorphism). If the process is stochastic and controlled by chance with 2D long-range interactions $\alpha = 1/2$ *the law of the great number* emerges, the space dimension *versus* time is two ($x^2 \propto t$) and the angle at infinity is 45°. Under the reserve of considering the reversibility of the time, $\frac{\partial}{\partial t} - D\frac{\partial}{\partial x} = \left(\frac{\partial^{1/2}}{\partial t^{1/2}} - \sqrt{D}\frac{\partial}{\partial x}\right)\left(\frac{\partial^{1/2}}{\partial t^{1/2}} + \sqrt{D}\frac{\partial}{\partial x}\right)$ the situation is the same and an artefact of algebra (90°=2x45°) allows a coming back to deterministic expression $\alpha \cong 1$. Based on these models with $\alpha = 1$ and $\alpha = 1/2$ the time is always immersed in the real space of number.

- The situation changes if we consider the whole field of definition of alpha and especially $1/2 < \alpha < 1$ [29,30,54]. In this case the information carried by the angle at infinity leads the entanglements between object and subject. This is due to the fact that velocity and energy lost their paradigmatic integer dimension, the related extensity of state variables and the invariance that requires the Noether's axioms [55]. These properties can be easily managed using non integer differential equation because the dimensionality of the main constant is no more a velocity or a diffusion constant (except for $\alpha = 2$) but a length to the power alpha over a standard time $x^{d_f} \propto t$. Even if the distance between object and subject stays represented by a density of action by unit of flow (resistance) the quadratic form of the energy is lost. This fundamental observation points out the artificial trick consisting of writing a constant with space time variable properties (anomalous diffusion, spectral physical dimension, etc.) conversely suggests to take into account another referential: 1b. The new referential is obtained by rotation of axis and by trapping the real axis upon the phase angle axe. The so called external representation, based on the conjecture that any quadratic form may be obtained by conjugating two fractional form (see below), restores the legitimacy to an over all 'energy'. This idea is now also suggested in the frame axiomatic thermodynamics [56]. In the case of the dynamics the time must hence be written using a complex number. The complex part takes in charge the information at infinity. Obviously the angle at the boundary of the first referential and the complex part of the time variable are related together. As the result an automorphism based on the opening of the torus at infinity leads the presence of a singular group associated to the object of the dynamics ($\omega \to 0$), which may be interpreted as an entanglement between the subject who drive the dynamics and the object passive with respect to it, as well as an entanglement between the past and the future. Time irreversibility (usually pointed out with the dissipation through the resistance) is split between the regression within the frequencies (spectral analysis) and a group of transformation at infinity re-establishing the legacy of the energy. This group being related to negentropic process might be called group of emergence, because it is the source of emerging autonomous structures for any open system exchanging energy with the outside. Let us observe that, in absence of this group at infinity the object in suspension in the space time and reduced to the asymptotic of its spectrum, may be call a 'tempo-object' [49]. More generally we suggest the reduction the analytical part of a Complex System to a 'tempo-object'. To reach its essence within the field of energy exchange with its environment the tempo-object must hence be completed by an additional part taking into account it complex component of the time and therefore of the group of transform at the boundary. The fractional dynamics being used as a model of complex issues, the duality of the representation using the time variable as a complex number to overcome the ambiguity involved at infinity by the tempo object and to trap the excess of energy within negentropic effect shows how a complex system may require some huge modification within our paradigmatic analytic posture and why it requires a hybrid point of view. This point of view makes smooth the standard distinction between the subject and the object (see for instance particle and wave as complementary infinite singularities in quantum mechanics [4]). The spectral treatment of the dynamics of complex systems, that is to say the use of a one-dimensional time, is possible but it cannot be exhaustive because it forgets the singular behaviour related to singularity at the boundary as well as the need for exchanging external energy with an environment in the absence of invariant parameters able to close the balance of energy. In the frame of non-integer dynamics the series stay diverging (tempo-object) and the renormalisation methods required to take into account the scaling properties of the object [50,57] stay enigmatic. In practice these methods play nevertheless the same role as the group at infinity (figure 1b).

If the mathematical problematic with regard to the analysis of the dynamics can be almost easily handled, the problematic of managing a complex media remains pending because we have to give physical meaning to the figure 1b. This issue explains the intuitive need of considering the local behaviour as Euclidean and, to do so, the need for the reference to quadratic forms even if the system responds to a non-extensive and hyperbolic behaviour (tree set of singularities). Let us give the example of batteries - a very complex system of storage of electric energy [51]. This question of engineering was considered for a long time, but only recently the analysis of the efficiency of batteries (generally characterized by fractal electrodes which fractal dimension is trapped all over their life), had understood why the currents required for utilizing the battery, are strongly related to the scaling properties of the electrode and why the negentropic process (the use of energy to produce work) is strongly correlated to the geometry of the heterogeneous structure. The battery use requires the optimization of both the storing of the energy (3D) and the power exchange (2D). One of the fundamental consequence of the characteristics of batteries is the fact that a Maxwell demon on a fractal surface of the electrode, is unable to measure the absolute flux of exchange required for application. This information is below his empirical horizon. This characteristic can be express within a general principle: *the local unit of time is controlled via the metric of the space and by the flow of exchange itself*. Without any knowledge about this flow we cannot define the numerical absolute value of the time constant of the process. The exchange with the environment must hence be externally normalized. The clock of the Maxwell demon is tuned by the flow to the power the fractal dimension of he geometry. Such a result shows that with respect to a process, the local time may be different of the external time unit. This general property is related to the property of the geometry at infinity and carries the same meaning that the group which must be associated with the tempo-object for reaching the essence of the complex object.

Due to the contact with the outside (open systems), that is to say, the presence of the invariant energy-momentum, the closing of the problematic onto the quadratic form of the energy (Figure 1b) requires the intervention of an additional term which is no more dissipative but organisational (long range correlations). The convolution of the harmonic behaviour with this new negentropic term is associated with the recovering of a pseudo homogeneity of the space-time, considered elsewhere in thermodynamics as a recovering of extensity [56]. Like a mirror effect it is a coupling between local and global properties and a balance between entropy and productive works (see above). The evolution of the energy balance into an auto-organisation appears as an effect associated to open boundary conditions, which has to be taken into account when tempo-object is considered. The absence of distinction between object and subject, which is the paradigm of the so-called objective knowledge, should therefore disappear for Complex Systems. Even if paradigm is questioned *mezzo voce* in quantum mechanics [4] via a uncertainty principle - to which Feynman's formulation of quantum mechanics gives precisely a geometrical meaning with 2D fractal path integral - and long distance entanglements in the field theory, this separation stays a pillar of the traditional analytic science [46,47]. The present hypothesis changes the time into a simple avatar of the energy requirement and in the absence of energy, or even of simple irreversibility, the standard time loses both its operational effectiveness and its psychological meaning. The consequence is that the time may hence be suppressed of the fundamental physics [43].

As the result even if usually it is asserted that the statistical effect is the origin of irreversibility, the statistical correlations -due to the emergence of the law of great number- usually evoked is probably only one reason among others but is clearly not the sole. The irreversibility of time, and especially the auto organization of complex systems should have a relationship with the general scaling effect even if this scaling effect is perfectly deterministic. Within a more rigorous way, we would like to assert that this affirmation could be backed upon the extraordinary mathematical tool, which is the Riemann zeta function [11-13,58-61].

4. The role of Riemann Zeta function and of the prime numbers

We recall that the Riemann zeta-function is given by the series $\zeta(s)=\sum_{1}^{\infty}\frac{1}{n^s}$ with $s=\alpha+i\theta$. It is shown that the zeroes of this function are the same as of $\chi(s)$ function with $\chi(s)=\left(1-2^{1-s}\right)\zeta(s)$ This function assures the validity of the analytic extension to **C\1**. The relationship $\pi^{-s/2}\Gamma(s/2)\zeta(s)=\pi^{-(1-s)/2}\Gamma((1-s)/2)\zeta(1-s)$ between $\zeta(s)$ et $\Gamma(s)$ which was discovered by Riemann in 1859 gave rise to his famous conjecture: *in the complex plane of parameters all zeros of zeta function that is to say* $\zeta(s)=0$ *are located on the line given by* $\alpha=1/2$ This relation comes from a mathematical relation between α and θ: $\forall\theta:\alpha=f(\theta)=const$ that cannot explained the specific value $\alpha=1/2$ With $F(s)=\left[(2\pi)^s/\pi\right]\frac{1-2^s}{1-2^{1-s}}\Gamma(1-s)\sin(\pi s/2)$ according to recent works [62] on gamma-function the relationship between $\zeta(s)$ et $\zeta(1-s)$ must be also be written by using the $\chi(s)$ function from the relation $\chi(s)=F(s)\chi(1-s)$ and $F(s)F(1-s)=1$ but due to the dual expression of s, the conjecture stays open even if the Riemann conjecture should be true if and only if the above functional relation has no other solution than the sole given by this conjecture.

Surprisingly the attempts to approach the understanding of the emergence of complex component of the time and therefore the key factor to strengthen the concept of hybrid structure of complex system opens the way to understand deeply the conjecture by the mean of a geometrical phase transition [29]. The main idea is to replace the variable α by an angle ψ, which will appear as the order parameter transition with respect to Riemann function symmetries. We can summarize this proposal within the following manner. Leaving aside the sum defining $\zeta(s)$ we can express the Riemann series within the following mathematical form ${}_0\zeta(s)\equiv\left\{1,\frac{1}{2^s},\frac{1}{3^s},\ldots,\frac{1}{n^s},\ldots\right\}$ The real part of this series suggests associating the Riemann series to the non-integer differential equation of α-order, or more precisely, to relate the ${}_0\zeta(s)$ series to Cole and Cole transfer function ${}^{\tau}Z_{\alpha}^{\rightarrow}(\omega)\propto\frac{1}{1+(i\omega\tau)^{\alpha}}$ [29,30,53] of a canonical function derived from a generalization of a first order equation. It can be shown that the Riemann series may then be considered as a discrete path 'integral' of $\alpha-$hyperbolic distance $1/\eta(\omega\tau)=\frac{1}{(\omega\tau)^{\alpha}}$ on the geodesic ${}^{\tau}Z_{\alpha}^{\rightarrow}(\omega)$ of the canonical non-integer dynamic in Fourier space

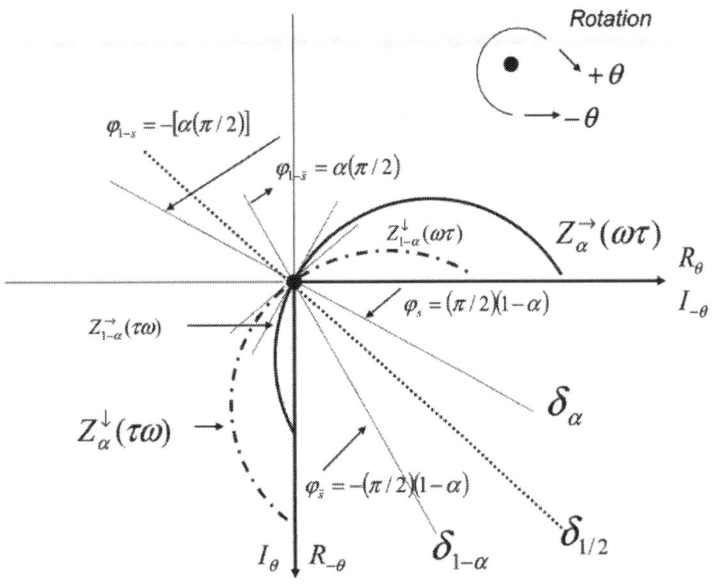

Figure 2. See explanation in the text

The homorphism between ${}^{\tau}Z^{\rightarrow}_{\alpha}(\omega)$and $\zeta(s)$ that can be generalized over the all set of Riemann function demonstrates the existence of scaling correlations between each term of the Riemann series. These correlations can be characterized by a phase angle at infinity $\varphi(\alpha)$ with $\varphi = \frac{\pi}{2}(1-\alpha)$. Due to $\zeta(s)$ and $\zeta(\bar{s})$symmetries as well symmetries of $\zeta(s)$ and $\zeta(1-s)$ the reference axis must be chosen upon $\frac{\pi}{4}$ axis and $\varphi(\alpha)$ can be replaced by $\psi(\alpha)$ with $\psi(\alpha) = \frac{\pi}{4}(2\alpha - 1)$ This relation authorizes the replacement of α by ψ in the Riemann series expression as well as in the sum; therefore ${}_0\zeta(s) = {}_0\zeta(\psi,\theta)$ With the same conventions for the rotation (Figure 2) let us observe (Figure 2) that ${}_0\zeta(\bar{s})$ can be associated with ${}^{\tau}Z^{\downarrow}_{\alpha}(\omega)$ and therefore ${}_0\zeta(\bar{s}) = {}_0\zeta(-\psi,-\theta)$ equality which leads the major following mathematical expression: $\zeta(\psi,\theta) = \zeta(-\psi,-\theta)$ We can therefore repeat the above arguing to show that ${}_0\zeta(1-s) = {}_0\zeta(\psi,-\theta)$ and ${}_0\zeta(1-\bar{s}) = {}_0\zeta(-\psi,\theta)$ therefore $\zeta(\psi,-\theta) = \zeta(-\psi,\theta)$ Let us observe that the use of $\psi(\alpha)$ as scaling invariant for all set definition of zeta functions leads a dependence between the $sign(\psi)$ and the $sign(\theta)$. This dependence is an identity for s and $\bar{s}$ and is naturally inverse for $1-s$ and $1-\bar{s}$. We can consider the Riemann Conjecture, now extended to the mathematical relation: $\forall\theta\ \zeta(\psi,\theta) = \zeta(-\psi,-\theta) = 0$ that is to say $\theta = f(\psi)$. This relation achieves a breaking of the symmetries into the set of zeta functions and creates an implicit relationship between both variable leading the mathematical relation $\forall\theta$ $\psi = const$. Nevertheless except if $\psi = 0$ this relation cannot fit any more the relations concerning the reciprocal sign relationship such as $sign(\psi) = sign(\theta)$ than $\psi(\alpha)$ must be equal to zero that is to say $\alpha = 1/2$. Obviously the same reasoning may be based from $\zeta(\psi,-\theta) = \zeta(-\psi,\theta) = 0$ Even if it is not a mathematical demonstration of the conjecture this analysis valid the Riemann Conjecture as a geometrical phase breaking of symmetries [29,30,53,62] but also unveils the role of the complementarity between $\zeta(s)$ and $\zeta(1-s)$in the emergence of stable states (related to prime number) in the spectrum of Riemann Zeta function.

As shown in figure 2, the homomorphism and the symmetries within rotation are based on the exchange of the space and time axis. This inversion leads the shift from $s \rightarrow \bar{s}$ Hence the long range geometrical correlations would be replaced by long-range correlations in time (history and anticipation). But taking $\psi(\alpha)$as reference of geometrical phase to express the set of zeta functions, the constraint $\alpha = 1/2$, appears related to standard ergodic properties of the fractional dynamics. Due to spatial mixing properties that may be expressed locally through an integer differential equation $\frac{\partial}{\partial t} - D\frac{\partial}{\partial x} = \left(\frac{\partial^{1/2}}{\partial t^{1/2}} - \sqrt{D}\frac{\partial}{\partial x}\right)\left(\frac{\partial^{1/2}}{\partial t^{1/2}} + \sqrt{D}\frac{\partial}{\partial x}\right)$the process is irreversible. Conversely for $\alpha \neq 1/2$ a phase $\psi(\alpha)$exists that cancels the commutativity of the dynamics and of the underlying $\alpha-$ Riemannian geometry. These properties associate our analysis to non-commutative geometry theory [44].

The fractional differential equation contains a long-range memory that prohibits time-space extensive properties and a local description able to be later involved within integration. These constraints are strongly related to the local irreversibility of the dynamics and its measurement [63] but nevertheless according to the fact that $\frac{\partial^{\alpha}}{\partial t^{\alpha}} * \frac{\partial^{1-\alpha}}{\partial t^{1-\alpha}} = \frac{d}{dt}$we may observe that, contingent to the right phase angle, the merging of transfer functions (Figure 2) may be useful for re-establishing pseudo extensive behaviour. Indeed above analysis leads the following assertion: *All quadratic form can be decomposed into to complementary fractional forms one of them being causal and the other a-causal*. As well, the merging of $\zeta(\psi,\theta) \cup \zeta(\psi,-\theta)$ and $\zeta(-\psi,-\theta) \cup \zeta(-\psi,\theta)$ (figure 2) should contribute to the definition of pseudo stable states associated with the 'teleonomic' like behaviour Complex Systems Issues from this analysis this assertion is the main conjecture concerning the specificity of complex systems. A certain form of causality may then be re established by considering the couple of the complementary component of the overall description (i) the objective spectral part for one side, and (ii) what it has called the 'group at infinity' that contains the long rang correlations. Let us observe that the zeroes of zeta function, we means the stable states, then disappears to be replaced by a goal (group at infinity) located below a local horizon of the physical action.

In summary, the analysis shows the existence of two type of irreversibility. The first with $\alpha = 1/2$ or$d_f = 2$ (related to the underlying 2D fractal geometry in the dynamics, with the existence of a velocity and the relevance of standard energy) is due to the loss of information within a process that may be locally time reversible and deterministic but which is globally dispersive. The second with $\alpha \neq 1/2$ or $d_f \leq 2$is characterized by a local irreversibility and unveiled by the existence of phase angle at infinity. This irreversibility leads negentropic effects able to lead emergence, auto-organisation and homeostasis. In both case the dynamics may be expressed with the same parameter t, the time, but if $\alpha \neq 1/2$ the time lost a part of its symmetries and may be expressed using a complex number. We conjecture that the specific associated irreversibility is then due to the exchange with the environment. In the case of $\alpha = 1/2$ the environment is no more than the object itself and the total dissipation of negentropic effect is therefore dissipated within entropic effect through balance energy exchanges. It is at this stage that the zeros of the Riemann function have to play a role in the differentiation between complicated systems (singular stable states) with respect to Complex Systems (dual stationary states). This balance is not in equilibrium if$d_f \neq 2$ and all specificity of complex systems [2] may appears through internal and external energy exchange process.

In spite of the limits of paradigmatic support [4] the aim of the scientific research looking for the spectrum of stable state looks like the research of the whole set of zeros of the zeta function [15, 55-

59]. We know that these zeros are related to the prime numbers. The ontological existence of these states is an indirect epistemological meaning of Riemann hypothesis. Our proposal to approach the conjecture gives the limits of this reductionist point of view. The analysis of the role of the phase upon the raising of a sign of rotation in the complex plan of parameters, that is to say the nature of the irreversibility of time, unveils the key role of hybridizing and coupling for systems strongly influenced by a memory. The image of such a system is enlighten by a general form of non-integer equation when $\alpha \neq 1/2$ and by the meaning of the merging symmetric Riemann functions By analogy, and taking into account the role of stable state as minimum of energy - that is zero value for the derivative of the energy - the properties of the Riemann zeta function suggest a radical breakthrough within the scientific paradigms able to defined a Complex Systems. Clearly to address the analytical problems, the scientist needs to deal with converging series to stable states. All means are good to go through divergences [50] and to reduce the description of the reality through his stable states, but this point of view can not be exclusive especially when interaction takes place with environment. In this situation zeta function suggests an analogy. Certain forms of stationary - we mean for living in an open system with a management of inside and outside energy - requires additional ingredient for managing the out stable sate flow of energy. The being of this flow must keep the divergences as diverging for its survival, and the state related to the minimum of energy must create a specific dynamics on time. Complex Systems behaves like a tempo object [49]. To solve its relationship with an instable environment it has to use transparency fences taking into account the correlations characterizing obviously the bulk, the properties of the boundary and the external environment. The analysis of the properties of the set of zeta function enlightens the methods to do so using the symmetries. Obviously these methods may take into account the existence of stable states somewhere but at the same time they must overcome the reductionism that could be associated to them. This method leads us to embrace the total fields of the problematic and not solely the state given by the zeros of Riemann function. The scientific meaning of the zeros appears as a step.

At this step, however, the reasoning points out an additional opportunity of progress given by the Goldbach hypothesis [64] applied upon the quadratic forms unveiled via the set of $\left\{{}^{\tau}Z^{\rightarrow}_{1/2}(\omega)\right\}$. We recall that the simplest Goldbach conjecture asserts that every even number is the sum of two prime numbers p. Today this hypothesis has not received any mathematical demonstration. However, it may be illuminated by the above approach about the degeneracy of the quadratic forms for $\alpha = 1/2$ and the associated stable states. Let us recall that the set of all primes numbers and Riemann zeta function are associated within the relation $\zeta(s) = \sum_{1}^{\infty} \frac{1}{n^s} = \prod_{p} 1/1 - \left(1/p^s\right)$. The homomorphism with the set $\left\{{}^{\tau}Z^{\rightarrow}_{1/2}(\omega)\right\}$ stays pending the role of the time constant. This time constant have to assure the incommensurability and the independence of the couple of terms involved in the procedure of merging ($\zeta(s) \cup \zeta(1-s)$) Due to the fact that we consider only the case $\alpha = 1/2$ and as noticed in chapter 5 stable states ($\omega \rightarrow 0$) are associated to un underlying geometry characterized by a space-time relationship based on diffusive equation that is to say $\eta^2 \propto \tau$ Combined with Pythagoras theorem, see figure 2, this relation leads the Goldbach relation between prime number:, $\tau^{\rightarrow} + \tau^{\downarrow} = 2\tau$. This is therefore a heuristic approach of Goldbach hypothesis, which appears strongly related to the spectrum of zeros of Riemann function.

5. Conclusion

The results obtained during last 5 years about the relationship between Zeta function, fractional differential equations, α – Riemann Manifold, irreversibility and complex systems can be summarized as follows. The Riemann's conjecture that we consider under the angle associated to the topology and to the differential equations of non-integer complex order, must be regarded as a geometrical phase transition leads by the order parameter which is the phase angle $\psi(\alpha)$ of the dynamics at infinity. This transition contributes to the stabilisation of all possible stable solutions of dynamic problem controlled by local Euclidean geometry of the space-time. That why some scientists look for the solution of the conjecture analysing the spectrum of stochastic matrix [27,28] or of fractal string structures [15]. However, due to the limited information contains at the boundary when $\alpha = 1/2$ (open systems with simple mirror effect at the boundary) these solutions only gain a physical Noetherian meaning (recovery of fundamental invariants such as energy relevance of the concept of standard velocity as a ratio between space and time etc.). The system may be complicated but never complex; a spectral analysis is sufficient to unveil the ontology of the stable states and associated objects. Due to the role of the major information at the boundary Complex Systems requires more advanced methods of approach. The simplest one requires the closing within a quadratic form using a couple of complementary operator to impose an asymmetry of time and to plunge this parameter in the Complex space. This time asymmetry is surprisingly related to the fact that the stable solutions of the zeta function require to take into account whether the angle theta is run positively or negatively (fundamental anisotropy of the complex plane is introduced by the group of tempo object). Here we find the properties of a non-commutative underlying $\alpha -$ geometry. The specific role of transcendence number $\sqrt{2}$ and intervention Goldbach's conjecture when $\alpha = 1/2$ are pointed out. These properties have probably very strong correlation with non-commutative geometry [44] but also with $\alpha -$ Gromov group [65, 68]

Under this perspective, we suggest to look the Complex Systems as a system that is able to survive only as hybrid open structure represented by a merging of a couple of system associated to $\zeta(s) \cup \zeta(1-s)$, the internal pseudo stable states then emerging being stable if and only if the boundary properties, that contain information and energy, are balance by a internal flow of energy related to auto organisation homeostasis or negentropic effects, or in other word by a complimentarily local and holistic joined attitude. Obviously in this case the standard paradigms of the science and especially the strict separation between the object and the subject being cancelled an objective posture may again be designed. We had called this new strategy with respect to Complex System a Sci**a**nce [9]. By the way Michel Bitbol in his book [9] about *the blinding near real: anti realism and quasi realism*, analyses in detail these problems already emerging in the quantum mechanics. But quantum mechanics stays inside the science because the Noether axioms keep their meanings. It is no more the case for instance for living structure and a cell for instance or a brain cannot be reduced to their chemical components. What is need in addition is still unknown but it is sure that it is related to their relationships with its boundary. We though that above analysis and especially the spectral concept of tempo object and group at infinity may help us to unveil a part of the mysteries still hidden by complexity of the Complex Systems

References

[1] Morin E 2005 *Introduction à la pensée complexe* (Paris: Seuil)
Morin E 2008 *On Complexity* Hampton Press
[2] RNSC 2010 *Complex Systems Roadmap* (Paris:Cargèse)
Perrier E 2013 *Panorama des Systèmes Complexes* (Paris: Belin) to be published
[3] Châtelet F 1999-2000 *Histoire de la philosophie* 8 tomes (Paris: Hachette-Pluriel)
[4] Bitbol M 1998 *L'aveuglante proximité du réel* (Paris: Flammarion)
[5] Kuhn T S 1962 *The Structure of Scientific Revolutions* (Chicago: University of Chicago Press)
[6] Nietzsche F 1974 *The Gay Science: With a Prelude in Rhymes and an Appendix of Songs* (New York: Vintage)
[7] Bouveresse J 1995-2010 *Lectures Collège de France 'Philosophie du langage et de la connaissance'* (Paris)
[8] Fagot-Largeault A 2008 *Lectures Collège de France 'Ontologie du devenir'* (Paris)

[9] Storck A and Le Méhauté A 2009 *La Revue des Sciences de Gestion*, N° 235
[10] Lyotard J F 1979 *La Condition postmoderne : rapport sur le savoir* (Paris: Minuit)
[11] Riemann B 1859 *Monastsber. Berlin Akad.* 671
Riemann B 1876 *Gesammelle mathematische Werke* (Hrsg. von H Weber und R Dedekind: Leipzig)
[12] Edwards H M 1974 *Riemann's zeta function* (New Yors: Academic Press)
[13] Ivic A 2003 *The Riemann Zeta-Function - Theory and Applications* (Dover Publications)
[14] Ellison X and Mendès France M 1975 *Les nombres premiers* (Paris:Hermann)
[15] Lapidus M L and van Frankenhuijsen M 1999 *Fractal geometry and number theory* (Birkhäuser)
Lapidus M L and van Frankenhuijsen M 2012 *Fractal geometry, Complex dimension, and zeta function* (Springer)
[16] Iliopoulos J 2007 *Pour La Science* N° 361 90
[17] Damour T and Polyakov A M 1994 *Nucl. Phys. B* **423** 532
Damour T, Piazza F and Veneziano G 2002 *Phys. Rev. Lett.* **89** 081601
Damour T and Donoghue J F 2010 *Phys Rev. D* **82** 084033
[18] Binétruy P 2006 *Supersymmetry* (Oxford University Press)
[19] Taylor J H and Weisberg J M 1982 *The Astrophysical Journal* **253** 908
[20] Mandelbrot B 2009 *Fractal, hazard et finance* (Paris: Flammarion)
[21] Bouchaud J P 2008 *Nature* **455** 1181
[22] *Géométrie Fractale, Dissipation, et Auto-Organisation: Distribution d'Energie dans les Systèmes Complexes* 1992 Proceeding Ecole d'été sur la croissance des systèmes biologiques Genève Septembre 1992
[23] Selberg A 1956 *J. Indian Math. Soc.* **20** 47
[24] Neudecker W and Williams W 1974 *Compositio Mathematica* **29** 197
[25] de Branges de Burcia L 1986 *Bull. of American Math. Society* **15** 1
[26] Keating J P and Snaith N C 2000 *Commun. Math. Physics* **214** 57
[27] Granville A 2010 *Prime Possibilities and Quantum Chaos* Mathematical Science Research Institute, 30 October 2010
http://www.msri.org/communications/emissary/pdfs/EmissarySpring02.pdf
[28] Baladi V, Bogomalny E, Colin Y, Combescure M, Le Drappier F, Vosoy A 2005 *Resonance periodic orbits spectrum and Zéta function in Quantum and Classical Chaos June 27 –July 5, 2005*
Faure F 2005 *Resonance periodic orbits spectrum and Zéta function in Quantum and Classical Chaos June 27 –July 5, 2005*
[29] Le Méhauté A R P, El Kaabouchi A and Nivanen L 2008 *Chaos, Solitons and Fractals* **36**(4) 659
[30] Le Méhauté A R P, El Kaabouchi A and Nivanen L 2010 *Computers & Mathematics with Applications* **59** 1610
[31] Nigmatullin R R 2012 *Nonlinear Dynamics* to be published
[32] Oldham K B and Spanier J 1974 *The fractional calculus and its applications* (New York: Academic Press)
[33] Nigmatullin R R 1992 *Theoretical and Mathematical Physics* **90** 242
[34] Samko S, Kilbas A A and Marichev O 2003 *Fractional Integrals and Derivatives: Theory and Applications* (Taylor & Francis Books)
[35] *Fractional Dérivation and Applications (FDA)*, Proceedings of FDA Bordeaux Meeting July 2005 Editors Le Méhauté A, Machado A, Sabatier J, Trigeassou J-C (E Book Lisbonne)
[36] Mandelbrot B 1977 *Fractal, Form Chance and Dimension* (San Francisco: Freeman)
Mandelbrot B 1982 *The fractal geometry of nature* (San Francisco: Freeman)
[37] Peitgen H O, Jürgens H and Saupe D 1992 *Chaos and fractals, new frontiers in science* (Springer-Verlag)

[38] Sapoval B 1997 *Universalités et fractales* (Paris: Flammarion)
[39] Havlin B-A 2000 *Diffusion and Reactions in Fractals and Disordered Systems* (Cambridge University Press)
[40] Oustaloup A 1983 *Systemes asservis d'ordre fractionnaire* (Paris: Masson)
Oustaloup A 1991 *La commande CRONE* (Paris: Hermes)
[41] Le Méhauté A 1991 *Fractal Geometry and Applications* (London: Penton Press)
Le Méhauté A, Nivanen L and R. Nigmatullin 1998 *Flèche du Temps et Géométrie Fractale*, (Paris: Hermes)
[42] Le Méhauté A R P and G. Crépy 1983 *Solid State Ionics* **9&10** 17
Le Méhauté A 1984 *J. Stat. Phys.* **36** 665
Le Méhauté A 1990 *New J. of Chem.* **14** 207
[43] Rovelli C 2012 *Et si le temps n'existait pas?* (Paris: Dunod)
[44] Connes A 1990 *Géométrie non Commutative* (Paris: InterEditions) ;
[45] Nigmatullin R R and Le Mehaute A 2005 *Journal of Non-Crystalline Solids* **351** 2888
[46] Laskin N 2000 *Phys. Lett.* **268A** 298
[47] Laskin N 2002 *Phys. Rev. E* **66** 056108
[48] Le Méhauté A, de Guibert A, Delaye A and Fillipi M 1982 *C. R. Acad. Sci. Ser. II* **294** 865
[49] Granel G 1968 *Le sens du temps et de la perception chez Husserl* (Paris: Gallimard)
[50] Laguës M and Lesne A 2008 *Invariances d'échelle* (Paris: Belin)
[51] Le Méhauté A and Dugast A 1983 *J. Power Sources* **9** 359
[52] Nottale L 1998 *La relativité dans tous ses états* (Paris: Hachette)
[53] Bedford T, Keane M and Series C 1991 *Ergodic theory, symbolic dynamics and hyperbolic spaces* (Oxford University Press)
[54] Le Méhauté A and Tayurskii D 2013 *Proceedings of the meeting on Fractional Derivation and Applications, Grenoble France 2013* to be published
[55] Kosmann-Scharzbach Y 2004 *Les theorems de Noether : Invariance et lois de conservation au 20ème siècle* (Paris: Ecole Polytechnique)
[56] Badiali J P and El Kaabouchi A 2013 *Entropy in non traditional systems*, to be published
[57] Connes A 2002 *Progress in Mathematical Physics (Poincaré Seminar 2002)* **30** 241
[58] Berry M V and Keating J P 1999 *SIAM Rev. Lett.* 236
[59] Berry M V 1984 *Proc. Roy. Soc. London A* **392** 45
[60] Elzalde E, Odintsov S D, Romeo A and Zerbini S 1994 *Zéta regularization techniques with applications* (Singapore: Word Scientific)
[61] Hawking S 1977 *Commun. Math. Physics* **55**(2) 133
[62] Nigmatullin R R and Le Méhauté A 2013 work in progress
[63] Tricot C 1999 *Courbes et dimension fractale* (Berlin : Springer-Verlag)
Tricot C 2008 *Géometrie et mesure fractale : une introduction* (Paris : Ellipses)
[64] Chudakov N G 1937 *Doklady Akademii Nauk SSSR* **17** 1937, p. 335–338
[65] Ghys E and de la Harpe P 1990 *Sur les groupes hyperboliques d'apres Mikhael Gromov* (Boston : Birkhäuser)
[66] Gromov M 1976 *Three remarks on geodesic dynamics and fundamental group* Preprint SUNY [reprinted in 2000 *L'Enseignement Mathématique* **46** 391
Gromov M 1988 *Essays in Group Theory* (Mathematical Sciences Research Institute Publications) **8** 75 (Berlin: Springer-Verlag)
[67] Gromov M 1991 *Rend. Semin. Mat. Fis. Milano* **61** 9
Gromov M 1993 *Lond. Math. Soc Lecture Notes* **182** 1
[68] Gromov M 2000 in *Visions in Mathematics, GAFA 2000 Special Volume, Part 1* 118 (Basel: Birkhäuser)

Conductivity in disordered structures: Verification of the generalized Jonscher's law on experimental data

I I Popov, R R Nigmatullin, A A Khamzin and I V Lounev
Institute of Physics, Kazan (Volga Region) Federal University, 420008, Kremlevskaya str.18, Kazan, Tatarstan, Russia

E-mail: ii.popov@yandex.ru

Abstract. The generalized Jonscher's law for complex conductivity (derived earlier) is tested on available experimental data. We suggest some criteria which are used for verification of the analytical expression describing the data related to complex conductivity. It is shown that the generalized Jonscher's law is suitable for description of the electrode polarization phenomenon.

1. Introduction

In paper [1] we derived the analytical expression generalizing the well-known Jonscher's law. This formula describes the contribution of the different charge carriers to the complex conductivity in the frequency region. In [1] the dependence of the complex conductivity against frequency was defined as the generalized Jonsher's law (GJ'sL). This dependence described well the complex conductivity data of the sodium nitrite embedded into the porous glass having mean pore size about 7 nm [2-4]. At derivation of the GJ'sL we did not use any specific model associated with concrete substance. The basic point was associated with the supposition that the substance has a self-similar structure [5-8]. This general supposition leads us to the idea that the GJ'sL should take place in a wide class of different heterogeneous substances. This fact, in turn, forced the authors of this paper to analyze attentively the scientific papers where the reliable data of the complex conductivity (or the complex dielectric permittivity (CDP)) were presented. The analysis of these frequency data could confirm or reject the "universality" of the GJ'sL put forward earlier as an alternative hypothesis. In this work we concentrate our attention only on analysis of experimental data of complex conductivity.

The basic aim of this paper is the verification of the GJ'sL on available experimental data that describe the behaviour of the complex conductivity against frequency.

2. The specific criteria of the generalized Jonscher's law that can be recognized in measured data

Preliminarily we reproduce the basic steps that led us to the analytical expression describing the GJ'sL [1]. Alongside with usual current associated with free carriers $\mathbf{j}_0 = \sigma_0 \mathbf{E}$ and polarization current $\mathbf{j}_p = \partial \mathbf{P} / \partial t$ we determine the so-called "fractal" current $\mathbf{j}_{frac} = \sigma_{frac}(\omega)\mathbf{E}$. In result the total current is determined by expression

$$\mathbf{j}_{tot} = \sigma_{tot}(\omega)\mathbf{E} = \mathbf{j}_0 + \mathbf{j}_p + \mathbf{j}_{frac} = \left(\sigma_0 + \frac{i\omega}{4\pi}(\varepsilon(\omega) - 1)\varepsilon_v + \sigma_{frac}(\omega) \right)\mathbf{E}, \quad (1)$$

here the value $\varepsilon_v = 8.887 \ 10^{-13}$ F/m - determines the dielectric permittivity of the vacuum. If one takes into account the contribution of the connected charges $\varepsilon_v(\varepsilon_\infty - 1)i\omega/4\pi$ then instead of (1) we obtain finally $\sigma_{tot}(\omega) = \sigma(\omega) + \varepsilon_v(\varepsilon_\infty - 1)i\omega/4\pi$ [9]

$$\sigma(\omega) = \sigma_0 + i\omega(\varepsilon(\omega) - \varepsilon_\infty) + \sigma_{frac}(\omega). \tag{2}$$

Here we normalized the expression $\left[\sigma(\omega) - \sigma_0 - \sigma_{frac}(\omega)\right]$ on the constant $\varepsilon_v/4\pi$. The fractal current $\mathbf{j}_{frac}$ figuring in (1) describes the slow/retarding dynamics of the carriers that is created by the fractal structure of the medium considered. It is defined by the following expression [1]

$$\mathbf{j}_{frac} = \sigma_{frac}(\omega)\mathbf{E} = {}_{-\infty}D_t^{\nu}\mathbf{P}_2, \ 0 < \nu < 1, \tag{3}$$

where the fractional derivative ${}_{-\infty}D_t^{\nu}$ [10] is taken from the delayed part of the polarization $\mathbf{P}_2(t)$. It is defined by expression [1]

$$P_2(t) = \frac{\chi E(t)}{1 + (i\omega\tau)^{\nu}}, \ E(t) = E_0 \exp(i \ t). \tag{4}$$

Here the value χ defines the dielectric susceptibility. Inserting (4) into (3) we obtain

$$\sigma_{frac}(\omega) = \frac{\chi\tau^{-\nu}}{1 + (i\omega\tau)^{-\nu}}, \ 0 < \nu \ \ 1. \tag{5}$$

In the result the total expression for complex conductivity accepts the form

$$\sigma(\omega) = \sigma_0 + i\omega(\varepsilon(\omega) - \varepsilon_\infty) + \frac{\chi\tau^{-\nu}}{1 + (i\omega\tau)^{-\nu}}. \tag{6}$$

In the case of small frequencies the fractal conductivity from (5) reproduces the conventional Jonscher's law

$$\sigma_{frac}(\omega) \underset{\omega \to 0}{\approx} \chi(i\omega)^{\nu}. \tag{7}$$

The limiting expression (7) coincides with results of the paper [11]. In this paper it was shown that the classical correction (7) (suggested in [12]) coincides with the low-frequency solution of the functional equation that describes the self-similar electric circuit. This circuit takes into account the influence of the fractal surface of electrodes.

Having expression (6) on can develop some important criterion for its checking. This criterion helps us to make the statement that the relationship associated with the classical correction in the form

$$\sigma_J(\omega) = \sigma_0 + i\omega(\varepsilon_J(\omega) - \varepsilon_\infty) + \tau^{\alpha-1}(i\omega)^{\alpha}, \tag{8}$$

is not realized in the data considered while the relationship (6) is the most probable and confirmed on the available data considered.

We consider only the real part of these expressions because this part is measured presumably in experiments. If the conductivity associated with polarization part is absent (the second terms in (6) and (8) are equaled zero) then for expression (8) we have usual power-law dependence. For the function (6) we have increasing function forming a plateau in high-frequency region. In the case of the presence of polarization part the real parts of both functions are increasing sharply at high-frequency region. Figure 1 demonstrates the dependencies of the real parts of expressions (6) and (8) in the absence/presence of polarization conductivity.

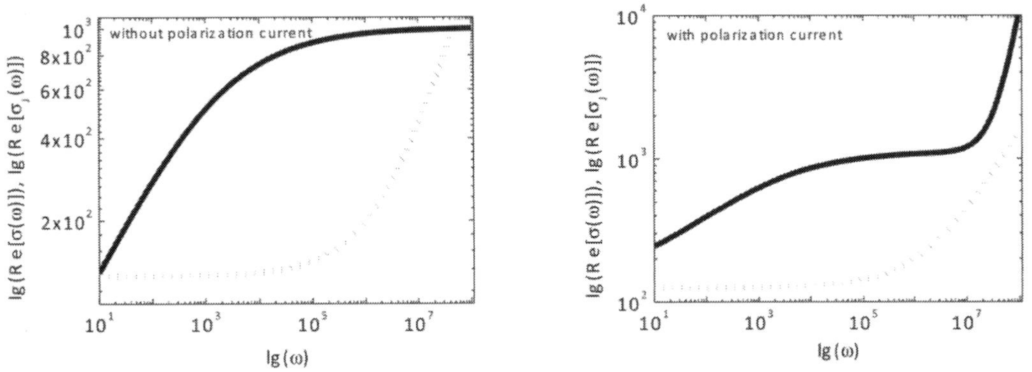

Figure 1. The model data for the real parts of the functions (6) and (8). As before the bold black line corresponds to function (6) the dotted-line corresponds to function (8). On the left we have two plots in the absence of the polarization current, on the right we have the plots when the contribution from the polarization current is taken into account. We used the following values of the fitting parameters: for function (6) $\sigma_0 = 10^1$, $\nu = 0.4$, $\chi = 10^3$, $\lg\tau = -3$; for function (8) $\sigma_0 = 10^{2.1}$, $\alpha = 0.65$, $\lg\tau = 5$.

From this figure one can notice that the functions (6) and (8) have different behavior in the typical frequency range 10-10^8 Hz.

3. Verification of the generalized Jonscher's law on real data

We think that the conditions of the fractal current appearance take place for example near the heterogeneous surface electrodes having self-similar (fractal) structure. This idea, as mentioned above, was realized in [11]. The same conditions can be realized also in materials having branching porous structure. So, one can expect that the dependence (6) should be closely associated with the electrode polarization phenomenon. In literature there are a lot of experimental papers that studied this phenomenon. So we have a good possibility to confirm/reject this new dependence (6) that can pretend on generalization of the conventional Jonscher's law for complex conductivity.

Let us start from data of the paper [13]. Experimental data together with its fit to expression (6) is presented by Fig. 2.

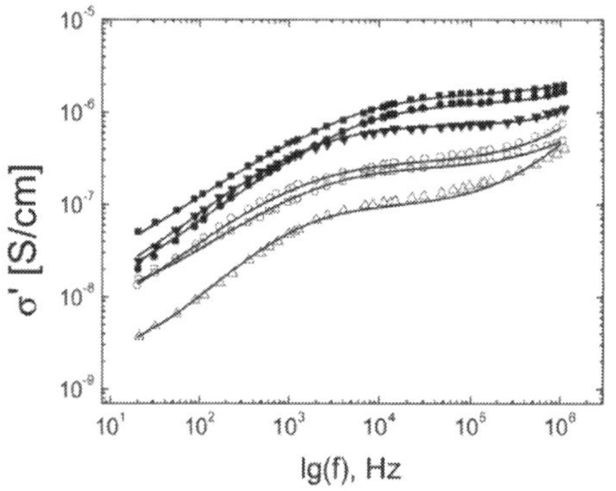

Figure 2. Frequency dependence of real part of ac conductivity σ' of the $(PEO)_{20}$-($LiClO4.3H_2O+NaClO_4.H_2O$)-x wt% MMT clay for nanocomposite electrolytes synthesized by melt compounded technique (see [13] for details). The solid lines correspond to the fit of data to modified expression (6).

As one can see the real part of conductivity has clearly expressed bend appearing at low-frequencies, which cannot be obtained from equation (8). In original paper [13] the authors can fitted only the high-frequency part of the curve in the real part of conductivity based on the conventional Jonscher's correction. Their fit was not given here. Solid fitting lines presented on figure 2 are realized in the correspondence with expression (6) which describes well all these data..

Let us consider other data that are given in paper [14]. In this paper the authors studied the same substance as it was used in [13] but in another experimental conditions. Experimental data (triangle points) are presented in Fig. 3.

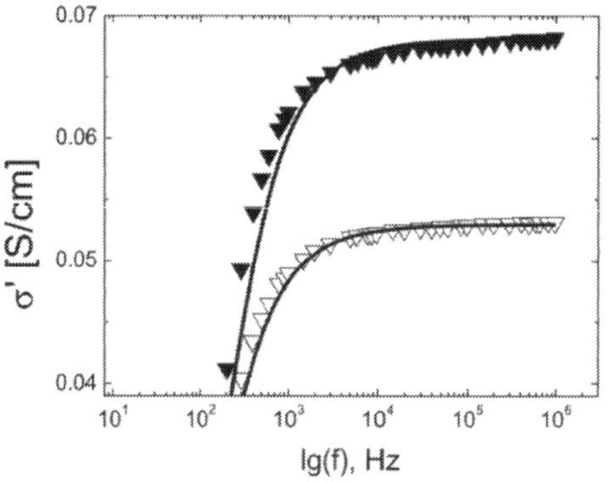

Figure 3. Frequency dependence of real part of ac conductivity σ' of PVA-PEO blend -MMT clay nanocomposites hydrocolloids at varying MMT clay concentration (wt.%) (see [14] for details). Again, the solid lines correspond to the fit to expression (6).

Here we consider the data corresponding to two values of clay concentrations (wt.%) because other data do not have any peculiarities from the curves presented here. As one can notice from these figures the bend is exist at low-frequencies as it is shown on model data on fig 1 and so one can assert that these data confirm the validity of the GJ'sL. As an additional confirmation on Fig. 3 we give also the fitting curves (solid lines) corresponding to expression (6).

Below we will demonstrate only the *qualitative* behavior of the curves that correspond to the electrode polarization phenomenon leaving aside the direct fit of the data. Figures 4 and 5 reproduce a typical behavior of the real parts of the complex conductivity and permittivity that were taken from paper [15] for some conducting glasses. The behavior of experimental data of conductivity on these figure is like as on figure 1. The behavior of dielectric permittivity one can explain in frame of this approach that was be done in [1]. So, one can conclude that for the most experimental data we observe again the realizability of the GJ'sL that is closely associated with electrode polarization phenomenon.

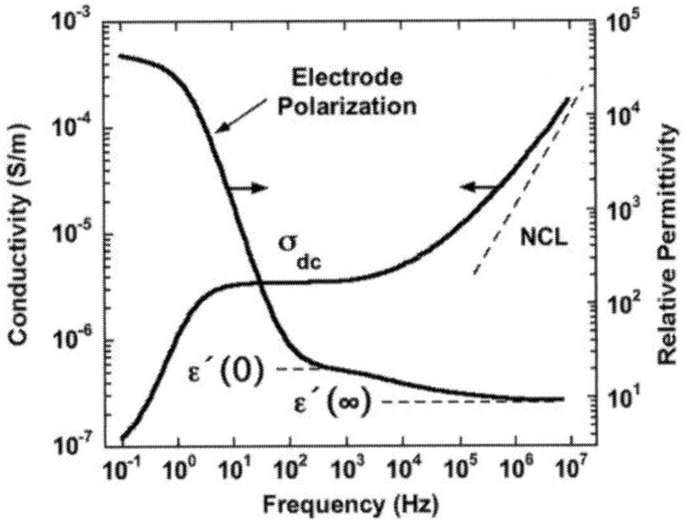

Figure 4. The typical behavior that is observed for the electrode polarization phenomenon for the real part of the conductivity and the real part of the dielectric permittivity at high temperatures for a lithium-phosphate glass (see [15] for details).

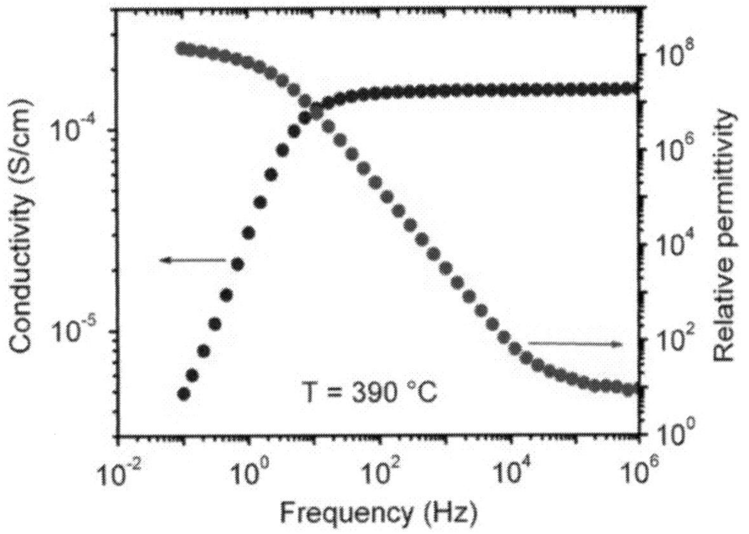

Figure 5. The typical electrode polarization phenomenon that is revealed itself in behavior of the real part of the conductivity and the real part of the dielectric permittivity at high temperatures for a Na–Ca–phosphosilicate glass (see [15] for details).

4. Basic conclusions

In spite of the fact that the GJ'sL is realized in many heterogeneous and disordered substances we should remark that this law is not universal. Not all data can be described by expression (6). Formula (6) always contains the character slope at low-frequency range that is not observed for all cases analyzed. The contrary examples are collected in papers [15, 16]. Probably, in such cases it is necessary to use models that have been suggested earlier. Concerning the electrode polarization effect the formula (6) proved their applicability and we can conclude that the GJ'sL was totally justified itself in description of this phenomenon.

References

[1] Popov I.I., Nigmatullin R.R., Koroleva E.Yu., Nabereznov A.A. 2012 *J. non. Cryst. Solids* **358**, 1
[2] Kutnjak Z., Vodopivec B., Blinc R., Fokin A.V.,. Kumzerov Yu. A, Vakhrushev S.B. 2005 *J. Chem. Phys.*, **123** 084708
[3] Naberezhnov A., Fokin A., Kumzerov Yu., Sotnikov A., Vakhrushev S., Dorner B. 2003 *Eur. Phys. Jour. E*, **12** 21
[4] Vakhrushev S.B., Koroleva E.Yu., Kumzerov Yu.A., Naberezhnov A.A., Fokin A.V., Korotkov L.N., Tovar E.V. 2006 *NanoTekhnika*, **1** 18. (in Russian)
[5] Le Mehaute, Nigmatullin R.R., Nivanen L. 1993 "*Fleches du Temps et Geometrie Fractale*" Paris, Hermez, (in French)
[6] Nigmatullin R.R. 2005 *Physica B: Physics of Condensed Matter*, **358** 201
[7] Nigmatullin R.R. 2006 *Physica A*, **363** 282
[8] Nigmatullin R.R., A. Le Mehaute 2005 *J. Non-Cryst. Solids*, **351** 288
[9] Dyre J.C., Shroder T.B. 2000 *Rev Mod. Phys.*,**72** (3) 873
[10] Samko S. G., Kilbas A. A., and Marychev O. I. 1993 *Fractional,Integrals and Derivatives. Theory and Applications* (Gordon and Breach, Amsterdam)
[11] Lui S.H. 1985 *Phys.Rev.Lett.* **55** 529
[12] Jonscher A.K. 1993 *Dielectric Relaxation in Solids*, London: Chelsea Dielectric press
[13] Choudhary S., Sengwa R.J. 2011 *Indian Journal of Pure & Applied Physics*, **49** 204
[14] Sengwa R.J., Choudhary S., Sankhla S. 2009 *Indian Journal of Engineering & Materials Sciences*, **16** 395
[15] Dyre J. C., Maass P., Roling B., Sidebottom D. L. 2009 *Rep. Prog. Phys.* **72** 046501
[16] Dyre J.C., Shroder T.B. 2000 *Rev Mod. Phys.*,**72**(3) 873

Two types of adsorbed water in natural montmorillonites at low temperatures by dielectric spectroscopy

M A Vasilyeva[1], Yu A Gusev[1] and V G Shtyrlin[2]

[1] Kazan Federal University, Institute of Physics, 420008, Kremlevskaya st. 18, Kazan, Russia

[2] Kazan Federal University, A.M. Butlerov Chemistry Institute, 420008, Kremlevskaya st.18, Kazan, Russia

E-mail: mariavasilyeva@mail.ru

Abstract. Dielectric spectroscopy was applied to natural clay mineral - montmorillonite with the exchangeable K^+ and Ni^{2+} cations where the effect of water adsorption in the samples on the dielectric response was examined in the temperature from -115°C to -75°C and frequency from 1 Hz to 1 MHz ranges. Two relaxation processes (1 and 2) were revealed. The process 1 was related to the adsorbed ordered water structures formed on the interior surface of interlayer channels while the relaxation process 2 was attributed to the disordered water structures built between two surface water monolayers.

1. Introduction

Montmorillonites are widely used in industrial processes in making wine, beer, oils, molding sand, ore pelletization, petroleum, pesticides, catalysts, adsorbents, cosmetics, ceramics, paintings, etc.

Broadband dielectric spectroscopy is certainly one of the most useful tools for investigating the molecular dynamics, the degree of intermolecular hydrogen bond interaction, cooperativity between guest and host molecules [1].

There are many studies on the dielectric properties of montmorillonite [2-9] but so far there are still many open questions in the problem of the interaction of water molecules and ions with the surface of the minerals.

The objectives of this study were to use dielectric spectroscopy in the temperature from -115°C to -75°C and frequency from 1 Hz to 1 MHz ranges at high resolutions to examine the effects of two various types of cations on the existence and properties of bound water in montmorillonite mineral.

2. Samples

Oglanlinsky montmorillonite (Turkmenistan) with the exchangeable cations K^+ and Ni^{2+} were considered in this study. All the samples were prepared in the laboratory of disperse systems of Institute of Colloid and Water Chemistry (Ukraine, Kiev) according to the procedure described previously [10]. All the samples were stored in ambient air humidity, thus one can assume that water molecules were absorbed from the air into the pores.

[1] To whom any correspondence should be addressed.

3. Experimental

Dielectric measurements in the frequency range of 1 Hz - 1 MHz were performed by BDS 80 based on an Alpha Impedance Analyzer (NOVOCONTROL) with automatic temperature control by QUATRO Cryosystem with a precision of 0.5°C. The accuracy of the complex dielectric permittivity was estimated to be better than 3%. The measurements were carried out in the following way: each of the samples was placed into the sample cell with a parallel plate capacitor configuration at room temperature and the measurements were then performed by cooling the samples from 20°C to -100°C. Then the samples were measured at intervals of 3°C upon heating them from -115°C to -75°C and then heating up to +300°C. The relative water content to the dry sample was determined by weighing the samples prior to and immediately after heating up to +300°C (see table 1).The inner diameter of the sample cell for dielectric measurements was 16 mm, the minimum spacing between electrodes was 0.5 mm.

4. Results and discussion

A typical view of the 3-D plots of the dielectric losses vs. frequency and temperature is shown in figure 1.

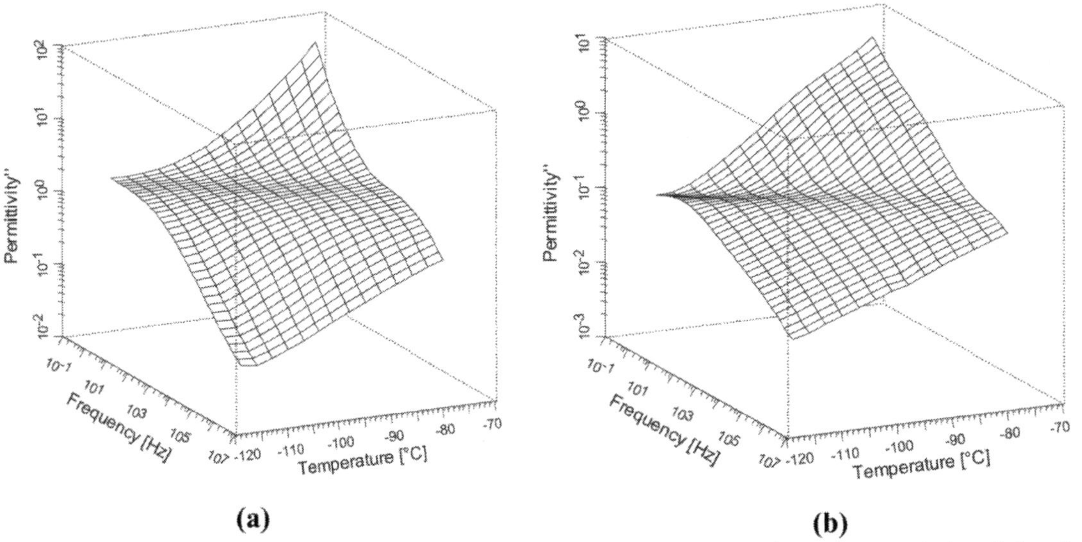

Figure 1. Three-dimensional plots of the frequency and temperature dependence of the dielectric losses for montmorillonite K^+ (a) and Ni^{2+} (b).

At low temperatures from -115°C to -75°C the spectra of montmorillonites consist of two relaxation processes marked 1 and 2 correspondingly. After annealing the samples at 300°C for 3 hours the relaxation processes are disappeared. This fact confirms our assertion that only water is the source of the dielectric relaxation of these materials.

For comparison we also measured the kaolinite samples with different exchangeable cations with the same protocol and did not find the low temperature process at all. We assume that the low temperature relaxation processes in montmorillonites can by associated with the interlayer water molecule relaxations because such water is absent in kaolinites [11-14].

For a quantitative analysis of the isothermal dielectric spectra at low temperatures a superposition of two Cole-Cole functions [15] in the form

$$\varepsilon^*(\omega) = \varepsilon_\infty + \sum_{j=1}^{2} \frac{\Delta\varepsilon_j}{1+\left(i\omega\tau_j\right)^{\alpha_j}} \qquad (1)$$

and Jonscher's empirical term, $B(i\omega)^{n-1}$ [16], has been fitted to the complex dielectric permittivity, $\varepsilon^*(\omega) = \varepsilon'(\omega) - i\varepsilon''(\omega)$. Here $\varepsilon'(\omega)$ and $\varepsilon''(\omega)$ are the real and loss parts of the complex permittivity, respectively; i is the imaginary unit and ω is the cyclic frequency, ε_∞ is the high frequency limit of the complex dielectric permittivity, $\Delta\varepsilon_j$ is the dielectric strength of the relaxation process, τ_j is a characteristic time scale and the exponent α_j is referred to as a measure of symmetrical broadening in the dielectric losses relaxation peak ($0<\alpha_j \leq 1$), n is a Jonscher parameter for the high frequency part of the respective relaxation process.

The temperature dependencies of relaxation times for the observed low temperature processes demonstrate Arrhenius behavior (figure 2). The calculated values of the activation energies of the relaxation processes are presented in table 1.

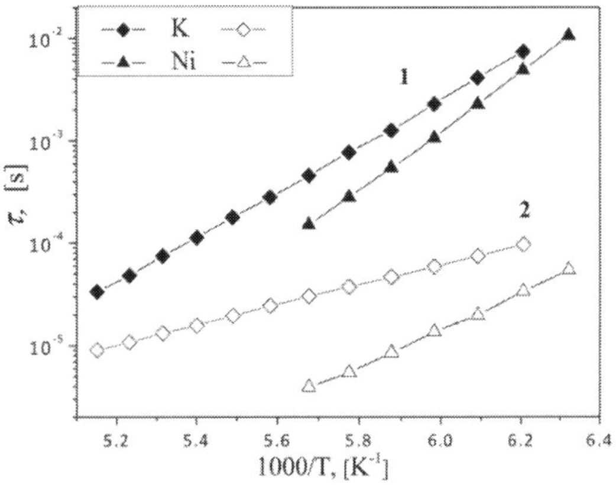

Figure 2. The temperature dependences of the relaxation times of the low temperature processes for montmorillonites. Solid symbols correspond to the 1 process; open symbols are related to the 2 process.

Table 1. The calculated values of the relative humidity (h) and activation energies of the relaxation processes (E_{a1} and E_{a2}) and the experimental values of the Gibbs free energy hydration ($\Delta_{hyd}G^o$) of the K^+ and Ni^{2+} ions [17].

	h, %	E_{a1}, kJ/mol	E_{a2}, kJ/mol	$\Delta_{hyd}G^o$, kJ/mol
Montmorillonite K	13.7	42	18	-295
Montmorillonite Ni	14.7	55	34	-1980

The activation energy values for the 1 process are in the same range that had been reported earlier for water absorption in porous silica glasses [18, 19]. In montmorillonites the water molecules responsible for the low temperature relaxations are located near the inner hydrophilic centers of interlayer channels (siloxane bonds and cations trapped in hexagonal rings) and around the interlayer cations [20]. Indeed during the hydration the water molecules interact with the exchangeable cations resulting in removing part of them from the surface with following joint motion as aqua complexes [13, 14, 21]. The slower relaxation process 1 coincides with the low temperature relaxation behavior in glass materials. Most probably, it corresponds to the adsorbed ordered water (ice-like) structures formed on the interior surface of interlayer channels while the faster relaxation process 2 reflects the disordered water structures built between two surface water monolayers. This explanation is in accordance with the Frank-Wen structural model of solvated ions [22]. According to this model a structural disordered aria of solvent (B) exists between first solvation sphere of ion (A) and bulk solution (C). In the water dielectric relaxation disordering effect of the B zone is manifested in the 2 process. Note that the presence of two layers of mobile water over the first layer of strongly adsorbed water was found by DS method for the water adsorption on α-Fe_2O_3 [23]. Disordered (weakly bound) water was revealed also in many porous materials and its content increases with decreasing the pore sizes [24].

In both processes (1 and 2) the activation energy depends on the type of interlayer cations presented in the structure of the clay and water content (see table 1). Influence of ions nature on the E_{a1} and E_{a2} values (table 1) can be explained by considering ion hydration. The E_{a1} and E_{a2} values change in the row of ions $K^+ < Ni^{2+}$ correlate with changing experimental values of the Gibbs free energy hydration ($\Delta_{hyd}G^o$) for this ions [17] (see table 1). It should be added that K^+ is the ion with "negative hydration" (life time of water molecules is less in immediate surrounding of K^+ than in bulk solution) [25, 26] and has the higher mobility in comparison with Ni^{2+}.

The detailed dielectric investigation of the hydrated clusters formation and their structures as a function of the exchangeable cations and water content in the montmorillonites is an open question for the continuation of the research.

5. Conclusions

Montmorillonites with the exchangeable K^+ and Ni^{2+} cations were examined by means of the dielectric spectroscopy in the temperature from -115°C to -75°C and frequency from 1 Hz to 1 MHz ranges. At the low temperature interval the two dielectric relaxation processes (1 and 2) were observed only for montmorillonites but not for kaolinites. The slower process 1 corresponds to the adsorbed ordered water (ice-like) structures formed on the interior surface of interlayer channels while the faster relaxation process 2 corresponds to the disordered water structures built between two surface water monolayers. The parameters of activation of the dielectric relaxation processes 1 and 2 correlate with values of the Gibbs free energy hydration of the exchangeable ions.

Acknowledgements
The authors thank Prof. Yu. Feldman, Dr. A. Greenbaum (Gutina), Dr. A. Puzenko, and Dr. P. Ben Ishai from Dielectric Spectroscopy Laboratory at Department of Applied Physics (The Hebrew University of Jerusalem, Israel) for helpful discussions.

References

[1] Kremer F and Schönhals A 2002 *Broadband Dielectric Spectroscopy* (Berlin: Springer)
[2] Mamy J 1968 *Ann. Agronomiques* **19** 175
[3] Hoekstra P and Doyle W T 1971 *J. Colloid Interface Sci.* **36** 513
[4] Calvet R 1975 *Clays Clay Miner.* **23** 257
[5] Lockhart N C 1980 *J. Colloid Interface Sci.* **74** 509
[6] Sposito G and Prost R 1982 *Chem. Rev.* **82** 553
[7] Kaviratna P D, Pinnavaia T J and Schroeder P A 1996 *J. Phys. Chem Solids* **57** 1897
[8] Ishida T, Makino T and Wang C 2000 *Clays Clay Miner.* **48** 75
[9] Rotenberg B, Cadéne A, Dufrêche J-F, Durand-Vidal S, Badot J-C and Turq P 2005 *J. Phys. Chem.* B **109** 15548
[10] Tarasevich Yu I and Ovcharenko F D 1980 *Adsorption sur des Min´eraux Argileux* (Institut Francais du P´etrole, Rueil Malmaison)
[11] Pusch R and Kamland O 1996 *Engineering Geology* **41** 73
[12] Murray H H 1999 *Clay Clay Miner.* **34** 39

[13] Tarasevich Yu I 1988 *Structure and Surface Chemistry of Layer Silicates* (Kiev: Naukova Dumka)
[14] Tarasevich Yu I and Ovcharenko F D 1975 *Adsorption on Clayey Minerals* (Kiev: Naukova Dumka)
[15] Cole K S and Cole R H 1941 *J. Chem. Phys.* **9** 341
[16] Jonscher A K 1996 *Universal Relaxation Law* (London: Chelsea Dielectrics Press) p 415
[17] Marcus Y 1991 *J. Chem. Soc. Faraday Trans.* **87** 2995
[18] Gutina A, Antropova T, Rysiakiewicz-Pasek E, Virnik K and Feldman Yu 2003 *Microporous Mesoporous Mater.* **58** 237
[19] Ryabov Y, Gutina A, Arkhipov V and Feldman Yu 2001 *J. Phys. Chem.* B **105** 1845
[20] Salles F, Devautour-Vinot S, Bildstein O, Jullien M, Maurin G, Giuntini J-C, Douillard J-M and van Damme H 2008 *J. Phys. Chem.* C **112** 14001
[21] Mank V V 1977 *Radiospectroscopic Investigations of Water Interaction Mechanism with Surface of Natural and Synthetic Ion-Exchangers* Thesis, Kiev
[22] Frank H S and Wen W Y 1957 *Disc. Faraday Soc.* **24** 133
[23] Zettlemoyer A C, Micale F J and Klier K 1975 Water: A Comprehensive Treatise *Water in Disperse Systems* v 5 ed F Franks (New York – London: Plenum Press) pp 249-291
[24] Tarasevich Yu A 2011 *Surface Phenomena on Disperse Materials* (Kiev: Naukova Dumka)
[25] Samoilov O Ya 1957 *The Structure of Aqueous Electrolyte Solutions and the Ion Hydration* (Moscow: Akad. Nauk SSSR)
[26] Krestov G A 1991 *Thermodynamics of Solvation: Solution and Dissolution, Ions and Solvents, Structure and Energetics* (New York: Horwood)

Differences in behaviour of adsorbed water in kaolinites and montmorillonites in temperature range from -90°C to +140°C by dielectric spectroscopy

M A Vasilyeva[1], Yu A Gusev[1] and V G Shtyrlin[2]

[1] Kazan Federal University, Institute of Physics, 420008, Kremlevskaya st. 18, Kazan, Russia
[2] Kazan Federal University, A.M. Butlerov Chemistry Institute, 420008, Kremlevskaya st. 18, Kazan, Russia

E-mail: mariavasilyeva@mail.ru

Abstract. Two different types of natural layer aluminosilicates (clay minerals), montmorillonite with the exchangeable K^+ and Ni^{2+} cations and kaolinite with the exchangeable K^+ and Ba^{2+} cations, were investigated by dielectric spectroscopy. The different effects of water adsorption in montmorillonites and kaolinites on the dielectric response were observed in wide temperature (-90°C ÷ +140°C) and frequency (1 Hz ÷ 1 MHz) ranges. The influence of the nature of the hydration centers on adsorbed water dynamics and influence of ions nature on the activation energy values of the relaxation processes were discussed.

1. Introduction

Kaolinites and montmorillonites are widely applied in many fields such as pharmaceuticals, ecology, hydrogeology, catalysis, and etc., due to their high specific surface area, chemical and mechanical stabilities, variety of surface and structural properties, higher values of cation exchange capacities, etc.

The physical effects related with the interaction of bound water with clay minerals play an important role in numerous technological applications. The useful information about the mechanism of the water molecules interaction with clay-contained substances on the mesoscopic level can be obtained by means of dielectric spectroscopy method. In spite of many studies on the dielectric properties of montmorillonites and kaolinites [1-8] full understanding of the adsorption mechanisms in clay materials was not achieved up to date.

The main aim of this study was to use dielectric measurements in wide temperature (-90°C ÷ +140°C) and frequency (1 Hz ÷ 1 MHz) ranges to examine the effects of the clay structure and various types of cations on the existence and properties of bound water in montmorillonites and kaolinites.

2. Samples

As objects of study two types of natural layer aluminosilicates (clay minerals), Oglanlinsky montmorillonite (Turkmenistan) with the exchangeable cations $K^+$and Ni^{2+} and Glukhovetsky kaolinite (Ukraine) with the exchangeable cations K^+ and Ba^{2+}, were taken. All the samples of clay

[1] To whom any correspondence should be addressed.

minerals were prepared in the laboratory of disperse systems of Institute of Colloid and Water Chemistry (Ukraine, Kiev) according to the procedure described previously [9]. The samples were stored in ambient air humidity, thus one can assume that water molecules were absorbed from the air into the pores.

Montmorillonite and kaolinite are layer silicate minerals composed of tetrahedral (T: Si tetrahedra) and octahedral (O: Al octahedra) sheets. Montmorillonite is formed by TOT layers. In this mineral isomorphic substitution (i.e., replacement of ions in the crystals without change of the structure) results in a net negative charge which is balanced through the electrostatic adsorption of cations in the interlayer space and cations on or near the external surfaces. Kaolinite is formed by TO layers and isomorphic substitution is rather small. Depending on their structure, type of interlayer cation, activity of water, and the degree of compaction, clays can be hydrated by a variable amount of water molecules in the interlayer space and on external surfaces. In the case of montmorillonite water is located between particles and in the interlayer space. In kaolinite water and ions are located on the external surfaces (figure 1).

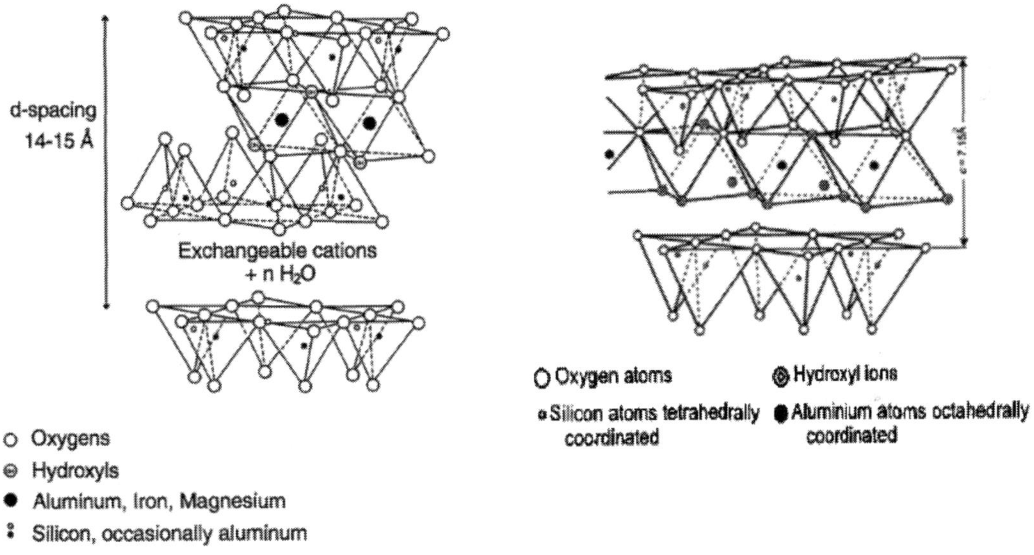

(a) **(b)**

Figure 1. Schematic structure of the montmorillonite (a) [10] and kaolinite (b) [11] clay minerals.

3. Experimental

For investigation broadband dielectric spectrometer BDS 80 based on an Alpha Impedance Analyzer (NOVOCONTROL) was used. Measurements were performed in the frequency range of 1 Hz ÷ 1 MHz with automatic temperature control by QUATRO Cryosystem with a precision of 0.5°C. The measurement protocol: each of the samples was placed into the sample cell with a parallel plate capacitor configuration at room temperature and the measurements were then made by cooling the samples from +20°C to -90°C. Then the samples were measured at intervals of 3°C upon heating them from -90°C to -58°C and at intervals of 6°C from -52°C to +140°C and then heating up to +300°C. The relative water content to the dry sample was obtained by weighing the samples prior to and immediately after heating up to +300°C (see table 1). The inner diameter of the sample cell for dielectric measurements was 16 mm and the minimum spacing between electrodes was 0.5 mm.

4. Results and discussion

A typical view of the 3-D plots of the dielectric losses vs. frequency and temperature is shown in figure 1.

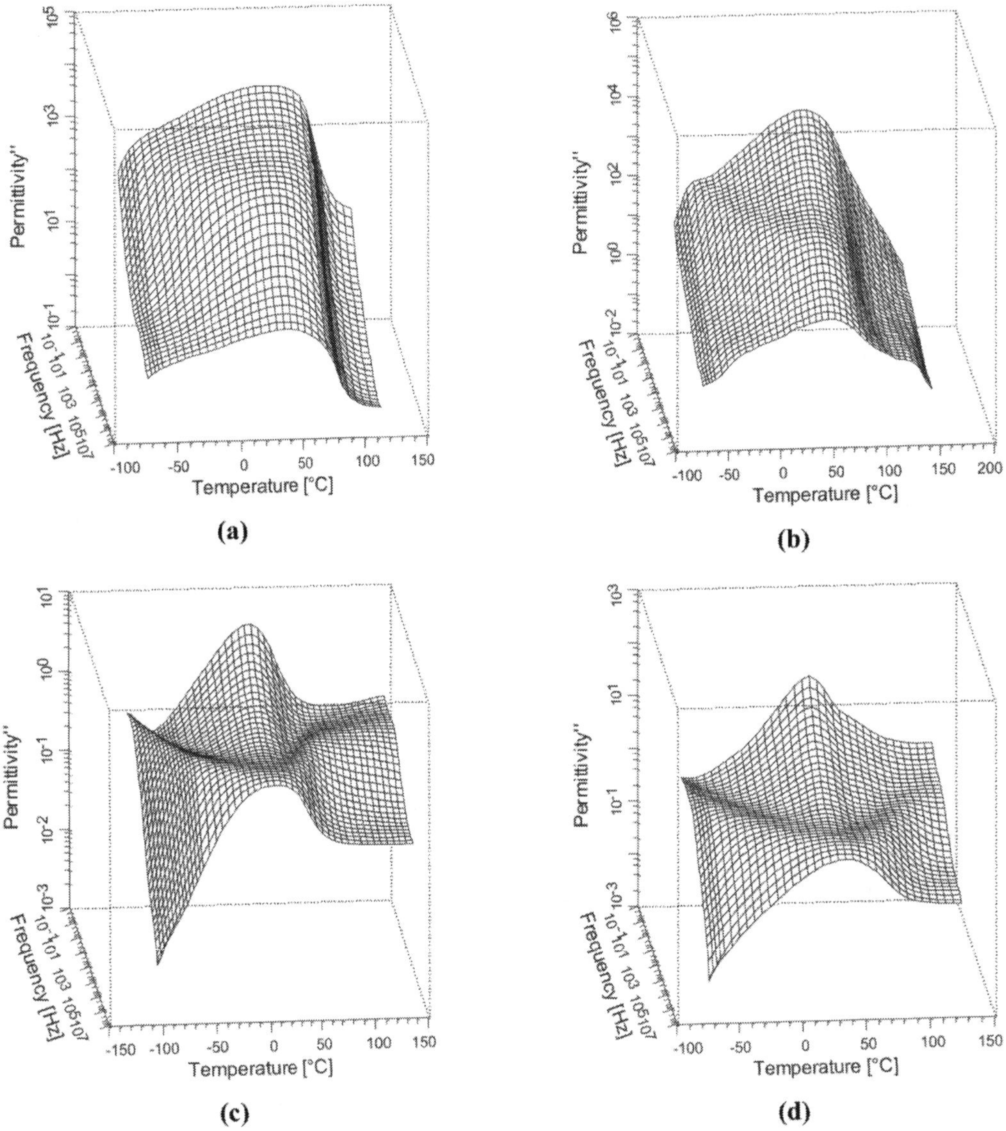

Figure 1. Three-dimensional plots of the frequency and temperature dependence of the dielectric losses for montmorillonite K (a), Ni (b) and kaolinite K (c) and K, Ba (d).

In temperatures from -90°C to +140°C a relaxation process with a specific saddle-like shape and distinctive kink point was observed twice for montmorillonite Ni (processes 1 and 2) at different frequency-temperature intervals while only one relaxation process 1 is observed in the case of kaolinite samples and montmorillonite K. Annealing the samples at 300°C for 3 hours leads to full disappearance of the relaxation processes. It means that the only water is the source of the dielectric relaxation.

Experimental data for the relaxation process 1 in kaolinites K and K, Ba were described by a single Cole-Cole function [12] with additional Jonscher's [13] empirical terms and conductivity term [14]. At the same time the process 1 in montmorillonites K and Ni in the full temperature range can't be well described by any known functions and we give for it only relaxation times corresponding to the maximum position of the relaxation peak (figure 2). Anomalous broadening of the spectral lines in the process 1 for montmorillonites may be explained by the exchange between several states but such exchange is not a fast enough and therefore described by very complex relations [7] being subject to further investigation.

The quantitative analysis of the measured dielectric spectra was performed based on the model described earlier [15, 16]. It considers the relaxation act as an occurrence of two simultaneous events: (a) the molecule reorientation and (b) presence of a defect site in its vicinity. In the framework of the model the temperature dependence of the relaxation time can be written in the form:

$$\tau = \tau_0 \exp\left\{\frac{H_a}{kT} + C\exp\left(-\frac{H_d}{kT}\right)\right\} \quad (1)$$

where H_a is the height of the potential barrier of the reorientation of water molecules; H_d is the energy of the defect formation; C is inversely proportional to the maximum possible defect concentration. The pre-exponential time constant τ_0 can vary with temperature.

The increasing temperature leads, on the one hand, to an increase in the probability to attain sufficient energy to break the bonds between the reorienting molecule and its neighbors. On the other hand, it leads to a decrease in the probability of finding a defect in the vicinity of the moving molecule. The interplay of these two terms yields the appearance of the minimum observed in the temperature dependence of the relaxation time.

The experimental data of the relaxation times were fitted to equation (1) by means of a least-squares fit procedure. The fitting curves are presented in figure 2 and the fitting results are listed in table 1.

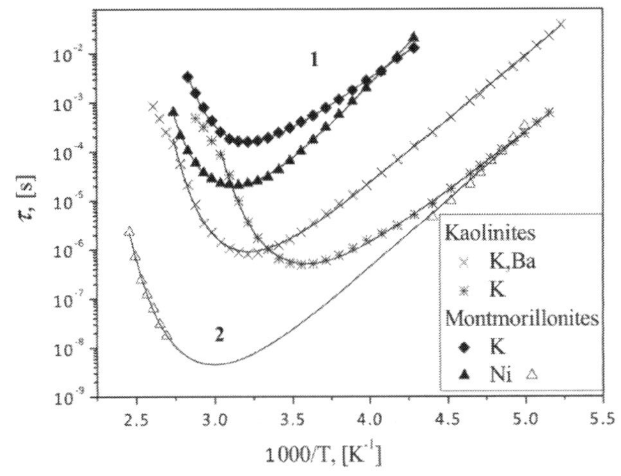

Figure 2. The temperature dependence of the relaxation times for montmorillonite and kaolinite samples (solid lines correspond to the fit by equation (1); two saddle-like processes 1 and 2 are marked by solid and open symbols respectively).

We associate the slower relaxation process 1 with intergranular water which is present both in kaolinites and montmorillonites. Thus the process 1 reflects the dynamics of the adsorbed water molecules at the outer surfaces of the powder particles.

The faster relaxation process 2 in montmorillonite Ni corresponds to the more disordered water. We suppose that the relaxation process 2 reflects mainly dynamics of the water molecules in interlayer space of the montmorillonite structure. Such type of water is absent in kaolinites. The apparent absence of the process 2 for montmorillonite K may be explained by strong disordering effect of K^+ ion on interlayer water that leads to shift of the process 2 in high-frequency range inaccessible for observation in the frequency range of the present experiment. Indeed, K^+ is characterized by "negative hydration" at which life time of water molecules is less in immediate environment of ion than in bulk solution [17, 18].

Table 1. The calculated values of the relative humidity (h) and the fitting parameters of the relaxation processes 1 and 2 according to equation (1).

	h, %	H_{a1}, kJ/mol	H_{d1}, kJ/mol	$\tau_{0(1)}$, s	H_{a2}, kJ/mol	H_{d2}, kJ/mol	$\tau_{0(2)}$, s
Montmorillonite K	13.7	43.2	36.0	$2\cdot10^{-12}$	-	-	-
Montmorillonite Ni	14.7	55.6	42.5	$1\cdot10^{-14}$	53.2	27.7	$3\cdot10^{-18}$
Kaolinite K	2.3	46.6	26.3	$2\cdot10^{-16}$	-	-	-
Kaolinite K, Ba	5.4	50.3	31.6	$6\cdot10^{-16}$	-	-	-

The values of the activation energies of the process 1 (H_{a1} and H_{d1}) for all clays under study are in a rather good agreement with the energies of molecular reorientation and defect formation in ice-like water [19] but depend on the properties of individual ions. Moreover, the values of H_{a1} and H_{a1} for kaolinites and montmorillonites (table 1) correlate with the Gibbs free energy hydration ($\Delta_{\text{hyd}}G^{\text{o}}$) of ions (-295, -1250, and -1980 kJ/mol for K^+, Ba^{2+}, and Ni^{2+} respectively [20]).

5. Conclusions

Montmorillonites with the exchangeable K^+ and Ni^{2+} cations and kaolinites with the exchangeable K^+ and Ba^{2+} cations were examined by means of the dielectric spectroscopy in wide temperature (-90°C ÷ +140°C) and frequency (1 Hz ÷ 1 MHz) ranges. Two types of relaxation processes (1 and 2) were revealed. The relaxation process 1 observed for montmorillonites K and Ni and kaolinites K and K, Ba was related to water on the outer surface of granules. The relaxation process 2 in montmorillonite Ni has been ascribed to interlayer water molecules. The activation parameters of the dielectric relaxation process 1 are in correlation with values of the Gibbs free energy hydration of the exchangeable ions.

Acknowledgements

The authors thank Prof. Yu. Feldman, Dr. A. Greenbaum (Gutina), Dr. A. Puzenko, and Dr. P. Ben Ishai from Dielectric Spectroscopy Laboratory at Department of Applied Physics (The Hebrew University of Jerusalem, Israel) for useful discussion.

References

[1] Hall P G and Rose M A 1978 *J. Chem. Soc. Faraday Trans.* 1 **74** 1221
[2] Lockhart N C 1980 *J. Colloid Interface Sci.* **74** 520
[3] Kaviratna P D, Pinnavaia T J and Schroeder P A 1996 *J. Phys. Chem Solids* **57** 1897
[4] Ishida T, Makino T and Wang C 2000 *Clays Clay Miner.* **48** 75
[5] Logsdon S D and Laird D A 2002 *J. Non-Cryst. Solids* **305** 243
[6] Logsdon S D and Laird D A 2004 *Clays Clay Miner.* **52** 411

[7] Rotenberg B, Cadéne A, Dufrêche J-F, Durand-Vidal S, Badot J-C and Turq P 2005 *J. Phys. Chem.* B **109** 15548
[8] Saltas V, Vallianatos F and Triantis D 2008 *J. Non-Cryst. Solids* **354** 5533
[9] Tarasevich Yu I and Ovcharenko F D 1980 *Adsorption sur des Min´eraux Argileux* (Institut Francais du P´etrole, Rueil Malmaison)
[10] Pusch R and Kamland O 1996 *Eng. Geol.* **41** 73
[11] Murray H H 1999 *Clay Clay Miner.* **34** 39
[12] Cole K S and Cole R H 1941 *J. Chem. Phys.* **9** 341
[13] Jonscher A K 1996 *Universal Relaxation Law* (London: Chelsea Dielectrics Press) p 415
[14] Froehlich H 1958 *Theory of Dielectrics* (Clarendon, Oxford)
[15] Macedo P B and Litovitz T A 1965 *J. Chem. Phys.* **42** 245
[16] Ryabov Y, Gutina A, Arkhipov V and Feldman Yu 2001 *J. Phys. Chem.* B **105** 1845
[17] Samoilov O Ya 1957 *The Structure of Aqueous Electrolyte Solutions and the Ion Hydration* (Moscow: Akad. Nauk SSSR)
[18] Krestov G A 1991 *Thermodynamics of Solvation: Solution and Dissolution, Ions and Solvents, Structure and Energetics* (New York: Horwood)
[19] Gutina A, Antropova T, Rysiakiewicz-Pasek E, Virnik K and Feldman Yu 2003 *Microporous Mesoporous Mater.* **58** 237
[20] Marcus Y 1991 *J. Chem. Soc. Faraday Trans.* **87** 2995

Prediction of thermodynamic properties of natural gases using Monte Carlo simulations

I Piyanzina, Yu Lysogorskiy, O Nedopekin

Institute of Physics, Kazan Federal University, Kazan, 42008, Russia

E-mail: i.piyanzina@gmail.com

Abstract. In this paper an applications of Monte-Carlo simulation in natural gas production is presented. We have investigated model of natural gas of the Bavlinskoye deposit located in the southeast of the Republic of Tatarstan. For this natural gas and for pure methane and ethane gases we have obtained thermal expansivity, isothermal compressibility, compressibility factor, heat capacity, Joule-Thompson coefficient and density at pressures up to 110 MPa at deposit temperature (463 K). Also we have obtained vapor pressures and liquid-vapor phase diagrams. Simulated properties for methane are in a good agreement with available experimental data.

1. Introduction

Today there are a number of problems in oil and gas industry related to the correct description of heavy hydrocarbons properties. Furthermore problems are encountered toxic and corrosive materials, unstable systems, microporous adsorbents, systems at extreme pressures. Natural gases are mixtures in which gaseous hydrocarbons (mainly methane) are most abundant, but they can contain significant amounts of liquid hydrocarbons, which affect strongly on the thermodynamic properties.

The properties of natural gases as well as other mixtures often can not be obtained from the superposition of their component properties that is why we should consider this system as complex.

Molecular simulation methods allow to predict all thermodynamic properties at all conditions and in the absence of phenomenological parameters. Molecular simulation is an emerging technique, which consists in performing a detailed simulation of microscopic systems involving typically a few hundreds or even thousands of molecules. On the basis of these simulations, appropriate statistical averages are performed to obtain the quantities that can be compared with experimental measurements.

2. Monte Carlo method

Molecular simulation includes methods of quantum chemistry and statistical thermodynamics. The important point is a choice of a statistical ensemble most suitable for the desired application [1]. In brief, the chosen statistical ensemble considers the possible states of a system compatible with some set of constraints, such as imposed temperature or imposed number of molecules. Practically, there are two main methods to simulate statistical ensembles. The first is molecular dynamics, which consists in solving the equations of motion, and the second is Monte Carlo (MC) simulation, in which a statistical method is used. According to the ergodic hypothesis they should give equal average values.

In the present work we used the method of Monte Carlo. The main point of this method is random transitions between states of the system with a probability determined by the given statistical ensemble. At each MC step randomly selected particle (or group of particles) is moved randomly. This

Journal of Physics: Conference Series **394** (2012) 012029 doi:10.1088/1742-6596/394/1/012029

causes the potential energy changes by a certain amount. Then the energy difference determines the transition probability from the "old" to the "new" state of the system. System parameters are calculated and are averaged over large number of steps. The potential energy of the molecular system consists of Lennard-Jones interaction and electrostatic interaction and intramolecular interactions associated with the stretching, bending and torsion of the molecules:

$$U=\sum_{i,j}U_{i,j}^{LJ}+\sum_{i,j}U_{i,j}^{el}+\sum_{k}U^{stretch}(l_k)+\sum_{m}U^{bend}(\theta_m)+\sum_{n}U^{tors}(\varphi_n). \quad (1)$$

The Monte Carlo simulations have wide range applications in different field of oil and gas industry.

- Method could be used to determine the unknown physical parameters of heavy hydrocarbons. In the case of the lack of experimental data for hydrocarbons, Monte Carlo simulations can be performed in order to obtain useful information for petroleum engineers such as saturation pressures, vaporization enthalpies and liquid densities at different temperatures.
- The method allows to determine thermophysical properties of natural gases. Molecular simulation can be used to predict the behaviour of such systems at all conditions, including very high pressures. It is also possible to determine the compressibility factor (Z = PV / RT), specific heat capacities, thermal expansion and isothermal compressibility with rather high accuracy.
- It should be noted, that in some natural deposits the pressure can reach 110 MPa and temperatures of 460 K or higher. Under these conditions, the natural gas can heat during the production process . The sign of the Joule-Thomson (JT) coefficient explains this feature and this fact should be considered during construction of wells. However, it is rather problematically to measure the Joule-Thomson coefficient in representative conditions due to experimental difficulties as well as small amount of gas sample available (a few hundreds of cm^3). Some estimations for JT coefficient from the equations of state are possible but usually with a large error, because these equations have not been validated for the extreme conditions, so that molecular simulation is a good candidate method.
- Furthermore molecular simulation is efficient tool to predict phase diagrams and phase densities, to determine solubility of gases, for example, in polymer coatings of pipes, it is also very useful to understand and predict the adsorption properties of zeolites, which are often used for separations in hydrocarbon processing.

3. Calculations

In this section we present calculations for the natural gas of the Bavlinskoye deposit located in the south-east of the Republic of Tatarstan. Similar calculations can be performed for any other natural gas. In addition, we also carried out calculations for methane and ethane in order to compare our results for these gases with known experimental data.

The calculations were performed by means of GIBBS code embedded in MedeA© interface. The periodic boundary conditions for simulation box were implied. We used Lennard-Jones (6-12) nonbond potentials [2-5] with the Lorentz-Berthelot mixing rule. The cutoff radius was one-half of the box length. The charges are being ignored, because the number of molecules described with electrostatic charges (CO_2) is rather small, and their electrostatic interaction energy is negligible. For a heavy alkanes the bending potential and the torsion potential were used as well. For more details about the Monte Carlo method as well as the calculation of interaction energy see, for example, Ref. [6]. In order to obtain the converging results, up to $3*10^7$ MC steps were used, whereas for the calculation of average values only last 50% of data were taken.

3.1. Densities

Percentage of components constituting the natural gas from the Bavlinskoye deposit is presented in Table 1. Firstly, we calculated the density of each component in the mixture at normal conditions in order to compare with known experimental data. The results are presented in the Figure 1. Computational error does not exceed ≈ 6.5%.

In the Figure 2 we can see density of methane and natural gas in a range of pressure from 30 to 110 MPa at a temperature of 463 K. The figure also shows that the experimental data from [7] for methane are in a good agreement with our calculations.

Table 1. Percentage of components constituting the natural gas from the Bavlinskoye deposit.

Component	CH_4	C_2H_6	C_3H_8	C_4H_{10}	C_5H_{12}	N_2	CO_2
Percentage	35	20.7	19.9	9.8	5.8	8.4	0.4

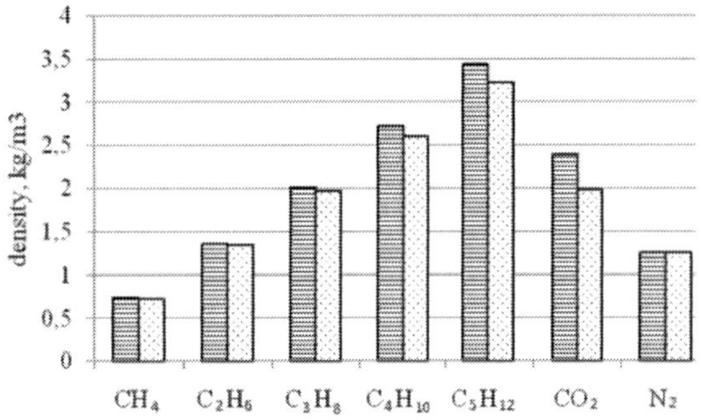

Figure 1. Density of each component of natural gas from the Bavlinskoye deposit at normal conditions (≡ - simulation, ∴ - experimental data [7]).

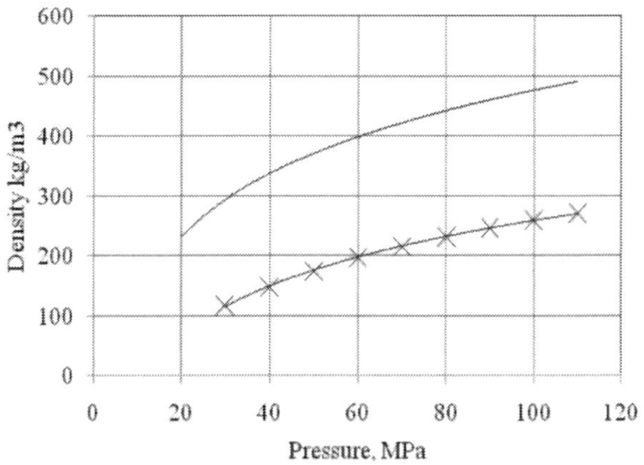

Figure 2. The calculated reassure dependence for the density of gases at T=463 K: the upper curve - natural gas of the Bavlinskoye deposit; the bottom curve - the methane (× - the experimental data [7]).

3.2. The Joule-Thompson coefficient and inversion pressure

Simulation results for Joule-Thomson coefficient are shown in Figure 3, where the computed Joule-Thomson coefficient of the natural gas from the Bavlinskoye deposit is plotted versus pressure at deposit temperature (463 K). We also present simulation results for methane and ethane. Those results could be partly confirmed by a measurement of the inversion pressure, *i.e.* the pressure corresponding to the change of sign of the Joule-Thomson coefficient (it can be obtained from volumetric measurements). Inversion pressures are collected in Table 2. For comparison, we also present results for natural gas of the San Joaquin Valley which were obtained in Ref. [6].

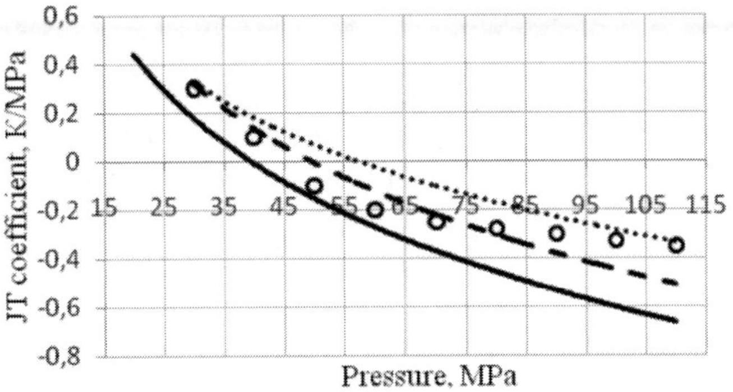

Figure 3. The Joule-Thompson coefficient for methane (•), for ethane (-), for natural gas of the Bavlinskoye deposit (solid line) and for natural gas of the San Joaquin Valley (°) [6] at 463 K.

Table 2. Simulation results and experimental data [7] for the inversion pressure of methane, ethane, natural gas of the Bavlinskoye deposit and of the San Joaquin Valley [6] (T=463K)

Gas	Simulation, P_{inv}, MPa	Experiment, P_{inv}, MPa
Methane	57	57
Ethane	52	50
Natural gas (Bavlinskoe deposit)	40	-
Natural gas (San Joaquin Valley)	42	42

3.3. *Thermodynamic derivative properties*

Thermodynamic derivative properties are the second order derivatives of the thermodynamic potential, with respect to other properties such as pressure, molar volume or enthalpy, which are first order derivatives of the thermodynamic potential. In the present work, we have calculated the thermal expansivity (Figure 4), the isothermal compressibility (Figure 5), compressibility factor (Figure 6) and the heat capacity (Figure 7) in a wide range of pressure up to 110 MPa at a temperature of 463 K.

3.4. *Phase behaviour*

The Figure 8 shows vapour pressure curves for methane and for natural gas of the Bavlinskoye deposit and Figure 9 presents the liquid-vapour phase diagram. For methane we present experimental data (they are marked by crosses). We obtained critical parameters for methane (T_c=190.6 K and P_c=4.6 MPa) and for natural gas ($T_c \approx 310$ K and $P_c \approx$ 8,6MPa). For methane results agree well with experimental data. For multicomponent mixtures of the natural gases the so-called pseudocritical pressure and temperature are the sums of the abundance of each component in the mixture (volume or moles) multiplied by the critical pressure and temperature of the same component respectively. According to this rule for natural gas of the Bavlinskoye deposit we obtained T_c = 284 K, P_c = 4.6 MPa. It should be pointed out that one has very smooth curve for methane and non-smooth curve for natural gas. This discrepancy and the non-smooth behaviour of the phase equilibrium curve for the mixture can be explained by the fact that each component of the mixture changes its phase under different conditions, in other words there is phase separation in the mixture.

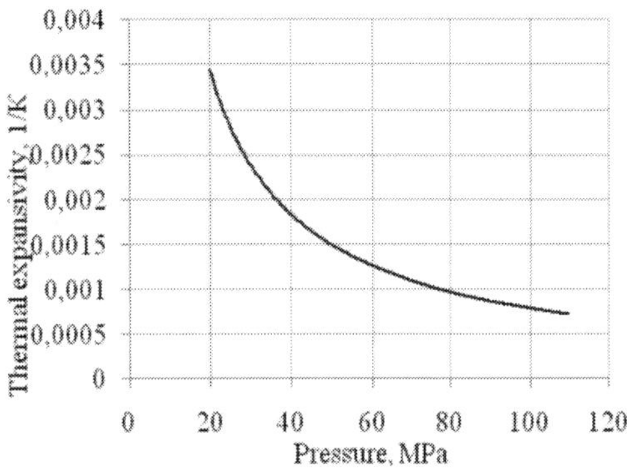

Figure 4. The simulated thermal expansivity of the natural gas of the Bavlinskoye deposit at 463K

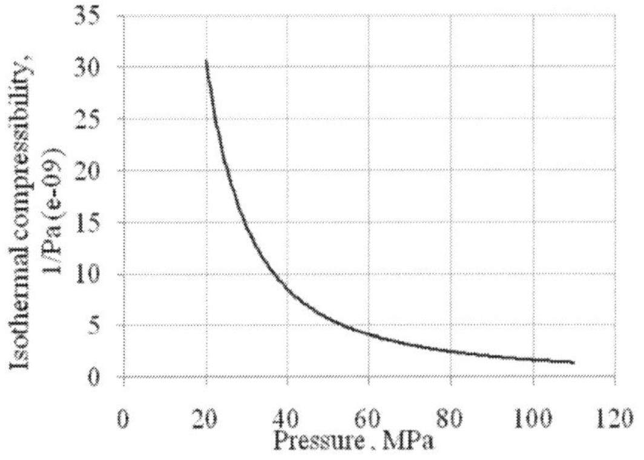

Figure 5. The simulated isothermal compressibility of the natural gas of the Bavlinskoye deposit at 463K.

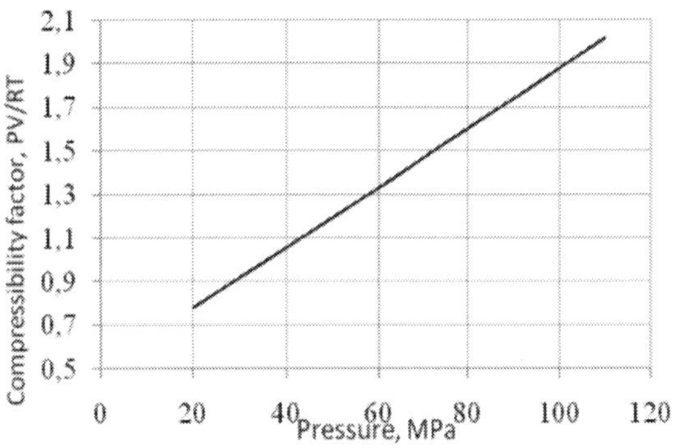

Figure 6. Monte Carlo simulation results for the compressibility factor of the natural gas of the Bavlinskoye deposit at 463K.

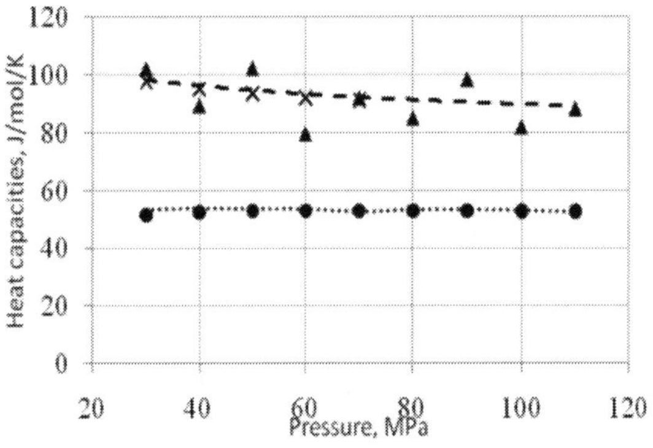

Figure 7. Heat capacities of methane, ethane and of the natural gas of the Bavlinskoye deposit (bottom curve is for methane along with experimental data (•) [7], upper curve is for ethane with experimental data (×) [7] and (▲) is for natural gas of the Bavlinskoye deposit).

The ratio of the mole fraction in the gas phase to the mole fraction in the liquid phase for the given component i of mixture is denoted as the K_i,. On the Figure 10 the dependence of K_i on the temperature is plotted. It could be seen that nitrogen, carbon dioxide and methane evaporate more intensively from the liquid mixture.

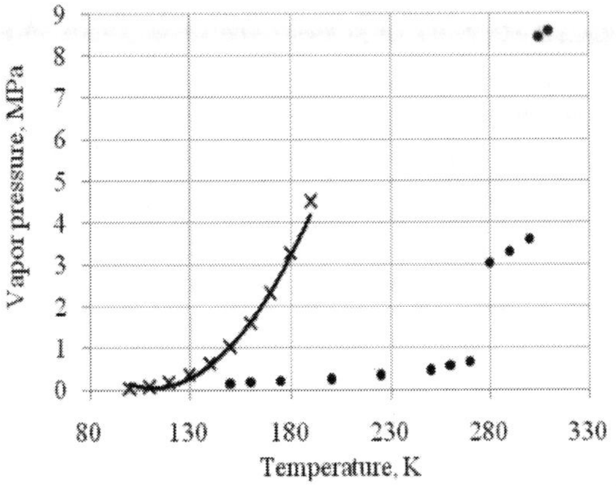

Figure 8. Simulated saturated vapor pressure of natural gas of the Bavlinskoye deposit (•) and of methane (solid line) with experimental data (×) [7].

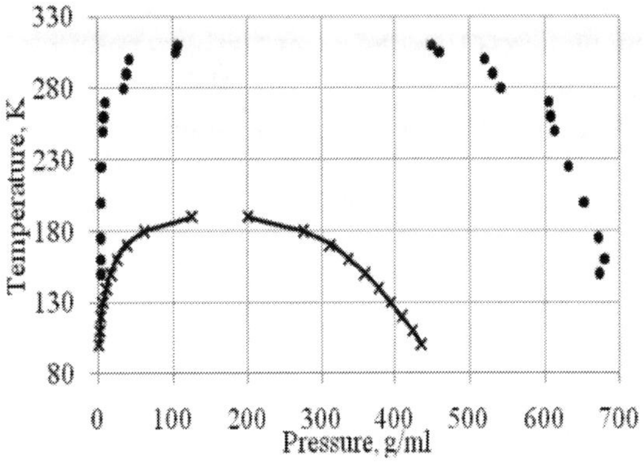

Figure 9. Liquid-Vapour phase diagram of natural gas of the Bavlinskoye deposit (•) and liquid-vapour phase diagram of methane (lower curve) with experimental data (×) [7].

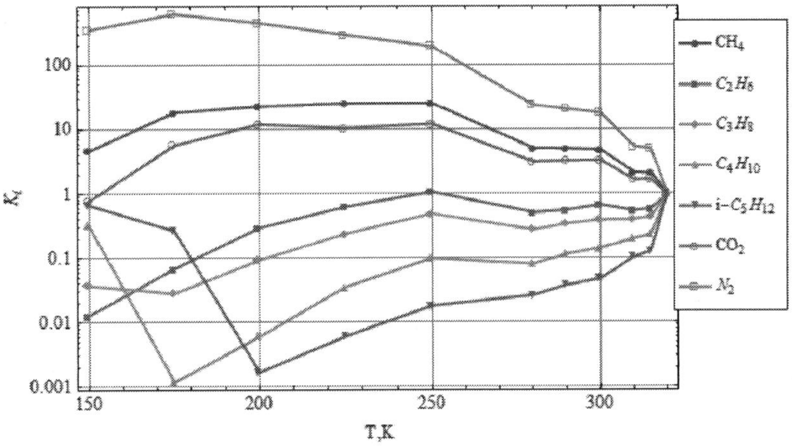

Figure 10. Temperature dependence of the K_i parameter for each component in the mixture (see text for details).

4. Conclusion

To sum up we presented the results of calculations of the basic thermodynamic parameters for the natural gas of the Bavlinskoye deposit and some parameters for methane and ethane: density, Joule-Thompson coefficient and inversion pressure. The known experimental data agree well with our calculations indicating the correct description of these systems by means of molecular simulation.

In a conclusion we should note that methods of molecular simulation in combination with other methods can significantly increase the efficiency of geological and geophysical studies. Computer simulation serves as a theoretical tool that allows to check the findings of existing analytical methods, theories or to supplement their predictions.

References

[1] Ungerer P, Lachet V and Tavitian B 2006 *Oil & Gas Science and Technology – Rev. IFP,* **61**(3) 387-403

[2] Boutard Y, Ungerer Ph, Teuler J M et al. 2005 *Fluid Phase Equilibria* **236** 25-41

[3] Möller D, Óprzynski J, Müller A and Fischer J. 1992 *Mol. Phys.* **75** 363

[4] Ungerer P, Beauvais C, Delhommelle J et al. 2000 *J. Chem. Phys.* **112** 5499
[5] Poling B E 2001 *The properties of gases and liquids* (New York: Mac Graw Hill)
[6] Lagache M, Ungerer Ph and Boutin A 2004 *Fluid Phase Equilibria* **220** 211
[7] National institute of Standards and Technology: Material Measurement Laboratory: Thermophysical Properties of Fluid Systems / Access mode: http://webbook.nist.gov/chemistry/fluid/

Rotational dynamics of copper(II) amino acid complexes by EPR and NMR relaxation methods

M S Bukharov[1,2], V G Shtyrlin[1], G V Mamin[2], A S Mukhtarov[1] and E M Gilyazetdinov[1]

[1] Kazan Federal University, A.M. Butlerov Chemistry Institute, 420008, Kremlevskaya St. 18, Kazan, Russian Federation

[2] Kazan Federal University, Institute of Physics, 420008, Kremlevskaya St. 18, Kazan, Russian Federation

E-mail: mikhail.bukharov@gmail.com

Abstract. Rotational dynamics of the copper(II) *bis*-complexes with glycine and *L*-aspartic acid has been studied by EPR and NMR relaxation methods in aqueous solutions at several temperatures. Dynamical parameters obtained by EPR were compared with nuclear magnetic relaxation dispersion (NMRD) results and were found to be in a good agreement. From EPR data dominating *trans* isomer for $Cu(Gly)_2$ and *cis* isomer for $Cu(L\text{-}Asp)_2^{2-}$ was found. On the basis of distance of closest approach of protons to central ion inferred from NMRD and crystal structure data the average slope angles of axial water molecule to equatorial plane were calculated and axial coordination of only one water molecule in the $Cu(L\text{-}Asp)_2^{2-}$ complex was established.

1. Introduction

Investigation of dynamical behaviour of molecules in solutions is very important fundamental problem of physics and chemistry of condensed matter. The reason of interest is lack of complete understanding of molecular dynamics depending on solvent nature, structure, and properties of molecules. Also investigations of rotational dynamics in solutions draw big attention because rotation of molecules and their fragments often determine the ways and effectiveness of chemical reactions.

In this work dynamical behaviour of the copper(II) amino acid *bis*-complexes with glycine and *L*-aspartic acid in aqueous medium was studied by EPR and NMR relaxation methods. Joint application of these two methods to study dynamics of particles in solution is already known for VO(II) and Gd(III) complexes [1, 2]. Copper(II) complexes with amino acids model many proteins and metal enzymes and serve as transport forms of copper in living cells so their investigation induces additional significant theoretical and practical interest in terms of understanding of living systems.

2. Materials and methods

Copper nitrate, glycine (GlyH) and *L*-aspartic acid (L-$AspH_2$) from Reanal were used to prepare solutions. The pH was adjusted to 6.90 and 7.29 respectively for copper(II) - glycine and copper(II) - *L*-aspartic acid systems by adding HNO_3 and KOH with an accuracy of 0.01 pH unit. Concentrations of copper(II) and both amino acids were 4.87 mM and 14.84 mM respectively. The salt background 1.0 M was created by addition of potassium nitrate recrystallized from water solution.

Measurements of the pH values were carried out with a Basic Titrino 794 automatic titrator from Metrohm. X-band EPR spectra were recorded on a Bruker ESP 300 spectrometer. Magnetic field was checked by an ER 035M magnetometer and microwave frequencies were measured with a Hewlett-Packard 5335A counter. NMRD measurements (0.1 – 20 MHz) in temperature range (278 – 300K) were carried out on a Stelar Spinmaster FFC2000 NMR relaxometer located in Polymerphysik laboratory of Institute of Physics at Technische Universität Ilmenau.

3. Results and discussion

3.1. EPR

EPR spectra of the studied compounds were registered in the temperature range 278-323 K. Simulations of the EPR spectra were performed with the modified computer program [3] that allows to determine the spin Hamiltonian parameters (g_o, A_o, A_o(N)), rotation correlation time (τ_R), and percentages of the complexes (P). EPR line positions were described by the following isotropic spin-Hamiltonian:

$$\hat{H} = g_0 \beta S_z H_z + A_0 \overline{IS} + \sum_i A_0^i \overline{I}^i \overline{S} \tag{1}$$

where g_0 is the isotropic g-factor, A_0 and A_0^i are the hyperfine (hf) and superhyperfine (shf) constants, and other symbols have their usual meaning. Line positions of single isotope molecule were calculated within the second-order perturbation approach. EPR spectra were fitted by assuming Lorentzian line shape with line width expressed by Wilson and Kivelson theory [4]:

$$\Delta H_{m_I} = \alpha' + \alpha'' + \beta m_I + \gamma m_I^2 + \delta m_I^3 \tag{2}$$

where coefficients α', β, γ, δ depend on anisotropy spin-Hamiltonian parameters, microwave frequency, rotational correlation time (τ_R) and α'' is the residual line width.

Examples of experimental and simulated EPR spectra of $Cu(L\text{-}Asp)_2^{2-}$ complex at different temperatures are presented on Fig.1. Simulations were performed taking into account two isomers with *cis* and *trans* arrangement of amino groups. Parameters of spectra simulated for 295K can be seen in Table 1.

Table 1. EPR spectra parameters of the $Cu(Gly)_2$ and $Cu(L\text{-}Asp)_2^{2-}$ complexes at 295K.

Complex	$g_0 \pm 0.0005$	A_0, G	τ_R[a], 10^{-11} s	A_N[b], G	$P \pm 0.05$
$Cu(Gly)_2$					
trans isomer	2.1288	64.4 ± 0.4	2.9 ± 0.1	9.3, 9.3	0.67
cis isomer	2.1280	75.8 ± 0.5	2.9 ± 0.1	10.7, 10.7	0.33
$Cu(L\text{-}Asp)_2^{2-}$					
trans isomer	2.1286	59.6 ± 0.3	6.2 ± 0.2	9.1, 9.1	0.39
cis isomer	2.1274	72.5 ± 0.4	6.2 ± 0.2	10.1, 10.1	0.61

[a] Rotational correlation times for two isomers are supposed to be the same.
[b] Errors in A_N values are not more than 1G.

As can be seen from Table 1 *trans* isomer dominates over *cis* isomer in the $Cu(Gly)_2$ complex and opposite is true for the $Cu(L\text{-}Asp)_2^{2-}$ complex. In our opinion the reason of this is axial coordination of a carboxyl group of ligand in the $Cu(L\text{-}Asp)_2^{2-}$ complex. In the case of *trans* isomer two carboxyl groups locate at one site of coordinated plane and compete with each other for the axial position because of their electrostatic repulsion. In *cis* isomer such repulsion doesn't take place (see Scheme 1). So *cis* isomer is more stable than *trans* isomer in the $Cu(L\text{-}Asp)_2^{2-}$ complex.

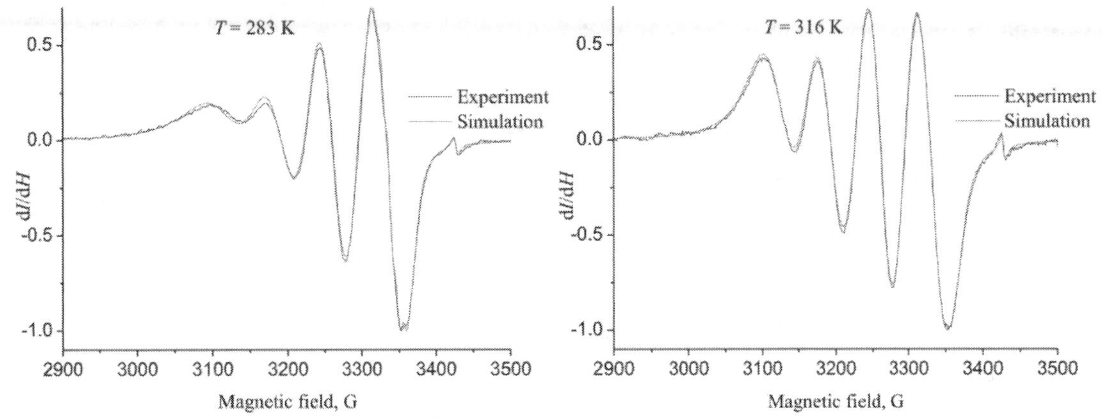

Figure 1. Experimental and simulated EPR spectra of copper(II) *bis* complex with aspartic acid ($Cu(L\text{-}Asp)_2^{2-}$) at different temperatures.

Assumed that, temperature dependences of rotational correlation times fulfill Arrhenius law ($\tau_R = \tau_0 \exp(E_a/RT)$) the activation energy of rotational motion of the complexes and τ_0 coefficient were found (see Table 2).

3.2. *NMRD*

In the $Cu(Gly)_2$ and $Cu(L\text{-}Asp)_2^{2-}$ complexes water molecules occupy axial positions and exchange rapidly with other molecules in solution [5]. In the case of $Cu(L\text{-}Asp)_2^{2-}$ complex one axial position can also be occupied by carboxyl group of the ligand (Scheme 1).

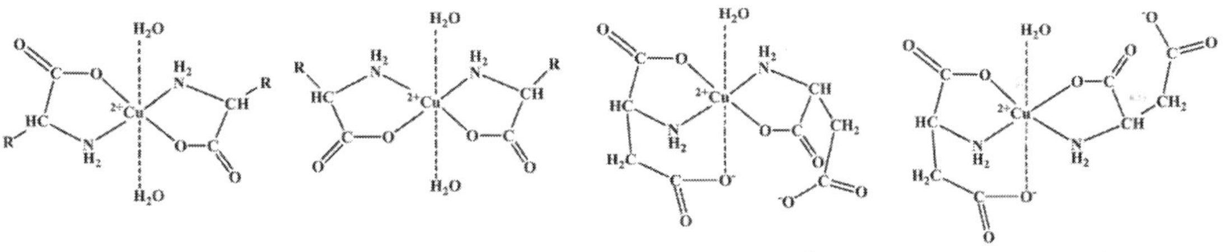

trans isomer — R = H for Gly^-, R = CH_2COO^- for Asp^{2-}; *cis* isomer; *trans* $Cu(L\text{-}Asp)_2^{2-}$ (1 ax. H_2O); *cis* $Cu(L\text{-}Asp)_2^{2-}$ (1 ax. H_2O)

Scheme 1

Proton relaxation of axial water molecules results in so called inner-sphere paramagnetic contribution in the observed relaxation times. The modified Solomon-Bloembergen-Morgan equation [6] taken into account *g*-tensor anisotropy of complexes was used to describe this contribution:

$$(T_{1p}^{-1})_{dd} = \frac{2}{15}\hbar^2\gamma_I^2\gamma_S^2 S(S+1)r^{-6}\left(A'\tau_c + B\frac{\tau_C}{1+\omega_S^2\tau_C^2} + A''\frac{\tau_C}{1+\omega_I^2\tau_C^2}\right) \qquad (3)$$

where γ_I and γ_S are nuclear and electron gyromagnetic ratios respectively, S is electron spin, ω_I and ω_S are the corresponding nuclear and electron Larmor frequencies, r is distance between protons and paramagnetic ion, coefficients A', A'', and B depend on *g*-tensor anisotropy and angle between electron-nuclear spin dipole and symmetry axis [6]. Correlation time τ_C equals τ_R for studied compounds [7].

To get acceptable fitting it was also necessary to account outer-sphere relaxation due to translational motion of water molecules near the complex coordination sphere. This contribution is described by the following equations [8]:

$$(T_{1p}^{-1})_{os} = \frac{32}{405}\pi \frac{N_A[M]\hbar^2\gamma_I^2\gamma_S^2 S(S+1)}{d(D_M + D_L)}\{7J(\omega_s) + 3J(\omega_I)\},$$

$$\text{where } J(\omega) = \frac{1+5z/8+z^2/8}{1+z+z^2/2+z^3/6+4z^4/81+z^5/81+z^6/648}; \qquad (4)$$

$$z = \sqrt{2}\left(\omega\tau_D + \frac{\tau_D}{\tau_S}\right)^{1/2}; \quad \tau_D = \frac{d^2}{D_M + D_L}.$$

In these equations d is the distance of closest approach of outer sphere water molecule protons to paramagnetic ion, τ_D is the diffusion correlation time, D_M and D_L are the diffusion coefficients for the solvent protons and the complex.

On Fig. 2 experimental (points) and simulated (lines) T_{1p}^{-1} proton relaxation profiles are presented for two solutions of copper(II) *bis*-complexes with glycine and *L*-aspartic acid. On Fig. 3 one can see comparison of inner-sphere and outer-sphere contributions for both systems at 284 K.

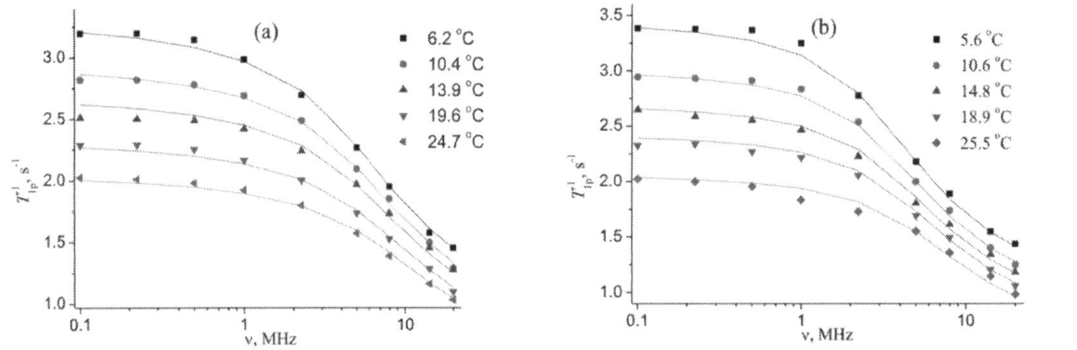

Figure 2. Experimental (dots) and simulated (lines) paramagnetic contributions in T_1^{-1} relaxation rates of solvent protons in $Cu(Gly)_2$ (a) and $Cu(L\text{-}Asp)_2^{2-}$ (b) water solutions at different temperatures.

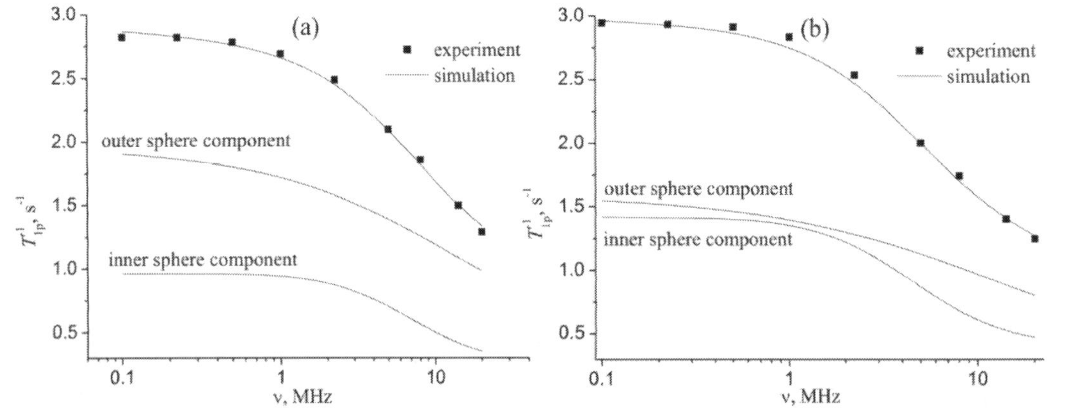

Figure 3. Inner-sphere and outer-sphere contributions for $Cu(Gly)_2$ (a) and $Cu(L\text{-}Asp)_2^{2-}$ (b) at 284K.

The parameters used to model NMR relaxation data are presented on the Table 2. In general there are six parameters for simulation. But it is not correct to vary all of them because they are correlated to each other and several solutions are possible. So in the case of $Cu(Gly)_2$ system we fixed rotational correlation times (namely τ_0 coefficient and activation energy of rotational motion) at the values found by EPR. Then by fitting relaxation data we could obtain distance *r* and parameters of outer sphere contribution.

Table 2. NMRD fitting parameters and parameters found by EPR.

	Inner-sphere component			Outer-sphere component		
Complex	τ_0, 10^{-14} s	E_a, kJ/mol	*r*, Å	$\tau_{0(os)}$, 10^{-14} s	$E_{a(os)}$, kJ/mol	*d*, Å
$Cu(Gly)_2$						
EPR	3.0±0.9	16.8±0.7				
NMRD	3.0[a]	16.8[a]	3.50±0.05	1.6±0.4	18.8±0.5	4.01±0.05
$Cu(L\text{-}Asp)_2^{2-}$						
EPR	3.2±0.2	18.6±0.2				
NMRD	3.2±0.4	17.5±0.6	3.51 (1 ax. H_2O) 3.15 (2 ax. H_2O)	1.6[b]	18.8[b]	4.29±0.05

[a] Values were fixed at those found from EPR.
[b] Taken as for $Cu(Gly)_2$ complex.

Using a such scheme in the case of $Cu(L\text{-}Asp)_2^{2-}$ complex results in non acceptable fitting of relaxation data. So to model experimental data of this system we fixed outer sphere relaxation parameters $\tau_{0(os)}$ and $E_{a(os)}$ at the values found for $Cu(Gly)_2$ system and obtained parameters for inner sphere component and distance *d*.

As can be seen from the Table 2 τ_0 coefficient found from NMR relaxation equals to one obtained from EPR. At the same time the value of activation energy is a little lower than in EPR. The distance of closest approach of protons to paramagnetic ion for outer sphere relaxation a little bigger for the $Cu(L\text{-}Asp)_2^{2-}$ complex then for $Cu(Gly)_2$ that is in agreement with bigger size of *L*-aspartic acid and presence of carboxylic groups screened paramagnetic ion from water molecules.

Comparison of rotational correlation times obtained by two methods is presented on Fig. 4. The difference of these values for the $Cu(L\text{-}Asp)_2^{2-}$ complex due to difference in activation energy is not understood yet and is the subject of the next studying.

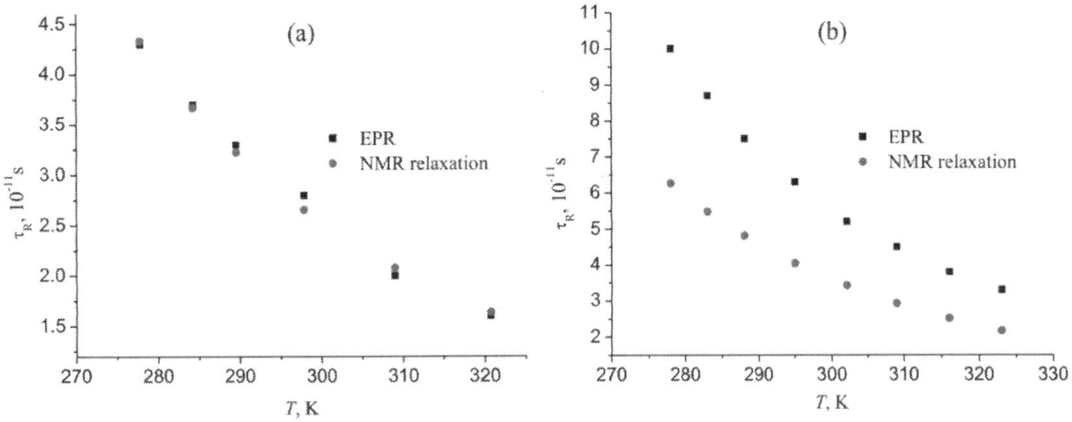

Figure 4. Temperature dependences of rotational correlation times for $Cu(Gly)_2$ (a) and $Cu(L\text{-}Asp)_2^{2-}$ complexes (b) obtained by EPR and NMR relaxation methods.

Let us take average length of Cu-O_{ax} bond equaled approximately 2.85 Å for the both complexes according to literature data [9,10]. Then using distance of closest approach r of axial water molecule protons found from NMR relaxation data one can calculate the average slope angle of water molecule to equatorial plane of complexes (Cu-O_{ax}-H angle). For the $Cu(Gly)_2$ complex this angle equals 127^0. It means that water molecule doesn't slope to equatorial plane of the complex. In the $Cu(L\text{-}Asp)_2^{2-}$ complex there are two possible ways of coordination: with two and one axial water molecule. In the first case we again obtain 127^0 and in the second case only 95^0. As this complex has negative charge the axial water molecule should indeed bend to equatorial plane by their hydrogen atoms. So we can conclude that in the $Cu(L\text{-}Asp)_2^{2-}$ complex only one axial position is occupied by water molecule while second one is taken up by carboxyl group of the ligand. This result is in agreement with above mentioned EPR data and higher stability constants of the $Cu(L\text{-}Asp)_2^{2-}$ ($\log\beta = 16.2$ [11]) relative to the $Cu(Gly)_2$ complex ($\log\beta = 15.2$ [11]).

4. Conclusions

Rather good agreement was found in rotational dynamics parameters of $Cu(Gly)_2$ and $Cu(L\text{-}Asp)_2^{2-}$ *bis*-complexes obtained by EPR and NMR relaxation methods. It was shown that combination of these two methods can be used to obtain reliable fitting parameters of proton relaxation profiles.

As opposed to $Cu(Gly)_2$ the *cis*-isomer of $Cu(L\text{-}Asp)_2^{2-}$ dominates in solution over *trans* isomer due to electrostatic repulsion of additional carboxyl groups of two $L\text{-}Asp^{2-}$ anions in the latter.

On the basis of T_{1p}^{-1} proton relaxation rates simulation and crystal structure data average slope angles of axial water molecules were estimated for the $Cu(Gly)_2$ and $Cu(L\text{-}Asp)_2^{2-}$ complexes.

Axial coordination of only one water molecule in the $Cu(L\text{-}Asp)_2^{2-}$ complex was confirmed by EPR and NMR relaxation methods.

5. Acknowledgment

We thank Prof. S. Stapf and Dr. C. Mattea for help in performing NMR relaxation experiments.

References

[1] Chen J W, Auteri F P, Budil D E, Belford R L and Clarkson R B 1994 *J. Phys. Chem.* **98** 13452
[2] Chen J W, Belford R L and Clarkson R B 1998 *J. Phys. Chem.* A **102** 2117
[3] Garipov R R, Shtyrlin V G, Safin D A, Zyavkina Yu I, Sokolov F D, Konkin A L, Aganov A V and Zakharov A V 2006 *J. Chem. Phys.* **320** 59
[4] Kivelson D 1960 *J. Chem. Phys.* **33** 1094
Wilson R and Kivelson D 1966 *J. Chem. Phys.* **44** 154
Wilson R and Kivelson D 1966 *J. Chem. Phys.* **44** 4445
[5] Zakharov A V and Shtyrlin V G 1985 *Rapid Ligand Exchange Reactions* (Kazan: KSU)
[6] Sternlicht H 1965 *J. Chem. Phys.* **42** 2250
[7] Poppel A A 1978 *Magnetic Relaxation Method for Analysis of Inorganic Substances* (Moscow: Khimia)
[8] Bennett H F, Brown R D, Koenig S H and Swartz H M 1987 *Magn. Reson. Med.* **4** 93
Polnaszek C F and Bryant R G 1984 *J. Chem. Phys.* **81** 4038
Hwang L-P and Freed J H 1975 *J. Chem. Phys.* **63** 4017
[9] Freeman H C and Snow M R 1964 *Acta Cryst.* **17** 1463
[10] Santana R C, Santos M G, Cunha R O, Ferreira K D, Carvalho J F and Calvo R. 2006 *J. Phys. Chem. Solids* **67** 745
[11] Perrin D D 1979 *Stability Constants of Metall-Ion Complexes. Pt. B: Organic Ligands* (Oxford: Pergamon Press)

New program for computation of the thermodynamic, spectral, and NMR relaxation parameters of coordination compounds in complex systems

A A Krutikov, V G Shtyrlin, A O Spiridonov, N Yu Serov, A N Il'yin, M S Bukharov and E M Gilyazetdinov

Kazan Federal University, A.M. Butlerov Chemistry Institute, 420008, Kremlevskaya St. 18, Kazan, Russian Federation

E-mail: akrutiko@ksu.ru

Abstract. A new approach providing a direct calculations of equilibrium constants, parameters of the chemical exchange reactions, and spectral characteristics of complexes on the basis of data of several methods including pH-potentiometry, multi-wavelength electronic spectroscopy, and NMR relaxation within a single computer program was proposed and realized in the program STALABS. Application of the STALABS program has been demonstrated on the example of investigation of the complex nickel(II) - *L*-histidine system by joint usage of the above three methods.

1. Introduction

Determination of thermodynamic and kinetic parameters of equilibria and spectral characteristics of coordination compounds in solutions of complex systems is an important problem of coordination and physical chemistry which cannot be resolved without a computer simulation. In most of the programs intended for these purposes only one additive physico-chemical response dependent on pH or component concentrations of the equilibrium system is modeled [1].

Earlier in the Kazan University two programs, CPESSP [2] and STABLAB [3], for computation of equilibria and physico-chemical properties of coordination compounds were created. The program CPESSP is able to calculate separately data of the methods of pH-potentiometry, spectrophotometry, or spin-lattice NMR relaxation in conditions of "rapid chemical exchange". The program STABLAB is used for the simultaneous calculation of the thermodynamic parameters of equilibria and kinetic characteristics of the proton and ligand exchange reactions from the parallel measurements of the spin-lattice (T_1) and spin-spin (T_2) relaxation times of the solvent nuclei.

The major aim of the present work is to develop a new approach for simultaneous calculation of equilibrium constants, parameters of the chemical exchange reactions, and spectral characteristics of complexes based on the data of several methods including pH-potentiometry, ion-selective potentiometry, multi-wavelength electronic spectroscopy, and NMR relaxation within a single computer program. The approach was implemented in the new program STALABS. The second aim of the work is to demonstrate the applicability of the STALABS program for investigation of the complex system nickel(II) - *L*-histidine.

2. Results and discussion

In the base of thermodynamic calculations of equilibrium system the Brinkley algorithm [4, 5] applied to the matrix algebra is put. A minimum set of components (B_l, $l = \overline{1, m_1}$), called a basis, is selected from all chemical forms (A_j, $j = \overline{1, M}$) participated in the reactions to determine the equilibrium concentrations of all the present species. So the reaction of any form A_j can be written as follows:

$$\sum_{l}^{m_1} \nu_{jl} B_l \xrightleftharpoons{\beta_j} A_j \tag{1}$$

where ν_{jl} is stoichiometric coefficient at the B_l component in j-th reaction and β_j is the equilibrium constant of this reaction. Equilibrium in the system with reactions of the type (1) corresponds to a minimum of the Gibbs potential which satisfies equations of the mass action law and mass balance:

$$[A_{ij}] = \exp(\theta_j + \sum_{l}^{m_1} \nu_{jl} \ln[B_{il}]) \text{ or } [A_{ij}] = \beta_j \prod_{l}^{m_1} [B_{il}]^{\nu_{jl}} \tag{2}$$

$$c_{il} = \sum_{j}^{M} \nu_{jl} [A_{ij}] \tag{3}$$

where $[A_{ij}]$ is equilibrium concentration of the species A_j, $\theta = \ln\beta$, $[B_{il}]$ and c_{il} are respectively equilibrium and total concentrations of the basis species in i-th experiment, $i = \overline{1, N}$, and N is the number of experiments.

For the calculation of the equilibrium concentrations of all the components of the system the first derivatives of the total concentrations of basis species by their equilibrium concentrations are used:

$$\frac{\partial c_{ik}}{\partial \ln[B_{il}]} = \sum_{j}^{M} \nu_{jk} \nu_{jl} [A_{ij}] \tag{4}$$

Symmetric Jacobi matrix **J** is composed from the values of these derivatives.

Minimization step is calculated by equation (5):

$$\mathbf{\Delta}_i = (\mathbf{J}_i)^{-1} \mathbf{r}_i \tag{5}$$

where $\mathbf{r}_i$ is the residuals vector with the $r_{ik} = c_{ik}^{\text{exp.}} - c_{ik}^{\text{calc.}}$ elements. The next values of the B_{il} components of the **B** vector on (x+1)-th iteration are given by equation (6):

$$\ln[\mathbf{B}]_i^{x+1} = \ln[\mathbf{B}]_i^{x} + \mathbf{\Delta}_i \tag{6}$$

The Bjerrum function (n_i) used in the pH-potentiometric data processing is calculated in the general form:

$$n_i^{\text{calc.}} = \frac{c_{i\text{H}} - [B_{i\text{H}}]}{c_{i\text{L}}} \tag{7}$$

Experimentally this function is calculated by the equation (8):

$$n_i^{\text{exp.}} = \frac{\dfrac{10^{\lg[B_{i\text{H}}]}(V_0 + \Delta V_i)}{a_\text{H}} - c_0^\text{T} \Delta V_i - \dfrac{10^{\lg K_w - \lg[B_{i\text{H}}]}(V_0 + \Delta V_i)}{a_\text{OH}}}{c_{i\text{L}} V_0} + \alpha \tag{8}$$

where V_0 is the initial volume of solution, ΔV_i is the added volume of titrant, c_0^T is the titrant concentration, a_H and a_OH are the activity coefficients of proton (H^+) and hydroxide ion (OH^-) respectively, and α is the ligand titration fraction.

To reduce the systematic errors in the program the joint or separate fitting of equilibrium constants with the following independent variables is provided: $c_{i\text{L}}$, c_0^T, α, a_H, and a_OH. Potentiometric methods are the simplest methods for calculation and analytical and numerical derivatives in this case match with great accuracy.

Spectrophotometric method is less accurate but more informative method for calculations of equilibria in solutions. The response function for the spectrophotometric method is written in the following matrix form:

$$\mathbf{D}^{\mathrm{M}} = \mathbf{C}^{\mathrm{M}}\mathbf{E} \tag{9}$$

where $\mathbf{D}^{\mathrm{M}}$ is matrix of the system absorption per unit concentration of the absorbing particles of dimension $N \times W$ (W is number of wavelengths), $\mathbf{C}^{\mathrm{M}}$ is matrix of the mass fractions of the absorbing particles of dimension $N \times S$ (S is number of the absorbing particles), and $\mathbf{E}$ is the molar extinction coefficients matrix of dimension $S \times W$.

Calculations of the molar extinction coefficients and equilibrium constants are performed separately by the variable separation method. The molar extinction coefficients are calculated as follows:

$$\mathbf{E} = (\mathbf{C}^{\mathrm{M}})^{-1}\mathbf{D}^{\mathrm{M}} \tag{10}$$

at $i = w$ or

$$\mathbf{E} = ((\mathbf{C}^{\mathrm{M}})^{\mathrm{T}}\mathbf{C}^{\mathrm{M}})^{-1}(\mathbf{C}^{\mathrm{M}})^{\mathrm{T}}\mathbf{D}^{\mathrm{M}} \tag{11}$$

at $i \neq w$. Such scheme of calculation is correct in the absence of species with negative molar extinction coefficients. Otherwise the method of non-negative linear least-squares should be used. Algorithm FNNLS [6] was applied to solve the problem of finding molar extinction coefficients with non-negative constraints by minimizing the system of equations for the residuals vector $\mathbf{r}$:

$$\mathbf{r} = (\mathbf{C}^{\mathrm{M}})^{\mathrm{T}}\mathbf{D}^{\mathrm{M}} - (\mathbf{C}^{\mathrm{M}})^{\mathrm{T}}\mathbf{C}^{\mathrm{M}}\mathbf{E} \tag{12}$$

Calculations of the analytical derivatives of the response function by the equilibrium constants were performed by the method described [7].

In the case of NMR relaxation method paramagnetic contributions (p) to the measured spin-lattice (T_1) and spin-spin (T_2) relaxation times of the nuclei of the solvent (A_0) from the M chemical forms $\{A_j, j = \overline{1, M}\}$ which are generated in solution and undergo exchange with the dominant state of solvent A_0 (diluted solutions) are given by additive relations [8]:

$$\frac{1}{T_{1p}} = \frac{1}{T_1} - \frac{1}{T_{1(A)}} = \sum_{j=1}^{M} K_{1j}[A_j] \tag{13}$$

$$\frac{1}{T_{2p}} = \frac{1}{T_2} - \frac{1}{T_{2(A)}} = \sum_{j=1}^{M} K_{2j}[A_j] \tag{14}$$

where $T_{1(A)}$ и $T_{2(A)}$ are spin-lattice and spin-spin relaxation times of the nuclei in pure solvent A_0; $[A_j]$ is equilibrium concentration of the A_j form, and K_{1j} и K_{2j} are the molar coefficients of spin-lattice and spin-spin relaxation respectively. For each A_j form with q nonequivalent positions of the nuclei including the outer coordination sphere the relaxation coefficients K_{1j} and K_{2j} take the generalized form [9, 10]:

$$K_1 = \sum_{n=0}^{q-1} \frac{P'_{(n)}}{\tau_{(n)} + T_{1(n)}} \tag{15}$$

$$K_2 = \sum_{n=0}^{q-1} \frac{P'_{(n)}}{\tau_{(n)}} \cdot \frac{T_{2(n)}^{-2} + T_{2(n)}^{-1} \times \tau_{(n)}^{-1} + \Delta\omega_{(n)}^{2}}{(T_{2(n)}^{-1} + \tau_{(n)}^{-1})^2 + \Delta\omega_{(n)}^{2}} \tag{16}$$

where n is number of the nonequivalent position, $P'_{(n)} = P_{(n)}/[A_j]$, $P_{(n)}$ is probability of presence of the solvent nuclei in n-th position of the A_j form; $\tau_{(n)}$ is the nuclei lifetime in corresponding position; $T_{1(n)}$ и $T_{2(n)}$ are spin-lattice and spin-spin relaxation times of the nuclei in this position, $\Delta\omega_{(n)}$ is difference of the Larmor precession frequency of nuclei between A_0 and n-th position of the A_j form (the j index of each element in the formulas (15) and (16) is omitted). Lifetime $\tau_{(n)}$ is defined by the solvent nuclei chemical exchange between the n-th position in the coordination sphere of the A_j form and free states in the bulk solution as following [11]:

$$\tau_{(n)}^{-1} = k_0^{(n)} + \sum_{m=1}^{M_{(n)}} k_m^{(n)}[A_{m(n)}] \tag{17}$$

where $k_0^{(n)}$ and $k_m^{(n)}$ are the rate constants of the chemical exchange reactions of first and second kinetics orders respectively without and with activation by certain chemical forms which are selected in subsets $\left\{A_{m(n)}, m=\overline{1, M_{(n)}}\right\}$ for each n value. When $\Delta\omega_{(n)} = 0$ equation (16) is similar to equation (15). If $T_{1(n)} >> \tau_{(n)}$ then the molar relaxation coefficients are calculated similarly to the molar extinction coefficients in spectrophotometry. The analytical derivatives of the response function by the equilibrium constants in the NMR relaxation method are approximately calculated as in spectrophotometry. Physico-chemical parameters of the NMR relaxation method are assumed to be independent of each other, the equilibrium concentrations, equilibrium constants, and other system parameters. Data calculations using numerical and analytical derivatives are in good agreement with each other.

Joint processing of the experimental data on the s properties measured by several methods with the simultaneous determination of physico-chemical parameters can be carried out by non-linear methods via minimization of the objective function.

$$\Phi(\eta,\theta) = \sum_{k=1}^{s}\sum_{i=1}^{N_k} r_{ki}^2 \quad (18)$$

where η is set of the spectral, NMR relaxation, and kinetics parameters, r_{ki} is normalized i-th residual for k-th property (Y_k) having the form $r_{ki} = (Y_{ki}^{\text{exp.}} - Y_{ki}^{\text{calc.}})\sqrt{w_{ki}}\left(\sqrt{N_k}\right)^{-1}$, N_k is number of experimental points by the k-th method data, w_{ki} is statistical weight giving by relation $w_{ki} = 1/(\sigma_0^2 Y_{ki}^{\text{exp.}} Y_{ki}^{\text{calc.}})$, σ_0 is the method error. The generalized residuals vector $\mathbf{r}$ of dimension $N = \sum_{k=1}^{s} N_k$ and the generalized Jacobian $\mathbf{J}$ of dimension $N \times d$ (d is number of fitted parameters) are introduced. Thus, to find the equilibrium constants by the several method data the following system of equations should be solved:

$$\mathbf{J}^{\mathrm{T}}\,\mathbf{J}\,\Delta\boldsymbol{\theta} = \mathbf{J}^{\mathrm{T}}\,\mathbf{r} \quad (19)$$

The correction vector to the approximated parameters, $\Delta\boldsymbol{\theta}$ (minimization step), results from equation (19). For these systems of equations the program provides two solution ways:

1) The subspace trust-region method based on the interior-reflective Newton method [12, 13]. This is a large-scale algorithm with variation of minimization step in each iteration by the approximate solution of a large linear system of equations using the preconditioned conjugate gradient method. This algorithm is able to optimize a large number of parameters in a small number of iterations and can converge from initial values far from the minima. It solves equilibrium problems better than many other classical minimization methods using first derivatives. It works with continuous functions and the number of data points must be greater than the number of parameters. Increment $\Delta\boldsymbol{\theta}_k = \boldsymbol{\theta}_{k+1} - \boldsymbol{\theta}_k$ is an approximate solution of the quadratic subproblem

$$\min\left\{ f_k(\Delta\boldsymbol{\theta}) = \mathbf{g}_k^{\mathrm{T}}\Delta\boldsymbol{\theta} + \frac{1}{2}\Delta\boldsymbol{\theta}^{\mathrm{T}}\mathbf{B}_k\Delta\boldsymbol{\theta} \right\}$$

where $\mathbf{g} = \mathbf{J}\,\mathbf{r}$, $\mathbf{B} = \mathbf{g}\,\mathbf{g}^{\mathrm{T}}$, and k is the iteration number. Then the following function is computed:

$$p_k^Y = \frac{(Y(\boldsymbol{\theta}_k + \Delta\boldsymbol{\theta}_k) - Y(\boldsymbol{\theta}_k))}{f_k(\Delta\boldsymbol{\theta}_k)}$$

If $p_k^Y > \mu$ (μ is the validity criterion) then $\boldsymbol{\theta}_{k+1} = \boldsymbol{\theta}_k + \Delta\boldsymbol{\theta}_k$, otherwise $\boldsymbol{\theta}_{k+1} = \boldsymbol{\theta}_k$ and calculation continues.

2) The method of the Jacobian matrix singular values decomposition [14, 15]. This method converges even when $\mathbf{J}^{\mathrm{T}}\mathbf{J}$ matrix is poor conditioned (having determinant close to zero). Poor conditionality occurs when initial parameters are far from real values or when two or more parameters are highly correlated. Firstly singular values decomposition of the weighted Jacobian matrix is performed:

$$\mathbf{J} = \mathbf{U}\,\boldsymbol{\Sigma}\,\mathbf{V} \quad (20)$$

where $\mathbf{U}$ is the unitary matrix of dimension $N \times N$, $\boldsymbol{\Sigma}$ is rectangular diagonal matrix of dimension $N \times d$ (d is number of fitted parameters), $\mathbf{V}$ is the unitary matrix of dimension $d \times d$. Then error

$\tau = \sigma_0 \max(\boldsymbol{\Sigma})$ is computed and vector **p** is composed from indexes of the **diag(Σ)** elements which are greater than τ. After that minimization step is calculated as follows:

$$\Delta\theta_i = \sum_{j=1}^{P} \frac{V_{ip_j}(\sum_{k}^{N} U_{kp_j} r_k)}{\Sigma_{p_j p_j}} \tag{21}$$

where $j = \overline{1,P}$ is index of the vector **p** element, $k = \overline{1,N}$ is number of experimental points, $i = \overline{1,D}$ is number of fitting parameters. This algorithm often converges in less iteration than the previous one when initial estimates are near the local minima, additionally it requires less computer time according to simplicity of calculation. Disadvantages of the algorithm are poor convergence by the initial approximations far from minima and forever loop in some cases.

Both of the minimization algorithms can work with analytical or numerical first derivatives.

The program calculates the following statistical information:

sum of weighted squared residual errors $SSE = \mathbf{r}^{\mathrm{T}}\,\mathbf{r}$,

confidence intervals $\boldsymbol{\delta\theta} = \sqrt{\dfrac{\mathbf{diag}(|(\mathbf{J}^{\mathrm{T}}\,\mathbf{J})^{-1}|)\cdot SSE}{N-D}}$,

and the Hamilton's *R*-factor.

Program was written in Qt Development Environment [16] with use of Eigen mathematical library [17]. Calculation data can be loaded in the form of formatted text or inserted into special dialog window.

Applicability of the STALABS program was tested on the example of investigation of the complex nickel(II) – *L*-histidine system by joint usage of the pH-potentiometry, multiwavelength spectrophotometry, and NMR relaxation methods. The computed species distribution diagram and reconstructed electronic absorption spectra of all the revealed species are given on Figures 1 and 2 as examples.

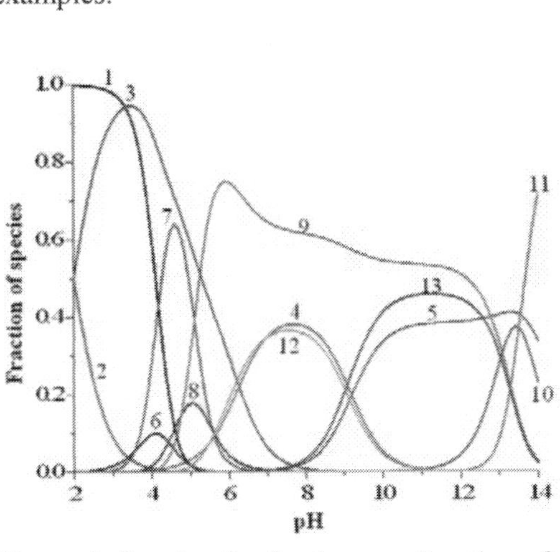

Figure 1. Species distribution as a function of pH for the Ni(II) – *L*-HisH system at a 1:4 metal:ligand ratio and T = 298 K with 1.0 M KNO_3 as background, $c_{Ni(II)}$ = 0.01 M; 1 – Ni^{2+}, 2 – $HisH_3^{2+}$, 3 – $HisH_2^{+}$, 4 – HisH, 5 – His^{-}, 6 – $Ni(HisH)^{2+}$, 7 – $Ni(His)^{+}$, 8 – $Ni(His)(HisH)^{+}$, 9 – $Ni(His)_2$, 10 – $Ni(His)(HisH_{-1})^{-}$, 11 – $Ni(HisH_{-1})_2^{2-}$, 12 – $Ni(His)_2(HisH)$, 13 – $Ni(His)_3^{-}$.

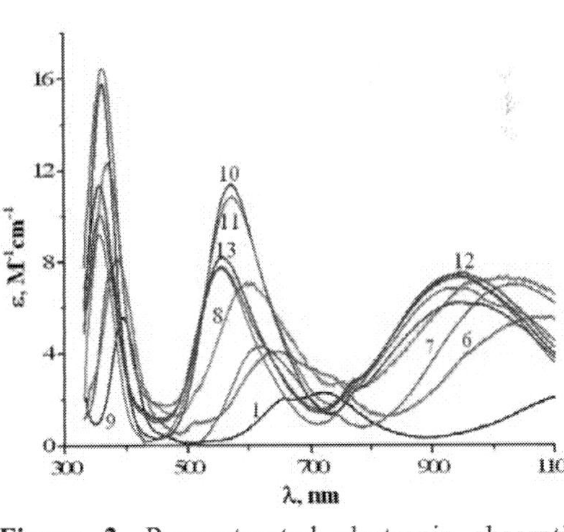

Figure 2. Reconstructed electronic absorption spectra of species in the Ni(II) – *L*-HisH system at T = 298 K with 1.0 M KNO_3 as background (numeration is given as on Figure 1).

Only four complexes ($Ni(HisH)^{2+}$, $Ni(His)^{+}$, $Ni(His)(HisH)^{+}$, and $Ni(His)_2$) have been described previously in the nickel(II) – *L*/*DL*-histidine systems [18]. The joint use of the above three methods in wide pH range (2-14) enables us to reveal and characterize four new complexes, $Ni(His)_2(HisH)$, $Ni(His)_3^-$, $Ni(His)(HisH_{-1})^-$, and $Ni(HisH_{-1})_2^{2-}$ (see Figure 1), and to establish the relaxation, chemical exchange kinetic, and spectral parameters for all the found species (Figure 2).

3. Conclusions

New computer program STALABS for a calculation of equilibrium constants, kinetic, relaxation, and spectral parameters of coordination compounds by the data of several physico-chemical methods including pH-potentiometry, multi-wavelength electronic spectroscopy, and spin-spin and spin-lattice NMR relaxation was elaborated.

Application of the STALABS program for investigation of the nickel(II) – *L*-histidine system by joint usage of the above methods allows one to achieve the most complete description of complex formation thermodynamics, ligand exchange reaction kinetics, and NMR relaxation and spectral characteristics of the eight complexes in this system.

References

[1] Gans P, Sabatini A and Vacca A 1996 *Talanta* **43** 1739
[2] Sal'nikov Yu I, Glebov A N and Devyatov F V 1989 *Polynuclear Complexes in Solutions* (Kazan: KSU)
[3] Ushanov V V, Shtyrlin V G, Nazmutdinova G A and Zakharov A V 1997 *Zh. Neorgan. Khim.* **42** 2019
[4] Brinkley S 1946 *J. Chem. Phys.* **14** 563
[5] Brinkley S 1947 *J. Chem. Phys.* **15** 107
[6] Bro R and de Jong S 1997 *J. Chemom.* **11** 393
[7] Gampp H, Maeder M, Meyer C J and Zuberbühler A D 1985 *Talanta* **32** 95
[8] Popel A A 1978 *Magnetic Relaxation Method for Analysis of Inorganic Substances* (Moscow: Khimia)
[9] Luz Z and Meiboom S 1964 *J. Chem. Phys.* **40** 2686
[10] Swift T J and Connick R E 1962 *J. Chem. Phys.* **37** 307
[11] Zakharov A V and Shtyrlin V G 1985 *Rapid Ligand Exchange Reactions* (Kazan: KSU)
[12] Coleman T F and Li Y 1996 *SIAM J. Opt.* **6** 418
[13] Coleman T F and Li Y 1994 *Math. Prog.* **67** 189
[14] Forsythe G E, Malcolm M A and Moler C B 1977 *Computer Methods for Mathematical Computation* (N. Y.: Prentice-Hall)
[15] Wilkinson J H and Reinsch C 1976 *Handbook for Automatic Computation Linear Algebra* (Heidelberg: Springer)
[16] http://qt.nokia.com
[17] http://eigen.tuxfamily.org
[18] Pettit L D 1984 *Pure & Appl. Chem.* **56** 247

Reconstruction of brain neuronal pathways in brain from the diffusion tensor MRI data.

A R Nasretdinov, K A Il`yasov and O V Nedopekin.
Institute of Physics, Kazan Federal University, Kazan 420008, Russia

E-mail: Azatto@gmail.com

Abstract. MRI based fiber tracking is a diagnostic method based on the diffusion tensor MRI data, which allows to find pathways of neuronal bundles in brain in vivo. In this work we propose a method of neuronal pathways reconstruction using A-star algorithm, with the possibility to assess its the effectiveness. One of the criteria is the probabilistic search parameter G, defined by a set of diffusion coefficients in a given volume element. The parameter G obtained trajectory correlated to its length has the meaning of entropy and allows to assess reliability of the found path. The proposed method was tested on simulated data with the characteristic behavior of trajectories of the complex variations, different cases of intersection of the beams passing through the intersection without a common voxels, and obtained characteristics of the corresponding probability.

1. Introduction

Magnetic resonance imaging (MRI) became now a powerful diagnostic tool in medicine. Apart visualization of body anatomy, MRI is able to provide information about metabolism, function and connectivity. Monitoring of brain activity with functional MRI opened a new age in neurosciences. Diffusion tensor MRI (DT MRI) [1] allows to track connectivity between different brain regions. The method is based on MRI measurements of local anisotropy of water molecules diffusion. Since myelin shears, covering axons are restricting diffusion of water molecules, the diffusion coefficients along and across axons are different. Assuming that direction of the eigenvector corresponding to the highest eigenvalue is coincides with the local direction of the axonal bundle it is possible to follow the pathway of the axonal bundle [2-4]. However since axon dimensions are much smaller that the resolution of MRI, such method detecting only major fiber track and gives ambitious results in regions of fibers crossing [5-7]. Partial volume averaging and noise contributes also to these errors.

Fiber tracking methods can be roughly divided in two classes – streamline and probabilistic methods [8]. Streamline line methods, like FACT [3] either detecting or not a single pathway fiber tract, while probabilistic methods [9,10] fiber gives different probability for different fiber track pathways. However assessing the reliability of the detected pathways is not possible by these methods. The purpose of this paper was to evaluate the reliability of the detected pathways in order to be able to compare probability of connection between different regions and in different persons.

2. Model description

If the classical A-star algorithm finds the shortest path, our modified algorithm finds the most likely path. To do this, we change one of the search criteria – G function. In this paper, a modified A-star algorithm was implemented in MatLab software package for the analysis of model data. We assumed that each voxel can contain multiple diffusion coefficients in their respective areas by which the cost of the transition at some neighboring voxel.

Define the concept of probability that, based on available data, in this way the real voxel direction coincides with the direction of k:

$$p_k = \frac{D_k}{\sum_{i=1}^{n} D_i}$$

where D_k and D_i is the diffusion coefficients in the directions of k and i.

The G cost of the transition from the parent to the current j-th cell in the direction of k is defined as:

$$G_k^j = -\ln \frac{D_k}{\sum_{i=1}^{n} D_i}$$

The G value for some arbitrary point j in the direction k will consist of the value of its parent cell G and cost of the transition from the current parent. The H function is determined by the Manhattan method, is estimated by the number of the path of elementary steps in the directions x and y, which is needed to reach the end point. Found the path must have a minimum value of G, among others. Function G increases with decreasing probability, that is, the lower the probability of movement in that direction, the more relevant the cost of transition. As we move to target the total value of G trajectory increases. This means that the further we move away from the starting area, the less likely to be in the current point.

3. Results

For simplicity here will be considered a two-dimensional model. Extension to three-dimensional case is straight forward. We define a certain curve in the form of data, represented by the set of directions and the corresponding diffusion coefficients.

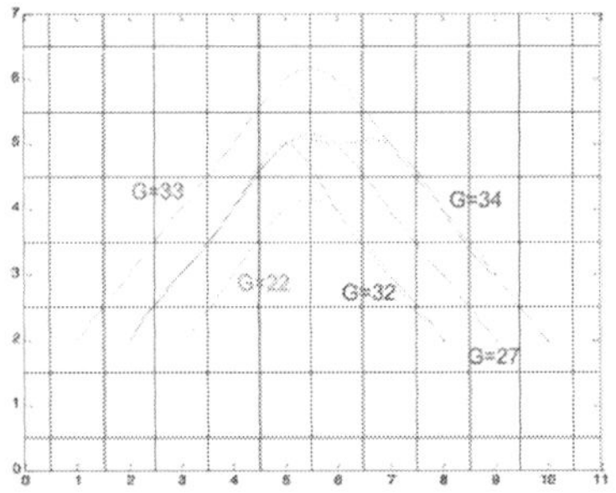

Figure 1. G-parameters of real and false trajectories.

We define a beam consisting of parallel trajectories with a characteristic curvature. Due to the presence of the isotropic component in each voxel are possible "jumps" between the paths, the results are presented in fig. 1. The smallest value of G corresponds to the highest probability. As you can see, the false path with G = 32 and the actual trajectory with G = 33 realized with almost equal probability. This is due to dependence of the parameter G of the distance, which corresponds to the fact of "the end point on, the less likely to reach it". Thus, the parameter G in its current form is not suitable for comparison reconstructed trajectories. To solve this problem we will consider the distance. We introduce a new parameter, which is determined from the following relationship:

$$\beta = \frac{G_{goal}}{h}$$

where G_{goal} is G value of the end point, it is also the path parameter, h - length of the traversed path.

The proposed parameter is essentially a kind of entropy, a measure of the probability of trajectory. We specify a parallel beam paths of the same length and calculate the corresponding β parameters (Fig. 2a).

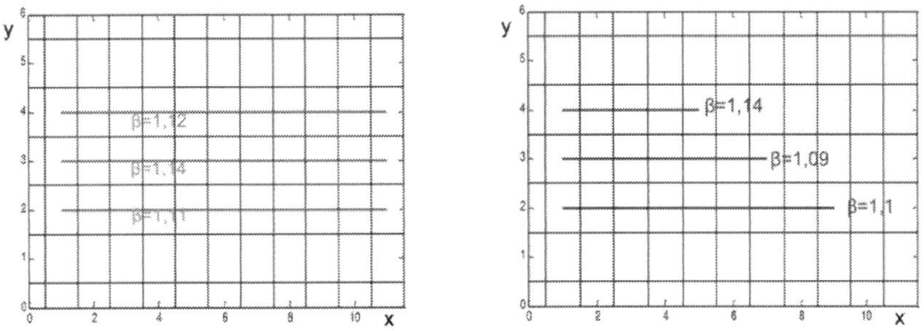

(a) (b)

Figure 2. Demonstration of independence of β parameter on the length of the trajectory.

The obtained data about the trajectories of the parameters are nearly equal, this means that the data paths are implemented with approximately the same probability. Change the location of relevant end points such that the lengths of the constructed paths were different (Fig. 2b). It is evident that the parameters for the resulting paths are also approximately equal. Thus, does not depend on the distance tracked and is the probability of the fact that, based on available data, the path is found true.

4. Dependence of β-parameter on the noise

For the model described early (Fig. 2a), we define the target areas on the neighboring trajectories and calculate the corresponding values.

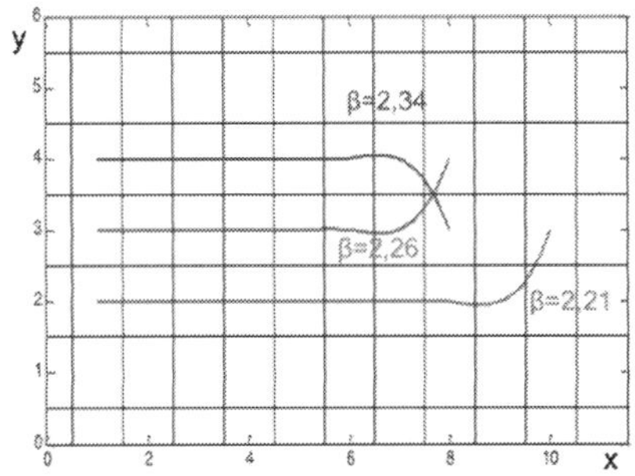

Figure 3. False paths and the corresponding β parameters.

Such transitions are possible, but significantly increases the value β . Comparing with previous results, we find that the implementation of the upper curve (Fig. 3) with a value of β= 2.34 is less likely than the top of the curve (Fig. 2a), for which β=1.12. Thus, the following values can be judged on how this or that found following the real trajectory.

With the increase of noise level jumps between the trajectories become more likely, that is, the cost of such a transition G decreases. Often this causes the accumulation of errors in choosing the direction of the search path tractography methods. When working with the model data is used the value of the signal to noise ratio of 10. Reducing this ratio, we can see how the value changes to "real" (lower value) and "imaginary" (with a high value) of the trajectories (Fig. 4a and Fig. 4b).

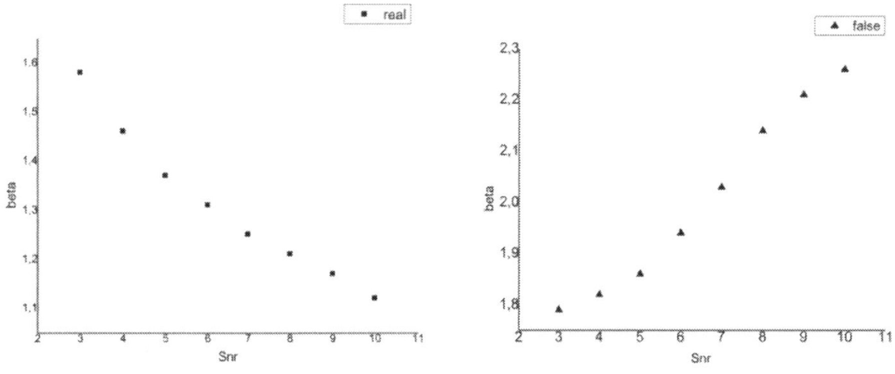

(a) (b)

Figure 4. The dependence of the β parameter for real and imaginary trajectory on the noise.

As can be seen, with increasing level of noise, setting the true trajectory begins to increase, reflecting the fact that it is becoming more difficult to fix the actual path. In second case, the noise decreases with increasing the value, thus increasing the probability of confusion with the actual path of the imaginary.

5. Changes in the degree of anisotropy

In living tissue, due to a number of reasons may be the destruction of the insulating shell pathways, as a consequence, the movement of water molecules is not limited to the barriers, and there are isotropic field. This situation can be modeled by first given a certain trajectory that describes the area of the anisotropy (Fig. 5), and then some of these areas are made isotropic.

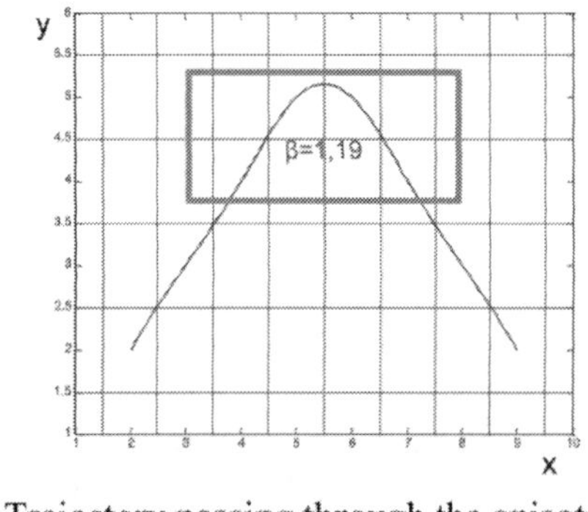

Figure 5. Trajectory passing through the anisotropic area.

Through intermediate isotropic field algorithm still builds the path, but it is an increase in the parameter β=2,72. Having by some a priori information about the behavior of the beam, we can ask a number of final and initial points along the way and keep track of how the value will vary along the length of the trajectory. Thus, we can not only detect the presence of an isotropic field, but also to localize it. The presence of the intersection of beams within a single voxel can also be reflected by the algorithm as the area during the passage through which the increase in the β parameter. This kind of analysis of real data provides information about the presence in the study area of any violations shell pathways, or the presence of tumors arise beam intersection. Consequently, the presence of changes in the system can be judged by the change trajectory, or alteration of the characteristic β parameter.

Conclusions

In this paper, we proposed a method to reconstruct pathways of the brain with the ability to assess the quality of reconstruction. Proposed likelihood parameter for reconstructed trajectories, allows to compare several pathways. Due to technical limitations, the minimum voxel size is much larger than the thickness of individual axons, thus making it difficult to resolve tracks in the areas of crossing. The proposed method provides a probabilistic assessment of the reconstructing trajectories, and gives possibility to compare them to the likelihood parameter.

References

[1] Basser P J, Mattiello J and LeBihan D 1994 *J Magn Reson* B **103** 247
[2] Bammer R, Acar B and Moseley M E 2003 *European journal of radiology* **45** 223
[3] Mori S, Crain B J, Chacko V P and van Zijl P C 1999 *Ann Neurol* **45** 265
[4] Mori S and van Zijl P C 2002 *NMR in biomedicine* **15** 468
[5] Jones D K 2003 *Magn Reson Med* **49** 7
[6] Kreher B W, Schneider J F, Mader I, Martin E, Hennig J and Il'yasov K A 2005 *Magn Reson Med* **54** 1216
[7] Moseley M E, Kucharczyk J, Mintorovitch J, Cohen Y, Kurhanewicz J, Derugin N, Asgari H and Norman D 1990 *Am J Neuroradiol* **11** 423
[8] Kreher B W Detektion von Hirnnervenfasern auf der Basis von diffusionsgewichteten Magnetresonanzdaten 2007 *Freiburg im Breisgau Albert-Ludwigs-Universität* 165
[9] Koch M A, Norris D G and Hund-Georgiadis M 2002 *NeuroImage* **16** 241
[10] Parker G J and Alexander D C sue 2005 *Philosophical transactions of the Royal Society of London* **360** 893

Entropy: A new measure of stock market volatility?

Sonia R. Bentes[1], and Rui Menezes[2]

[1]ISCAL, Av. Miguel Bombarda, 20, 1069-035, Lisbon, Portugal

[2]ISCTE, Av. das Forças Armadas, 1649-025, Lisbon, Portugal

E-mail: smbentes@iscal.ipl.pt

Abstract. When uncertainty dominates understanding stock market volatility is vital. There are a number of reasons for that. On one hand, substantial changes in volatility of financial market returns are capable of having significant negative effects on risk averse investors. In addition, such changes can also impact on consumption patterns, corporate capital investment decisions and macroeconomic variables. Arguably, volatility is one of the most important concepts in the whole finance theory. In the traditional approach this phenomenon has been addressed based on the concept of standard-deviation (or variance) from which all the famous ARCH type models – Autoregressive Conditional Heteroskedasticity Models– depart. In this context, volatility is often used to describe dispersion from an expected value, price or model. The variability of traded prices from their sample mean is only an example. Although as a measure of uncertainty and risk standard-deviation is very popular since it is simple and easy to calculate it has long been recognized that it is not fully satisfactory. The main reason for that lies in the fact that it is severely affected by extreme values. This may suggest that this is not a closed issue. Bearing on the above we might conclude that many other questions might arise while addressing this subject. One of outstanding importance, from which more sophisticated analysis can be carried out, is how to evaluate volatility, after all? If the standard-deviation has some drawbacks shall we still rely on it? Shall we look for an alternative measure? In searching for this shall we consider the insight of other domains of knowledge? In this paper we specifically address if the concept of entropy, originally developed in physics by Clausius in the XIX century, which can constitute an effective alternative. Basically, what we try to understand is, which are the potentialities of entropy compared to the standard deviation. But why entropy? The answer lies on the fact that there is already some research on the domain of Econophysics, which points out that as a measure of disorder, distance from equilibrium or even ignorance, entropy might present some advantages. However another question arises: since there is several measures of entropy which one since there are several measures of entropy, which one shall be used? As a starting point we discuss the potentialities of Shannon entropy and Tsallis entropy. The main difference between them is that both Renyi and Tsallis are adequate for anomalous systems while Shannon has revealed optimal for equilibrium systems.

1. Introduction

Financial market volatility is central to theory and practice in finance domain. Although its actuality this is not however an entirely new issue and has emerged in a systematic way when [1] first argued that the observed stock market volatility were inconsistent with the predictions of the present value models, quite popular in the past. Furthermore, [2] found out that the intemporal variation appeared to be inexplicably high and could not be rationalized even in models with a stochastic discount factor.

Even thought some authors questioned about the conclusion of excessive volatility, like [3], latter tests accounting for dividend nonstationarity and small sample bias continued to lend support to Shiller's initial claim (see [4-5]).

In this brief overview we have tried to shed some light on the theme and to unfold some of its major implications. Nevertheless, given the impracticability of analyzing the volatility as a whole we focus on its particular aspect of measurement. Here, however we face an obstacle: since volatility is not observed, there has been no agreement on how to measure it, thus emerging a plethora of techniques. Another conclusion that appeared to have raised is that volatility is volatile.

The main contribution of this paper is to compare two different approaches: one based on the statistical measure of the standard deviation or variance and the other one centered on the concept of entropy. In this regard, we particularly focus on the concept of Tsallis entropy which constitutes a possible generalization of the Boltzmann-Gibbs or Shannon entropy. These measures were both generated in the domain of physics, although the latter is also attributed to the Information Theory, and their application to financial phenomena falls in the domain of the so-called econophysics. In an analogy with terms like biophysics, geophysics and astrophysics this word was originally introduced by [6] in an attempt to legitimize the study of economics by physicists. One argument is that there were found some regularities between these two areas. Other, refers to the benefits of the experimental method commonly used in physics which departs from the observed data without imposing any prior model. Also, it is worthy to note the evidence of common research interests between these two areas. As [7] pointed out an active domain of research in physics is the characterization of the process of prices changes, *i.e.*, volatility. In our particular investigation we apply the concept of entropy to capture the presence of nonlinear dynamics in seven stock market indexes since the standard deviation evidence some limitations. The empirical analysis is conducted with data from different countries for comparative purposes.

The remainder of the paper is organized as follows: Section 2 describes the most commonly used measure of volatility - the standard deviation. Section 3 presents two different measures of entropy: the Shannon entropy and a possible generalization of it - the Tsallis entropy. Section 4 exhibits the empirical findings and Section 5 draws the conclusions.

2. Volatility and Entropy Measures: Some Concepts

In this Section we define various measures of volatility. We begin with the standard deviation and then analyze the Tsallis entropy and a special case of it - the Shannon entropy. Before proceeding further on we shall first clarify the term volatility. According to a wide range of research, volatility is popular as a synonymous of risk and uncertainty.

Based on the possibility that volatility could be not constant over time, *i.e.*, "volatility is volatile", some authors have divided the various techniques in two different categories: time invariant (or independent) and time variant (or dependent) measures. In the first group we include the techniques studied in this paper since they do not depend on time. The other one clearly exceeds the scope of our investigation and it is related with, for example, the ARCH (Autoregressive Conditional Heteroscedastic) models, and their subsequent derivations.

2.1. A traditional measure of volatility

A conventional way of measuring volatility is to compute the returns r_t of an asset:

$$r_t = \ln P_t - \ln P_{t-1}, \tag{1}$$

where P_t and P_{t-1} denote the prices at time t and $t-1$, respectively, and then estimate the corresponding standard deviation over some historical period T, given by

$$\hat{\sigma} = \sqrt{\frac{\sum_{t=1}^{T}\left(r_t - \langle r\rangle\right)^2}{T-1}}, \tag{2}$$

with $\langle r\rangle$ representing the sample average return, $\langle r\rangle = \sum r_t / T$.

Although this measure has some advantages since it is simple to estimate and has the ability to capture the probability of occurring extreme events it also shows some drawbacks. One is that it could lead to an abrupt change in volatility once shocks fall out of the measurement sample. And, if shocks are still included in a relatively long measurement sample period, then an abnormally large observation will imply that the forecast will remain in an artificial high level even thought the market is subsequently tranquil. Finally, it only captures linear relationships, ignoring all kind of nonlinear dynamics among data. In the light of this, some more sophisticated measures has emerged in a way to improve the understanding of volatility. Regarding to this, a measure that appears to be particularly relevant is the concept of entropy, which constitutes our major aim in this study.

However, it is worthy to note that in spite of all the flaws that have been recognized by a wide body of research the standard deviation is still the most popular measure of volatility being used as a benchmark for comparing the forecast ability of more complex models.

3. Theoretical framework: the concept of entropy

An alternative way to study stock market volatility is by applying concepts of physics which significant literature has already proven to be helpful in describing financial or economic problems, such as, the concept of entropy. Although there are many different understandings of this notion the most commonly used in literature is as a measure of ignorance, disorder, uncertainty or even lack of information (see [8]).Later, in a subsequent investigation Shannon [9] provided a new insight into this matter showing that entropy wasn't only restricted to thermodynamics but could instead be applied in any context where probabilities can be defined.

3.1. Shannon entropy

For a given a probability distribution $p_i \equiv p(X = i)$, $(i = 1,\ldots,n)$ of a given random variable X, Shannon entropy $S(X)$, for the discrete case can be defined as

$$S(X) = -\sum_{i=1}^{n} p_i \log p_i, \tag{3}$$

where $0\log 0$ is defined as 0 and the normalized associated probabilities $\sum_{i=1}^{N} p_i = 1$. Shannon entropy has been most successful in the treatment of equilibrium systems in which short/space/temporal interactions with ergodicity and independence dominate. However, there are many anomalous systems in nature that do not verify the simplifying assumption of ergodicity and independence. To overcome this kind of drawback Tsallis [11] derived a new measure of entropy, known as Tsallis entropy.

3.2. Tsallis entropy

For any non-negative real number q and considering the probability distribution $p_i \equiv p(X = i)$, ($i = 1,\ldots,n$) of a given random variable X, Tsallis entropy (Tsallis, [10]) denoted by $S_q(X)$ for the discrete case, is defined as

$$S_q(X) = k\frac{1-\sum_{i=1}^{W} p_i^q}{q-1}, \tag{4}$$

where the q-exponential function is defined by

$$y = \left[1+(1-q)x\right]^{1/(1-q)} \equiv e_q^x(x) \quad , \left(x \geq 0, q \in {}^\circ\right), \tag{5}$$

whose inverse is the q-logarithm function

$$y = \frac{x^{1-q}-1}{1-q} \equiv \ln_q(x) \quad \left(x \geq 0, q \in {}^\circ\right). \tag{6}$$

The entropic index q characterizes the statistics we are dealing with; as $q \to 1$, $S_q(X)$ recovers $S(X)$ since the q-logarithm uniformly converges to a natural logarithm as $q \to 1$. This index may be regarded as a biasing parameter since $q < 1$ privileges rare events and $q > 1$ privileges common events ([11]). A concrete consequence of this is that while the Shannon entropy yields exponential equilibrium distributions, Tsallis entropy yields power-law distributions.

4. Empirical Analysis

This Section explores the empirical relevance of the theoretical results obtained by both perspectives. To do so we gathered data from several different countries in order to detect whether some similarities among all of them can be found. This is especially relevant in the context of globalization, where we living in, which also constitutes an active area of research.

4.1. Data

In our empirical research the data set compounds the daily returns of the FTSE 100 (U.K), SP 500 (U.S.A.), CAC 40 (France), MIB 30 (Italy), NIKKEI 225 (Japan), IBEX 35 (Spain) and PSI 20 (Portugal) extending from 8 January 1990 to 7 April 2006. Each index contains 4240 observations, which is large enough to make our analysis meaningful. These data were collected on a daily basis without considering the re-investment of dividends and were computed in accordance to Eq. (1) where the closing prices were the inputs.

4.2. Standard deviations results

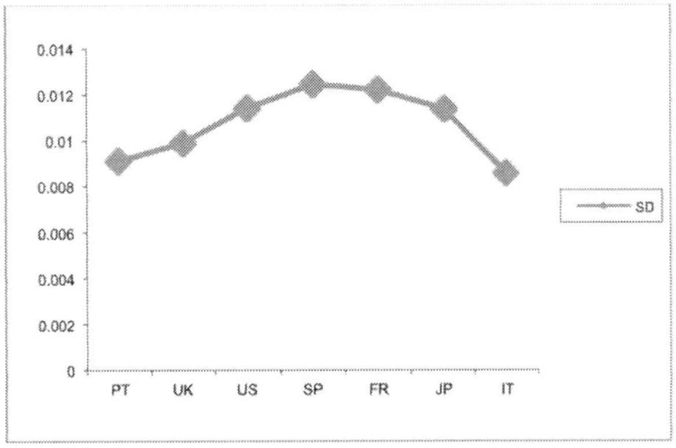

Fig. 1. Standard deviation of stock index returns

We observe that the higher volatility indexes are: first the IBEX 35, second CAC 40 and third, NIKKEI 225. Accordingly, PSI 20 and FTSE 100 are the lower ones. The ranking also shows that all values are close to zero which may suggest that all of them exhibit low volatility in spite of their particular values. However, these outcomes should not be regarded in a strict sense since they exhibit some drawbacks, already mentioned in Section 2 of this research. Indeed, they are influenced by abnormally high observations and are not able to capture nonlinear dynamics.

4.3. Entropy results

In the domain of the econophysics approach we have computed the Tsallis and Shannon entropies, which are depicted in Fig. 2.

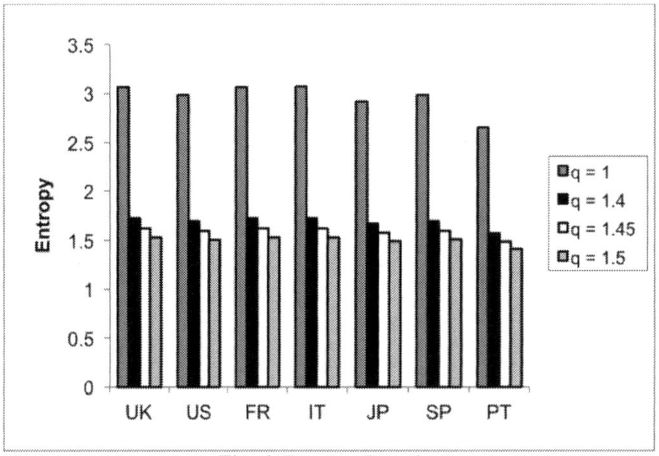

Fig. 2 Entropy Results

All entropies were estimated with histograms based on equidistant cells. For the calculation of Tsallis entropy we have set values at 1.4, 1.45 and 1.5 for the index q, which is consistent with the finding that when considering financial data their values lie within the range $q \simeq 1.4\text{-}1.5$ (see [11]). It is worthy to note that MIB 30 index (Italy) attained the highest levels along with CAC 40 and NIKKEI 225. As for the others the results are somewhat mixed since they vary according to technique adopted.

In the overall, it appears that the use of entropy as a measure of uncertainty allows better insights over the identification of volatile markets, by distinguishing them more sharply, than simply using the standard deviation. This leads us to the conclusion that entropy is more general and better suited for describing stock market volatility with its generality arising directly from the fact that it can be computed from metric and non-metric data. Apart from that the major advantages of entropy when compared to the standard deviation can be summarized as follows [12]: (i) it incorporates much more information than the latter; (ii) it is not dependent upon any particular distribution; in other words, it is distribution free, thus avoiding the introduction of errors through the fitting of the distribution of returns to a normal-like distribution. This is especially true when dealing with non-symmetric distributions with generally non-normal additional moments, which seems to be the case in some phenomena in Finance; (iii) Since entropy is independent of the mean for all types of distributions, it satisfies the first order conditions and (iv) finally, due to its common understanding of mean uncertainty, it also serves as a measure of dispersion.

Nevertheless, some disadvantages have also to be weighted when considering the use of any kind of entropy. First one has to do with its inherent complexity when compared to the simple standard deviation. Second, is related to the amount of statistical bias in theses measures due to the degrees of freedom allowed in an experiment.

5. Conclusions

In this paper we have investigated the volatility of seven indexes: CAC 40, MIB 30, NIKKEI 225, PSI 20, IBEX 35, FTSE 100 and SP 500. Our major goal was to compare two different perspectives: one based on the standard deviation and other supported by the concept of entropy. For our purpose two variants of this notion were regarded: the Tsallis and the Shannon entropies.

In particular, the results from both entropies have shown nonlinear dynamics in the volatility of all indexes and must be understood in complementarity since they point out to the same conclusions. However, most of the outcomes are not in accordance with the statistics produced by the standard deviation, which emphasizes that this method is not able to capture the overall behaviour of dispersion. This is especially relevant for the decision making process in which all the information is regarded as necessary and useful. Nonetheless, in spite of all the divergences encountered there is still something in common that relates to the fact that regardless of the method applied the CAC 40 and NIKKEI 225 returns are always the most volatile ones.

In this study we specially address the concept of entropy as an alternative to the standard deviation since it can captures the uncertainty and disorder in a time series without imposing any constraints on the theoretical probability distribution, which constitutes its major advantage. Additionally, in order to capture global serial dependence one should use a specific measure such as, for example, mutual information.

References

[1] Shiller R J 1981 *American Economic Review* **71** 421
[2] Grossman S J and Shiller R J 1981 *American Economic Review* **71** 222
[3] Flavin M A 1983 *Journal of Political Economy* **91** 929.
[4] Kleidon A W 1986 *Journal of Political Economy* **94** 953
[5] Mankiw N G, Romer D and Shapiro M D 1985 *Journal of Finance* **40** 667
[6] Stanley H E, Afanasyev V, Amaral L A N, Buldryrev S V, Goldberger A L, Havlin S, Maass P, Mantegna R N, Pemg C-K, Prince P A, Salinger M A, Stanley M H R and Viswananthan G M 1996 *Physica A* **224** 302
[7] Mantegna R N and StanleyH E 2004 *An Introduction to Econophysics : correlations and Complexity in Finance* (Cambridge University Press, Cambridge)
[8] Golan A 2002 *Journal of Economertics* **107** 1
[9] Shannon C E 1948 *The Bell System Technical Journal* **27** 379
[10] Tsallis C 1988 *Journal of Statistical Physics* **52** 479
[11] Tsallis C, Anteneodo C, Borland L and Ososrio R 2003 *Physica A* **324** 89
[12] Philippatos G C and Wilson C J 1972 *Computational Statistics and Data Analysis* **32** 1

Reserves Represented by Random Walks

J A Filipe, M A M Ferreira, and M Andrade
Instituto Universitário de Lisboa (ISCTE-IUL), BRU-IUL, Lisboa, Portugal

Email : jose.filipe@iscte.pt

Abstract. The reserves problem is studied through models based on Random Walks. Random walks are a classical particular case in the analysis of stochastic processes. They do not appear only to study reserves evolution models. They are also used to build more complex systems and as analysis instruments, in a theoretical feature, of other kind of systems. In this work by studying the reserves, the main objective is to see and guarantee that pensions funds get sustainable. Being the use of these models considering this goal a classical approach in the study of pensions funds, this work concluded about the problematic of reserves. A concrete example is presented.

1. Introduction

The problem of Gambler's ruin, which reserves behave according to a simple random walk, is presented in a lot of textbooks about the stochastic processes theory in relation with Markov Chains, Random Walks, Martingales and even in other contexts. Very clear approaches to this subject are Billingsley [1] and Feller [2] that solve the problem through the classic first step analysis in order to obtain a difference equation. This proceeding is followed in this work. In alternative, Grimmett and Stirzaker [3] and Karlin and Taylor [4] present also resolutions through the Martingales Theory, as an applications' example of the Martingales Stopping Time Theorem.

In the next section the gambler's ruin problem is presented. In the following an approach based on the general random walk, enlarging the former one, is outlined. The work finishes with conclusions and a brief list of references.

2. Gambler's Ruin

So consider a gambler disposing of an initial capital of x monetary unit that intends to play a sequence of games till his/her fortune reaches a value of k monetary unit. Suppose that x and k are integer numbers that satisfy the conditions $x > 0$ and $k > x$. In each game, the gambler either wins 1 monetary unit with probability p or loses 1 monetary unit with probability $q = 1 - p$. What is the probability that the gambler ruins before attaining his/her target? That is, which is the probability of losing the x monetary unit before accumulating wins in the amount of $k - x$ monetary unit?

Call $X_n, n = 1, 2, \dots$ the result of the nth game. Evidently $X_1, X_2, \dots$ are independent and identically distributed random variables with probability function:

$$P(X_n = 1) = p, P(X_n = -1) = q = 1 - p.$$

So, the reserves, that is: the fortune, of the player after the nth game correspond to the simple random walk:

$$S_0 = x, S_n = S_{n-1} + X_n, n = 1, 2, \dots .$$

It is intended to determine the gambler's ruin probability. Call this probability $\rho_k(x)$. It corresponds to the probability that $S_n = 0$ and $0 < S_i < k, i = 0, 1, \dots, n - 1$ for $n = 1$ or $n =$

2 or .... If $\rho_k(x)$ is conditioned to the result of the first game it is obtained, as a consequence of the Total Probability Law,

$$\rho_k(x) = p\rho_k(x+1) + q\rho_k(x-1), \qquad 0 < x < k \qquad (2.1).$$

Considering conveniently$0 \le x \le k$, this difference equation is easy to solve with the support of the evident border conditions

$$\rho_k(0) = 1,\ \rho_k(k) = 0 \qquad (2.2).$$

Write (2.1) as

$$\rho_k(x) - \rho_k(x-1) = \frac{p}{q}\left(\rho_k(x+1) - \rho_k(x)\right), \qquad 0 < x < k \qquad (2.3).$$

For $x = k - 1$ and considering (2.2),

$$\rho_k(k-2) = \rho_k(k-1)\left(1 + \frac{p}{q}\right).$$

Based on this equation, when $x = k - 2$, (2.3) becomes:

$$\rho_k(k-3) = \rho_k(k-1)\left(1 + \frac{p}{q} + \left(\frac{p}{q}\right)^2\right).$$

Going on with this proceeding it is obtained the general expression

$$\rho_k(k-y) = \rho_k(k-1)\left(1 + \frac{p}{q} + \left(\frac{p}{q}\right)^2 + \cdots + \left(\frac{p}{q}\right)^{y-1}\right), 0 < y \le k \qquad (2.4).$$

Considering again (2.2), after (2.4), with $y = k$ it is obtained

$$\rho_k(k-1) = 1\Big/\left(1 + \frac{p}{q} + \left(\frac{p}{q}\right)^2 + \cdots + \left(\frac{p}{q}\right)^{k-1}\right) \qquad (2.5).$$

Finally, substituting (2.5) in (2.4) and performing the change of variable $y = k - x$ it is obtained the solution of the difference equation (2.1) with the border conditions (2.2):

$$\rho_k(k-1) = \begin{cases} \dfrac{1-(p/q)^{k-x}}{1-(p/q)^k}, \text{if } p \neq \dfrac{1}{2} \\ \dfrac{k-x}{k}, \qquad \text{if } p = \dfrac{1}{2} \end{cases} \qquad (2.6).$$

Call N_a the first passage time by a of the random walk S_n:

$$N_a = \min\{n \ge 0 : S_n = a\}.$$

In consequence it is possible to write $\rho_k(x) = P(N_0 < N_k | S_0 = x)$. And it is pertinent to take in (2.6) the limit as k converges to ∞ to evaluate $\rho(x)$, the ruin probability of a gambler infinitely ambitious. In the context of the simple random walk S_n, $\rho(x) = P(N_0 < \infty | S_0 = x)$. After (2.6)

$$\rho(x) = \lim_{k\to\infty} \rho_k(x) = \begin{cases} (q/p)^x, \text{if } p > \dfrac{1}{2} \\ 1, \qquad \text{if } p \le \dfrac{1}{2} \end{cases} \qquad (2.7).$$

Note that $\mu = E[X_n] = 2p - 1$. It is relevant to see, after (2.7), that the ruin probability is 1 for the simple random walk at which the mean of the step is $\mu \le 0 \Leftrightarrow p \le \frac{1}{2}$.

3. The General Random Walk

Suppose that the contributions (pensions) received (paid), by time unit, for a fund may be described as a sequence of random variables $\xi_1, \xi_2, \ldots$ $(\eta_1, \eta_2, \ldots)$. State that $\xi_n(\eta_n)$ is the value of the contributions (pensions) received (paid) by the fund during the nth time unit and so $X_n = \xi_n - \eta_n$ is the reserves variation occurred in the fund at the nth time unit. Supposing that $X_1, X_2, \ldots$ is a sequence of non degenerated random variables, independent and identically distributed, so the stochastic process defined recursively as:

$$\tilde{S}_0 = x, \tilde{S}_n = \tilde{S}_{n-1} + X_n, \qquad \text{with } n = 1, 2, \ldots,$$

is a general random walk that represents the evolution of the fund reserves, since the initial level x till the value $\tilde{S}_n$ after n time units.

It is intended to study the game reserves exhaustion probability, that is the fund ruin. For x and k real numbers fulfilling $x > 0$ and $k > x$, it is considered first the evaluation of $\rho_k(x)$, the probability that the fund reserves decrease from an initial value x to a value in $(-\infty, 0]$ before reaching a value in $[k, +\infty)$. Then, calculating the limit, as in the former section, it is considered the evaluation of $\rho(x)$, the eventual fund ruin probability, admitting so that the random walk, that represents its reserves, evolves with no restrictions at the right of 0.

The method exposed is recognized in the stochastic processes literature as Wald's Approximation. The expositions of Grimmett and Stirzaker [3] and Cox and Miller [5], about this subject are closely followed in this work. It will be considered the process $S_n = \tilde{S}_n - x$, that is, the random walk

$$S_0 = 0, S_n = S_{n-1} + X_n, n = 1, 2, \ldots.$$

instead of the $\tilde{S}_n$ process.

So, when evaluating $\rho_k(x)$, what in fact is being considered is the probability that the process S_n is visiting the set $(-\infty, -x]$ before visiting the set $[k - x, +\infty)$. And when evaluating $\rho(x)$ what is being considered is only the probability that the process S_n goes down from the initial value 0 till a level lesser or equal than $-x$.

Begin considering the non-null value θ for which the X_1 moments generator function assumes the value 1. It is assumed that such a θ exists, that is, θ satisfies

$$E\left[e^{\theta X_1}\right] = 1, \theta \neq 0 \qquad (3.1).$$

Define the process:

$$M_n = e^{\theta S_n}, n = 0, 1, 2, \ldots.$$

It is obvious that $E[|M_n|] < \infty$ and that, after (3.1)

$$E[M_{n+1}|X_1, X_2, \ldots, X_n] = E\left[e^{\theta(S_n + X_{n+1})}|X_1, X_2, \ldots, X_n\right] = e^{\theta S_n} E\left[e^{\theta X_{n+1}}|X_1, X_2, \ldots, X_n\right] = M_n.$$

So, the process M_n is a Martingale in relation to the sequence of random variables $X_1, X_2, \ldots$. Consider now N the S_n first passage time to outside the interval $(-x, k - x)$:

$$N = \min\{n \geq 0: S_n \leq -x \text{ or } S_n \geq k - x\}.$$

It is easy to check that the random variable N is a stopping time – or a Markov time – for which the following conditions are fulfilled:

$$E[N] < \infty \text{ and}$$
$$E[|M_{n+1} - M_n||X_1, X_2, \ldots, X_n] \leq 2e^{|\theta|a}, \text{for } n < N$$

and $a = -x$ or $a = k - x$.

In relation with this subject see Grimmett and Stirzaker [3]. Under these conditions, it is possible to apply the Martingales Stopping Time Theorem and, in consequence:

$$E[M_n] = E[M_0] = 1 \qquad (3.2).$$

Also

$$E[M_n] = E\left[e^{\theta S_N} | S_N \leq -x\right] P(S_N \leq -x) + E\left[e^{\theta S_N} | S_N \geq k - x\right] P(S_N \geq k - x) \ (3.3).$$

Performing the approximations

$$E\left[e^{\theta S_N} | S_N \leq -x\right] \cong e^{-\theta x}$$

and

$$E\left[e^{\theta S_N} | S_N \geq k - x\right] \cong e^{\theta(k-x)},$$

and considering that $P(S_N \leq -x) = \rho_k(x) = 1 - P(S_N \geq k - x)$, after (3.2) and (3.3), it is obtained

$$\rho_k(x) \cong \frac{1 - e^{\theta(k-x)}}{e^{-\theta x} - e^{\theta(k-x)}}, \text{when } E[X_1] \neq 0 \quad (3.4).$$

This is the Classic Approximation for the Ruin Probability in the conditions stated in (3.1). Note that to admit a non-null solution θ for the equation $E\left[e^{\theta X_1}\right] = 1$ implies in fact to assume that $E[X_1] \neq 0$.

Out of these considerations is the situation for which the only solution of the equation $E\left[e^{\theta X_1}\right] = 1$ is precisely $\theta = 0$; it means, situation at which $E[X_1] = 0$. This case may be dealt through the following passage to the limit:

$$\rho_k(x) \cong \lim_{\theta \to 0} \frac{1 - e^{\theta(k-x)}}{e^{-\theta x} - e^{\theta(k-x)}} = \frac{k - x}{k}, \qquad \text{when } E[X_1] = 0 \qquad (3.5).$$

As for $\rho(x)$, the probability that the process S_n decreases eventually from the initial value 0 to a level lesser or equal than $-x$, is also got from (3.4), now for a different passage to the limit:

$$\rho(x) \cong \lim_{k \to \infty} \frac{1 - e^{\theta(k-x)}}{e^{-\theta x} - e^{\theta(k-x)}} = e^{\theta x}, \qquad \text{if } \theta < 0 \Leftrightarrow E[X_1] > 0 \qquad (3.6).$$

As it may be deduced from the former section results about the simple random walk, it is legitimate to accept $\rho(x) = 1$ when $\theta \geq 0 \Leftrightarrow E[X_1] \leq 0$.

4. Example

Suppose that $X_1, X_2, \ldots$ constitute a sequence of independent random variables with normal distribution with mean μ and standard deviation σ. That is, admit that X_n, the fund reserves variation at the nth time unit, has normal distribution with those parameters. In this case, the moments generator function is:

$$E\left[e^{\theta X_1}\right] = \frac{1}{\sqrt{2\pi}\sigma}\int_{-\infty}^{+\infty} e^{\theta x - \frac{(x-\mu)^2}{2\sigma^2}}dx = e^{\theta\mu + \frac{\theta^2\sigma^2}{2}}.$$

The solution of the equation (3.1) is given by

$$\theta = \frac{-2\mu}{\sigma^2}, \mu \neq 0.$$

The ruin probability $\rho_k(x)$ is obtained substituting this result in (3.4):

$$\rho_k(x) \cong \frac{1 - e^{-\frac{2\mu(k-x)}{\sigma^2}}}{e^{\frac{2\mu x}{\sigma^2}} - e^{-\frac{2\mu(k-x)}{\sigma^2}}}, \text{when } \mu \neq 0 \quad (4.1).$$

It is evident that this particularization does not influence the approximation to $\rho_k(x)$ when $\mu = 0$ that, as it was seen, is given by (3.5). After (3.6),

$$\rho(x) \cong e^{-\frac{2\mu x}{\sigma^2}}, \qquad \text{when } \mu > 0 \qquad (4.2).$$

5. Conclusions

The simple and general random walks are classic stochastic processes broadly studied. They do not appear only as reserves evolution models. Are also used to build more complex systems and as analysis instruments, in a theoretical feature, of other kind of systems.

In the approach presented some methodologies applied in the study of this kind of processes are highlighted: Difference Equations and Martingales Theory.

It is important to note that, with this approach the reserves systems are treated as if they were physical systems. It is not obvious that the direct application of these principles to financial reserves funds, ignoring their own valuation and devaluation dynamics as time goes by, is legitimate.

Those models, and the consequent stability systems appreciation done with basis on the evaluation of the probability of the reserves exhaustion or ruin, seem valid only in scenarios where constant prices are considered. The integration of factors associated to the time depreciation process of the value of the money in the modulation of financial reserves, although complicating eventually the mathematical models involved, seems important.

Acknowledgments

This work was financially supported by FCT through the Strategic Project PEst-OE/EGE/UI0315/2011.

References

[1] Billingsley P 1986 *Probability and Measure* 2nd Ed. John Wiley & Sons, New York

[2] Feller W 1968 *An Introduction to Probability Theory and its Applications* Vol. 1, 3rd Ed. John Wiley & Sons, New York

[3] Grimmett G R and Stirzaker D R 1992 *Probability and Random Processes* 2nd Ed. Clarendon Press, Oxford

[4] Karlin S and Taylor H M 1975 *A First Course in Stochastic Processes* 2nd Ed. Academic Press, New York

[5] Cox D R and Miller H D 1965 *The Theory of Stochastic Processes* Chapman & Hall, London

[6] Andrade M 2010 *Journal of Mathematics and Technology* **1** 96
[7] Ferreira M A M 2010 *Journal of Mathematics and Technology* **1** 91
[8] Ferreira M A M and Andrade M 2011 *International Journal of Latest Trends in Finance and Economic Sciences* **1**(2) 91
[9] Ferreira M A M, Andrade M, Filipe J A and Coelho, M. P., 2011 *International Journal of Latest Trends in Finance and Economic Sciences* **1**(4) 171
[10] Figueira J 2003 *Aplicação dos Processos de Difusão e da Teoria do Renovamento num Estudo de Reservas Aleatórias*, PhD Thesis, ISCTE
[11] Figueira J and Ferreira M A M 1999 *Portuguese Review of Financial Markets* **2** 75

Studying Pensions Funds Through an Infinite Servers Nodes Network: A Theoretical Problem

M A M Ferreira, M Andrade and J A Filipe
Instituto Universitário de Lisboa (ISCTE-IUL), BRU-IUL, Lisboa, Portugal

E-mail : manuel.ferreira@iscte.pt

Abstract. This study intends to present a representation of a pensions fund through a stochastic network with two infinite servers nodes. With this representation it is allowed to deduce an equilibrium condition of the system with basis on the identity of the random rates expected values, for which the contributions arrive to the fund and the pensions are paid by the fund. In our study a stochastic network is constructed where traffic is represented. This network allows to study the equilibrium in the system and it is admissible to get a balance to a pensions fund. A specific case is studied. When the arrivals from outside at nodes A and B are according to a Poisson process, with rates λ_A and λ_B, respectively, the system may be seen as a two nodes network where the first node is a $M/G/\infty$ queue and second a $M_t/G/\infty$ queue. For this case in the long term the conditions of equilibrium are as follows: $m_A\lambda_A\alpha_A = m_B(\rho\lambda_A + \lambda_B)\alpha_B$. In this formula it is established a relationship among the two nodes. Several examples are given in the study.

1. Introduction

Consider two nodes, service centres, A and B both with infinite servers. The traffic through arches a to e is as it is schematized in Figure 1.

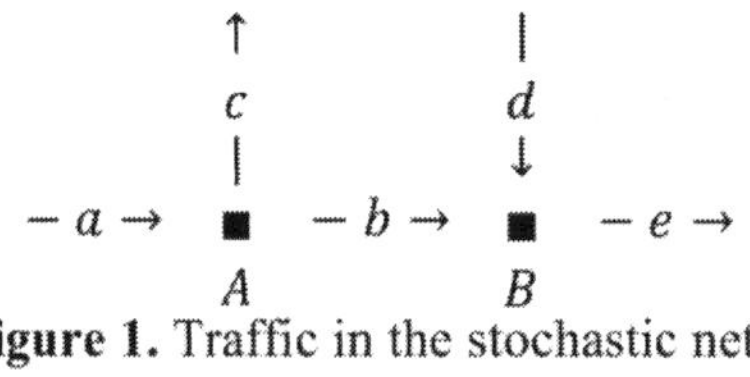

Figure 1. Traffic in the stochastic network

The users arrive to node A by arch a at rate λ_A. And the service time at this node is a positive random variable with distribution function (d.f.) G_A and finite mean α_A. After node A the users go to node B through b with probability p. Or just abandon the system through arch c with probability $1 - p$.

The users coming directly from outside through d at rate λ_B have also access to the service supplied at B, according to a positive random variable with density function G_B and finite mean α_B. The system is abandoned by these users through arch e.

In [1] this system is suggested as a representation of a pensions fund. So at node A arrive individuals that pay, during the service time, their contributions to the fund. The pensioners are at node B, which service represents their pensions payment by the fund. This representation reflects also the functions of the common social security funds and that is why it accepts the access of pensioners that have not formerly participated, at node A, in the building of the fund.

The target of this study is, having this representation in mind, to obtain results about the transient behavior of the system from the point of view of its equilibrium and autonomy.

2. The Fund Equilibrium

Let $N_A(t)$ and $N_B(t)$ be the random variables (r.v.) that represent the number of individuals by time t at nodes A and B, respectively. Consider also the sets of the independent and identically distributed random variables:

$$X_{A_1}(t), X_{A_2}(t), X_{A_3}(t), \dots, \left(X_{B_1}(t), X_{B_2}(t), X_{B_3}(t), \dots\right)$$

which designate the unitary contributions, pensions by time t, with mean $m_A(t)$ and $m_B(t)$.

The system is in equilibrium when the expected values of the rates at which the contributions are being received and the pensions are being paid by the fund are identical:

$$E\left[\sum_{i=1}^{N_A(t)} X_{A_i}(t)\right] = E\left[\sum_{j=1}^{N_B(t)} X_{B_j}(t)\right].$$

That is, by Wald's equation:

$$m_A(t)E[N_A(t)] = m_B(t)E[N_B(t)] \qquad (1).$$

Eq. (1) just stays that at each instant the mean value of the unitary pension should be proportional to the mean value of the unitary contribution, with the ratio between the averages of the numbers of contributors and pensioners as proportionality factor. Being $t = 0$ the origin time, its solution corresponds, for $t > 0$, to the following pairs:

$$\left(m_A(t); m_B(t)\right) = \left(m_A(t); \frac{m_A(t)E[N_A(t)]}{E[N_B(t)]}\right),$$

where $m_A(t)$ is independent of the equilibrium.

If the mean value of the unitary pension is initially 1, and grows continuously with an interest rate r,

$$m_B(t) = e^{rt}$$
$$m_A(t) = e^{rt}(E[N_B(t)]/E[N_A(t)]).$$

It is elementary, after Eq. (1),

$$E[N_A(t)] < E[N_B(t)] \Rightarrow m_A(t) > m_B(t).$$

So, in equilibrium, the mean value of the unitary pension is smaller than the mean value of unitary contribution whenever the number of pensioners at B is bigger than the number of contributors at A.

3. The Case of Poisson Arrivals

If the arrivals from outside at nodes A and B are according to a Poisson process, with rates λ_A and λ_B, respectively, the system may be seen as a two nodes network where the first node is a $M/G/\infty$ queue and second a $M_t/G/\infty$ queue, see for instance Ferreira and Andrade [2]. So, $N_A(t)$ is Poisson distributed with parameter, see Ross [3]

$$\lambda_A \int_0^t (1 - G_A(v))\, dv.$$

The output of the first node is a non-homogeneous Poisson process with intensity function $\lambda_A G_A(t)$ and, consequently, the global arrivals rate at node B is $p\lambda_A G_A(t) + \lambda_B$. Under this conditions $N_B(t)$ is Poisson distributed with parameter, see Harrison and Lemoine [4]:

$$\int_0^t (p\lambda_A G_A(v) + \lambda_B)(1 - G_B(t - v))dv.$$

And Eq. (1) is written like this

$$m_A(t)\lambda_A \int_0^t (1 - G_A(v))\, dv = m_B(t) \int_0^t (p\lambda_A G_A(v) + \lambda_B)(1 - G_B(t - v))dv \quad (2).$$

When $t \to \infty$ the equilibrium conditions assumes the following form where $m_i = \lim_{t\to\infty} m_i(t)\,, i = A, B$:

$$m_A \lambda_A \alpha_A = m_B(\rho\lambda_A + \lambda_B)\alpha_B \qquad (3).$$

If the service times at nodes A and B have density functions concentrated in the intervals $[0, a]$ and $[0, b]$, $m_A\lambda_A\alpha_A = m_B(\rho\lambda_A + \lambda_B)\alpha_B$ for $t \geq a + b$.

4. Example

In this section some concrete examples of service times distributions will be considered.

4.1. Uniformly Distributed Service Times

If the service times are uniformly distributed, supposing that $\alpha_B < \alpha_A$, it is obtained for Eq. (2) in $0 \leq t < 2\alpha_A + 2\alpha_B$, not to repeat what has just been mentioned:

i) $m_A(t)\lambda_A\left(t - \frac{t^2}{4\alpha_A}\right) = m_B(t)\lambda_B\left(t - \frac{t^2}{4\alpha_B}\right) + \quad m_B(t)p\lambda_A\left(\frac{t^2}{4\alpha_A} - \frac{t^3}{24\alpha_A\alpha_B}\right), \quad$ if $0 \leq \frac{t}{2} < \alpha_B$

ii) $m_A(t)\lambda_A\left(t - \frac{t^2}{4\alpha_A}\right) = m_B(t)\lambda_B\alpha_B + \ m_B(t)p\lambda_A\left(-\frac{\alpha_B^2}{3\alpha_A} - \frac{t\alpha_B}{2\alpha_A}\right), \alpha_B \leq \frac{t}{2} < \alpha_A$

iii) $m_A(t)\lambda_A\alpha_A =$
$m_B(t)\lambda_B\alpha_B + \quad m_B(t)p\lambda_A\left(-\alpha_A - \frac{\alpha_A^2}{12\alpha_B} + t - \frac{(t-\alpha_A)^2}{4\alpha_B} + \frac{(t-2\alpha_B)^3}{24\alpha_A\alpha_B}\right), \quad$ if $\alpha_A \leq \frac{t}{2} <$
$\alpha_A + \alpha_B$.

4.2. Exponentially Distributed Service Times

If the service times are exponentially distributed the equilibrium distribution is given by:

i) $m_A(t)\lambda_A\alpha_A\left(1 - e^{-\frac{t}{\alpha_A}}\right) = m_B(t)(p\lambda_A + \quad \lambda_B)\alpha_B\left(1 - e^{-\frac{t}{\alpha_B}}\right) - m_B(t)\frac{\rho\lambda_A\alpha_A\alpha_B}{\alpha_A - \alpha_B}\left(e^{-\frac{t}{\alpha_A}} -\right.$
$\left. e^{-\frac{t}{\alpha_B}}\right)$, if $\alpha_A \neq \alpha_B$

$ii)$ $m_A(t)\lambda_A\alpha_A\left(1-e^{-\frac{t}{\alpha_A}}\right)=m_B(t)\,(p\lambda_A+\lambda_B)\alpha_A\left(1-e^{-\frac{t}{\alpha_A}}\right)-m_B(t)p\lambda_A te^{-\frac{t}{\alpha_A}},$
if $\alpha_A=\alpha_B$

4.3. Service Times with a Particular Distribution Function

Solving Eq. (2) in the way presented above becomes quite difficult with other standard distributions for the service times. So now it will be considered a collection of density functions, see Ferreira and Andrade [5] and Ferreira and Andrade [6], for the service times given by

$$G_i(v)=1-\frac{(1-e^{-\rho_i})(\gamma_i+\beta_i)}{\gamma_i e^{-\rho_i}(e^{(\gamma_i+\beta_i)v}-1)+\gamma_i}, v\geq 0, \gamma_i>0, \rho_i>0, -\gamma_i\leq\beta_i\leq\frac{\gamma_i}{e^{-\rho_i}-1}, i=A,B.$$

The mean distribution is $\alpha_i=\rho_i/\gamma_i$. In this case Eq. (2) becomes

$$m_A(t)\frac{\lambda_A}{\gamma_A}ln\frac{e^{(\gamma_A+\beta_A)t}}{e^{-\rho_A}(e^{(\gamma_A+\beta_A)t}-1)+1}$$
$$=m_B(t)\frac{p\lambda_A+\lambda_B}{\gamma_B}ln\frac{e^{(\gamma_B+\beta_B)t}}{e^{-\rho_B}(e^{(\gamma_B+\beta_B)t}-1)+1}-m_B(t)p\lambda_A I(t)$$

where

$$I(t)=\int_0^t\frac{(1-e^{-\rho_A})(\gamma_A+\beta_A)}{\gamma_A e^{-\rho_A}(e^{(\gamma_A+\beta_A)v}-1)+\gamma_A}\times$$
$$\frac{(1-e^{-\rho_B})(\gamma_B+\beta_B)}{\gamma_B e^{-\rho_B}(e^{(\gamma_B+\beta_B)(t-v)}-1)+\gamma_B}dv.$$

$I(t)$ is non-negative and not bigger than

$$\frac{(\gamma_A+\beta_A)(\gamma_A+\beta_A)^t}{\gamma_A+\gamma_B}.$$

4.4. Approximations

The Eq. (2) solution seems to be significantly more complex in circumstances different from those that have been mentioned. For instance, if the service times follow a LogNornal, Gamma or Weibull distributions. In some cases, only the numerical solution can eventually be stained.

For appropriate values of *t*, the following approximations concerning the equilibrium conditions are suggested:

$$\frac{m_B(t)}{m_A(t)}\cong\frac{\lambda_A\alpha_A}{(p\lambda_A+\lambda_B)\alpha_B}\qquad(4);$$

$$\frac{m_B(t)}{m_A(t)}\cong\frac{\lambda_A}{\lambda_B}\qquad(5).$$

Eq. (4) seems reasonable for values of *t* big enough and Eq. (5) is preferred for *t* close to zero. For details see Figueira and Ferreira [7].

5. Observations

Some values of the parameters p and λ_B have a special influence in the system behaviour. One may consider the suppression of the arch b when $p = 0$, of the arch c when $p = 1$ or of the arch d for $\lambda_B = 0$. Under those circumstances the traffic in those arches can be neglected.

It may be admitted that the ratio $m_B(t)/m_A(t)$ remains constant. This corresponds to the assumption that all the users of the system face identical conditions of effort and benefit, independently of the moment they join the system. Eq. (3) supplies a natural candidate for the value of that constant: $\lambda_A\alpha_A/(p\lambda_A + \lambda_B)\alpha_B$. In such situation Eq. (2) should include an "excess" functions $h(t)$:

$$h(t) = m_B(t)\frac{\lambda_A\alpha_A}{(p\lambda_A+\lambda_B)\alpha_B}\int_0^t (p\lambda_A G_A(v) + \lambda_B)\big(1 - G_B(t-v)\big)\,dv - m_A(t)\lambda_A\int_0^t \big(1 - G_A(v)\big)dv.$$

The function $h(t)$ is also interpreted in the sense of the expected value of a random variable depending on t. This approach can be generalized in a natural way to some other predefined function $m_B(t)/m_A(t)$.

Assuming that the system is initially empty appears to be a strong restriction of the analysis performed. When someone meets the system already in operation and does not known when it did start, the results that have been mentioned seem to have a lesser utility. In such case, there re-evaluation or finding a estimation procedure for the initial time are determinant for practical purposes.

Acknowledgments

This work was financially supported by FCT through the Strategic Project PEst-OE/EGE/UI0315/2011.

References

[1] Carvalho P 1993 *Planos e Fundos de Pensões* Texto Editora, Lisboa
[2] Ferreira M A M and Andrade M 2011 *International Journal of Academic Research*, **3** (1-II) 427
[3] Ross S 1996 *Stochastic Processes* 2nd Edition, Wiley, New York
[4] Harrison J and Lemoine A 1981 *Journal of Applied Probability* **18** 561
[5] Ferreira M A M and Andrade M 2009 *International Journal of Academic Research* **1** 84
[6] Ferreira M A M and Andrade M 2010 *Journal of Mathematics and Technology* **1** 58
[7] Figueira J and Ferreira M A M 1999 *Portuguese Review of Financial Markets* **2** 75
[8] Andrade M 2010 *Journal of Mathematics and Technology* **1** 96
[9] Andrade M, Ferreira M A M, Filipe J A and Coelho M 2012 *International Journal of Academic Research* **4** 205
[10] Ferreira M A M 2010 *Journal of Mathematics and Technology* **1** 91
[11] Ferreira M A M and Andrade M 2011 *International Journal of Latest Trends in Finance and Economic Sciences* **1**(2) 88
[12] Figueira J 2003 *Aplicação dos Processos de Difusão e da Teoria do Renovamento num Estudo de Reservas Aleatórias*, PhD Thesis, ISCTE

Journal of Physics: Conference Series **394** (2012) 012036 doi:10.1088/1742-6596/394/1/012036

Representation of Reserves Through a Brownian Motion Model

M Andrade, M A M Ferreira and J A Filipe
Instituto Universitário de Lisboa (ISCTE-IUL), BRU-IUL, Lisboa, Portugal

E-mail : marina.andrade@iscte.pt

Abstract. The Brownian Motion is commonly used as an approximation for some Random Walks and also for the Classic Risk Process. As the Random Walks and the Classic Risk Process are used frequently as stochastic models to represent reserves, it is natural to consider the Brownian Motion with the same purpose. In this study a model, based on the Brownian Motion, is presented to represent reserves. The Brownian Motion is used in this study to estimate the ruin probability of a fund. This kind of models is considered often in the study of pensions funds.

1. Introduction

The Brownian Motion is often used as an approximation for some Random Walks and also for the Classic Risk Process. About the convergence of Random Walks to the Brownian Motion see, for instance, [1], page 23. And on the convergence of the Classic Risk Process to the Brownian Motion see, for instance [2], page16 and [3]. As the Random Walks and the Classic Risk Process are used frequently as stochastic models to represent reserves, it is natural to consider the Brownian Motion with the same purpose.

Consider the following Brownian Motion definition, see [1], page 18:

Definition 1.1

The Brownian Motion with drift μ and diffusion coefficient σ^2 is a stochastic process $X(t)$ with continuous paths and independent Gaussian increments with mean and variance of an increment, $X(t+s)-X(t)$, given by μs and $\sigma^2 s$, respectively.

Note

- If $X(0)=x$, the Brownian Motion is assumed to begin in x,
- A Brownian Motion beginning at the origin, with drift 0 and diffusion coefficient 1, is named *standard* Brownian Motion and it is referred in general by $B(t)$,
- So the general Brownian Motion $X(t)$ with the beginning in x, drift μ and diffusion coefficient σ^2 may be written as

$$X(t)=x+\mu t+\sigma B(t) \qquad (1).$$

The computation of $\rho_k(x)$, the ruin probability of a system which reserves behave in accordance with a general Brownian Motion $X(t)$ will be performed in the next section. It is the probability that $X(t)$, with the beginning in x, reach 0 before arriving at k, $x>0$ and $k>x$. And it is considered also, then, the eventual system ruin probability $\rho(x)$, or the probability that $X(t)$ general Brownian Motion, with the beginning at x, reach anytime 0.

The problem may be approached in different ways. See, for instance, [1], page 24, for an approach of the problem based on the approximation to the Brownian Motion through the simple Random Walk. Or [4], page 360, for an approach to the problem through the application of the Martingales Stopping Theorem. The one presented here stands on the deduction of an ordinary differential equation for $\rho_k(x)$, relatively easy to solve, see, for instance, [5], page 193, or [6], page 373.

2. The Ruin Probability

Suppose initially $\mu \neq 0$. Consider h a time interval small enough such that the probability that $X(t)$ reach 0 before k is negligible. Thus by the Total Probability Theorem:

$$\rho_k(x) = E\left(\rho_k\big(X(h)\big)\right) + o(h) \qquad (2)$$

where $o(h)$ is the usual Landau notation for a function $f(h)$ such that $\lim_{h\to 0}\frac{f(h)}{h} = 0$. Writing now $\Delta X = X(h) - x$ and expanding $\rho_k\big(X(h)\big)$ as a Taylor's Series around x:

$$\rho_k\big(X(h)\big) = \rho_k(x + \Delta \mathrm{X}) = \sum_{n=0}^{\infty} \frac{(\Delta X)^n \rho_k^{(n)}(x)}{n!} \qquad (3).$$

Note that as the increments $X(h) - X(0) = X(h) - x = \Delta X$ are Gaussian Random variables with mean μh and variance $\sigma^2 h$:

- $E(\Delta X) = \mu h$,
- $E((\Delta X)^2) = \sigma^2 h + o(h)$,
- $E((\Delta X)^\nu) = o(h)$, for $\nu > 2$.

With these results in mind, applying operator E to both sides of equation (3) and substituting in (2) it is obtained

$$\mu h \rho_k'(x) + \frac{1}{2}\sigma^2 h \rho_k''(x) = o(h) \qquad (4).$$

Performing the division of both sides by h and making h to converge to 0 it is obtained, finally

$$\mu \rho_k'(x) + \frac{1}{2}\sigma^2 \rho_k''(x) = 0 \qquad (5),$$

with the border conditions $\rho_k(0) = 1$ and $\rho_k(k) = 0$.

The expression (5) is a second order ordinary differential equation that may be reduced to the following first order ordinary differential equation

$$\mu\rho_k(x) + \frac{1}{2}\sigma^2\rho_k'(x) = C_1 \qquad (6).$$

The general solution of (6) is

$$\rho_k(x) = C_1\frac{1}{\mu} + C_2\frac{2}{\sigma^2}e^{-2\mu x/\sigma^2} \qquad (7),$$

resulting from the border conditions $C_1 = \mu - \frac{\mu}{1-e^{-2\mu k/\sigma^2}}$ and $C_2 = \frac{\sigma^2}{2\left(1-e^{-2\mu k/\sigma^2}\right)}$.

So finally

$$\rho_k(x) = \frac{1-e^{-2\mu(k-x)/\sigma^2}}{e^{2\mu x/\sigma^2} - e^{-2\mu(k-x)/\sigma^2}}, \quad \mu \neq 0 \qquad (8).$$

Making μ converge to 0 in (8),

$$\rho_k(x) = \frac{k-x}{k}, \qquad \mu = 0 \qquad (9).$$

To estimate now the probability $\rho(x)$, of the eventual ruin of a system which reserves behave in accordance with the general Brownian Motion $X(t)$, make k converge to infinite in (8) and (9) obtaining:

$$\rho_k(x) = \begin{cases} e^{-2\mu x/\sigma^2}, & if\ \mu > 0 \\ 1, & if\ \mu \leq 0. \end{cases}$$

3. Concluding Remarks

It was exposed how to estimate the probability of the eventual ruin of a system which reserves behave in accordance with the general Brownian Motion. It is a very simple and intuitive way of doing it.

But this kind of models, and the consequent appreciations of the stability of the systems, done based on the evaluation of the probability of the exhaustion of the reserves, or ruin, are valid only in scenarios at which are considered constant prices. The integration of factors associated to the temporal depreciation process of the money value, in the modeling of financial reserves, although make eventually more complex the mathematical models involved, seems desirable.

Acknowledgments

This work was financially supported by FCT through the Strategic Project PEst-OE/EGE/UI0315/2011.

References

[1] Bhattacharya R N and Waymire E C 1990 *Stochastic Processes with Applications* John Wiley & Sons, New York
[2] Grandell J 1991 *Aspects of Risk Theory* Springer-Verlag, New York
[3] Iglehart D L 1969 *Journal of Applied Probability* **6** 285
[4] Karlin S and Taylor H M 1975 *A First Course in Stochastic Processes* 2nd Ed. Academic Press, New York
[5] Karlin S and Taylor H M 1981 *A Second Course in Stochastic Processes* 2nd Ed. Academic Press, New York
[6] Ross S 1996 *Stochastic Processes* John Wiley & Sons, New York
[7] Andrade M 2010 *Journal of Mathematics and Technology* **1**(1), 96
[8] Figueira J 2003 *Aplicação dos Processos de Difusão e da Teoria do Renovamento num Estudo de Reservas Aleatórias* PhD Thesis, ISCTE
[9] Figueira J and Ferreira M A M 1999 *Portuguese Review of Financial Markets* **2**(1), 75
[10] Ferreira M A M and Andrade M 2011 *International Journal of Latest Trends in Finance and Economic Sciences*, **1**(2), 91

AUTHOR INDEX

Institute of Physics
Dirac House, Temple Back
Bristol BS1 6BE UK

CURRAN ASSOCIATES INC
proceedings.com